Pesticide Formulation

Recent Developments and Their Applications in Developing Countries

Editors
Wade Van Valkenburg
B. Sugavanam
Sushil K. Khetan

UNIDO

UNITED NATIONS INDUSTRIAL DEVELOPMENT ORGANISATION
VIENNA

PUBLISHING FOR ONE WORLD

NEW AGE INTERNATIONAL (P) LIMITED, PUBLISHERS
New Delhi • Bangalore • Chennai • Cochin • Guwahati • Hyderabad
Jalandhar • Kolkata • Lucknow • Mumbai • Ranchi
Visit us at www.newagepublishers.com

Published by New Age International (P) Ltd., Publishers
First Edition: 1998
Reprint: 2008

Branches:

- 36, Malikarjuna Temple Street, Opp. ICWA, Basavanagudi, **Bangalore**. © (080) 26677815
- 26, Damodaran Street, T. Nagar, **Chennai**. © (044) 24353401
- Hemsen Complex, Mohd. Shah Road, Paltan Bazar, Near Starline Hotel, **Guwahati**. © (0361) 2543669
- No. 105, 1st Floor, Madhiray Kaveri Tower, 3-2-19, Azam Jahi Road, Nimboliadda, **Hyderabad**. © (040) 24652456
- RDB Chambers (Formerly Lotus Cinema) 106A, 1st Floor, S.N. Banerjee Road, **Kolkata**. © (033) 22275247
- 18, Madan Mohan Malviya Marg, **Lucknow**. © (0522) 2209578
- 142C, Victor House, Ground Floor, N.M. Joshi Marg, Lower Parel, **Mumbai**. © (022) 24927869
- 22, Golden House, Daryaganj, **New Delhi**. © (011) 23262370, 23262368

ISBN : 978-81-224-1069-3

Rs. 595.00

C-08-05-2513

2 3 4 5 6 7 8 9 10

Printed in India at Glorious Printers, Delhi.
Typeset at Printek India, Delhi.

PUBLISHING FOR ONE WORLD
NEW AGE INTERNATIONAL (P) LIMITED, PUBLISHERS
4835/24, Ansari Road, Daryaganj, New Delhi-110002
Visit us at **www.newagepublishers.com**

J. Wade Van Valkenburg Jr. (1925-1996)

Wade Van Valkenburg was best known for his singular contribution of elevating pesticide formulation research from a proprietary industrial activity to a fully-fledged scientific discipline. His early career at Dow Chemical Company in the 1950s and 1960s, where he led pesticide formulation research and supervised physical chemistry research, gave him the requisite exposure and opportunity. He published several books and book chapters. Of these, *Pesticide Formulations* in 1973 was a watershed for its first comprehensive treatment of the subject.

He received his education at Kalamazoo College, the University of Wisconsin and the University of Michigan, where he earned a Ph.D. degree in colloid and surface chemistry. He also received a law degree from the William Mitchell College of Law and practised patent law. He worked with 3M Company in the Magnetic Media Division. He was the president of Accessible Technologies Inc. and also provided consultancy on intellectual property protection. He was active member of several professional bodies, including the American Chemical Society.

He became a UNIDO advisor on pesticide formulation technology in India in the late 1980s for a project aiming at development and the introduction of technologies for safer formulation. The idea for this book grew out of the experiences gained during this project. His participation brought experience and requisite thrust. He did the spadework by compiling the manuscripts and doing the first round of editing. Unfortunately, he was not to see the fruits of his labour and succumbed to liver cancer in February 1996.

This book is dedicated to his memory.

Contents

Foreword

World food security is a must if the world is to live in peace and be able to feed the more than 10 billion people expected to inhabit this planet by the first quarter of the twenty-first century. A dramatic increase in agricultural production will be necessary, especially in the developing countries and countries in economic transition, since the bulk of the population will be living in less-well-off countries and will depend on agriculture for its livelihood. Crop production and crop protection during the growing season and during storage and transport of harvested crops are therefore of paramount importance.

Among the various inputs into agriculture, pesticides play a vital role in protecting crops from insects, disease and weeds. In addition, they are widely used in the control of vector-born disease and in keeping animals free of insects. However, with the continued use of pesticides, especially those that are highly toxic and persistent, as well as their misuse, especially in developing countries and countries in economic transition, it has become clear that all is not well: workers in the production and user sectors are being exposed, the environment is being contaminated and above all the problem of disposing of empty containers, accumulated pesticides and their wastes is continuing and growing in many places. Integrated pest management (IPM) policies have now been adopted by many governments, but the issues associated with the use of pesticides, both within the definition of IPM and outside, will dominate well into the next millenium.

While the manufacture of active ingredients is beyond the capabilities of most developing countries, more and more of these countries are opting for pesticide formulation since it gives them the option of changing pesticides as and when needed and also of moving towards user- and environment-friendly pesticide formulations. The United Nations Industrial Development Organization (UNIDO), in line with its mandate, provides assistance related to formulation technology, quality control, quality assurance, planning and technology selection and integrated safety, waste minimization and waste disposal. It is allocating a large share of its technical cooperation to Africa, where there is the greatest need for food security, and in future the use of pesticides there will have to be based on minimum risk/maximum benefits. This book seeks to give the pesticide industry better access to advances in pesticide technology and to encourage it to seek the advice needed to make the industry more responsible for the safety of the public and the environment.

Agenda 21 gives a clear mandate to international agencies, governmental and non-governmental organizations and the industry to coordinate their efforts to achieve overall safety in the use of agrochemicals through information collection and dissemination, capacity-building and greater inter-ministerial cooperation. This book, along with others on the topic, aims to assist in decision making at different levels so that the best use is made of synthetic organic pesticides in combination with other, non-chemical means of achieving food security.

Carlos Magariños
Director-General
United Nations Industrial Development Organization

June, 1998

Preface

In the early 1970s, two books were published on pesticide formulation. The first, *Industrial Production and Formulation of Pesticides in Developing Countries,* was published by UNIDO in 1972 and consisted of two volumes. The second, *Pesticide Formulation,* was edited by J. Wade Van Valkenburg and published in 1973; it was geared to an audience in both developed and developing countries. The two books laid the foundation for many other books written on the subject since that time.

With the cost of inventing new pesticides having escalated during the 1980s owing to rising registration costs and more stringent toxicological and environmental regulations, we saw many multinational companies ending their activities and/or merging to be more effective and profitable. At the same time, developing countries were carrying out an ever-larger share of pesticide formulation since few of them (some of the exceptions were China, India, Mexico and Argentina) could afford to manufacture active ingredients. The end of the cold war and the break-up of the Soviet Union brought a new dimension to the pesticide industry: most of the economies in transition opened up their markets to pesticides from the developed countries. Despite the advances in pesticide technology over the past two decades, developing countries still depend on the old generation of pesticides, which are toxic and persistent and, even worse, use archaic equipment for their application. This dilutes the benefits of agrochemicals on two fronts, production and application, and unnecessarily exposes people and the environment to hazards that could have been reduced to a minimum or even eliminated.

Having witnessed the development of technology for pesticide formulation over the last two decades, UNIDO decided it was time to update its 1983 publication *Pesticide Formulation in Developing Countries,* which was itself a major revision of the earlier UNIDO publication referred to above. We, the editors, are honoured to have been entrusted with this task. It has been a rewarding experience to have been in contact with the many experts from around the world who contributed to this book. Most of them have also been involved in United Nations activities in the area of pesticides and have wide experience in industry, research and development and policy issues. There are many others who could have made significant contributions but space time and the budget do not allow developing this book into a stand-alone treatise.

The book is mainly intended to make interested persons in developing countries aware of developments in pesticides formulation technology and should be used in conjunction with other publications on the topic. It achieves a good balance between authors from both developing and developed countries and attempt to highlight the science and technology where there have been substantial technical advances. Pesticide technology is rapidly changing. The original technology for organic synthesis, screening, field tests, analytical tests etc. is still there, but in a much more sophisticated form. These technological advances, together with the demands and concerns of society, constitute the substance of this book. Exciting new developments in two areas - transgenic crops incorporating *in situ Bacillus thuringiensis* genes and herbicide-resistant varieties - will have a great impact on the pesticides market but are outside the scope of this book.

The evolving technology places great emphasis on (a) synergistic mixtures that can reduce the amount of pesticide use, (b) integrated pest management, (c) safety, (d) quality, (e) ecotoxicology and (f) waste management. Events that arouse public concern are causing governments to increase their level of control over the pesticides industry. While regulations, registration, legal liability and the like are mentioned, a fuller discussion of these issues is beyond the scope of the book. In 1992, while this book was being prepared, the United Nations Conference on Environment and Development (UNCED) took place at Rio de Janeiro. Agenda 21, adopted by UNCED, gave the world at large a mandate for the sound management of toxic chemicals and hazardous wastes and the preservation of biodiversity. These concerns are addressed in the final chapter of this book, "Environmental toxicological considerations in pesticide formulation", which discusses the sound management of pesticides from cradle to grave, by industry, governments and the public, a task that will continue into the new millenium.

We hope that this book will create an awareness in developing countries of the issues involved in pesticides and that it will impart knowledge to the people there who are involved with pesticides, encouraging them to seek guidance in reducing the risks involved in pesticides production and application, in line with the mandate of Agenda 21.

EXPLANATORY NOTES

Dollars ($) refer to United States Dollars

AED	atomic emission detection
AI	active ingredient
B.t.	*Bacillus thuringiensis*
BR	briquette
CDA	controlled droplet application
CIPAC	Collaborative International Pesticide Analytical Council
CS	capsule suspension
CSMA	Chemical Specialties Manufacturers Association
DF	dry flowable
DIN	Deutsches Institut für Normung (German Institute for Standardization)
DL	dustless dust
DMI	demethylation inhibitor
DP	dustable powder
DSC	differential scanning calorimetry
DTA	differential thermal analysis
EBIs	ergosterol biosynthesis inhibitors
EC	emulsifiable concentrate
EICD	electroconductivity detector
EPA	United States Environmental Protection Agency
EW	oil-in-water emulsion
FAO	Food and Agriculture Organization of the United Nations
FID	flame ionization detector
FTIR	Fourier transform infra-red
GEMS	Global Environmental Monitoring System
GIFAP	International Group of Associations of Manufacturers of Agrochemical Products, renamed Global Crop Protection Federation (GCPF)
GL	gel
GLC	gas-liquid chromatography
GR	granule
HDPE	high-density polyethylene

HLB	hydrophilic-lipophilic balance
HPLC	high-performance liquid chromatography
IOBC	International Institute for the Biological Control of Noxious Animals and Plants
IPM	integrated pest management
IR	infra-red
IRM	insecticide resistance management
IRPTC	International Register of Potentially Toxic Chemicals
IRRI	International Rice Research Institute
ISO	International Organization for Standardization
LDPE	low-density polyethylene
LLDPE	linear low-density polyethylene
ME	microemulsion
MG	microgranule
MSD	mass spectral detector
NOEL	no-effect level
OECD	Organization for Economic Cooperation and Development
OF	oil-miscible flowable concentrate
O-P	organophosphorus
PA	polyamide
PEO	polyethylene oxide
PET	Polyethylene terephthalate
PP	polypropylene
PVA	polyvinyl alcohol
QA	quality assurance
QSAR	quantitative structure activity relationship
SC	suspension concentrate
SEWS	surveillance and early warning system
SFE	supercritical fluid extractor
SOP	standard operating procedures
SP	water-soluble powder
TCD	thermal conductivity detector
TCDD	3,3,4,4-tetrachlorodibenzo-p-dioxin

TGA	thermal gravimetric analysis
TK	technical concentrate
TLV	threshold limit value
UL	ultra-low volume (ULV) liquid
UNEP	United Nations Environment Programme
UNIDO	United Nations Industrial Development Organization
VLV	very-low volume (applications)
WG	water-dispersible granules
WHO	World Health Organization
WP	wettable powder
WTO	World Trade Organization

GENERAL PERSPECTIVES

CHAPTER

1

Development of Crop Protection Agents - Invention to Sales

B. Sugavanam

UNIDO, Vienna International Centre
A-1400 Vienna, Austria

and

L.G. Copping

34, Saxon Way, Saffron Walden, Essex,
United Kingdom of Great Britain
and Northern Ireland

The discovery and development of a new crop protection agent is a long and multidisciplinary process that involves good and effective interaction between groups of scientists. As our knowledge base increases and as chemistry becomes more complex, the restrictions imposed upon agrochemical companies increase and the number of chemicals that have to be screened to produce a new product increases. This chapter addresses the challenges faced in the search for new pesticides and reviews the various approaches taken by different organizations in establishing screens, preparing compounds, proving field efficacy and assessing marketing viability.

1 IDENTIFICATION OF TARGETS

The costs associated with the discovery of a new agrochemical are huge. It is essential that research is funded by the current sales of a company. Usually a forward-thinking organization invests between 5 and 10% of its sales in research designed to find new products. If sales are to be maximized it is of fundamental importance that each new introduction repay more than its research costs. Currently a new pesticide, from discovery to market, may cost $60 million or more. If this compound is an insecticide its probable life expectancy is of the order of 7-8 years. It will be less if it is an acaricide and slightly more (usually) if it is a fungicide. Herbicides, on the other hand, will be expected to maintain market share for 12 years or more. The reasons for these differences are twofold. First, resistance problems are very common with insecticides and acaricides and also with the newer fungicides with specific biochemical modes of action. Secondly, new introductions are always challenging older compounds for a share of existing markets. Clearly, if a company wishes to remain in business, it is essential that the cost of development, plus lost interest on the invested capital, plus profits must be recouped within the lifetime of the new product. In the aggregate, this means $60 million plus appropriate interest plus profit has to be returned in 5 years for an acaricide, 7 years for an insecticide, 9 years for a fungicide and 12 years for a herbicide. If it is assumed that a new product at maturity will achieve a 20% penetration into each market for which it is sold, it becomes apparent that it is essential that new products are targeted to big markets.

The principal markets are as follows:

Herbicides

- Small-grain cereals
- Corn (maize)
- Soybean
- Rice
- Industrial weed control

Fungicides

- Small-grain cereals
- Vines
- Rice
- Potatoes

Insecticides
- Cotton
- Corn (maize)
- Rice

Acaricides
- Cotton
- Top fruit and citrus

Vegetables are excluded from these lists because they are fragmented markets made up of a large number of different crops all affected by different pests, weeds and diseases; together they represent a large market but individually each market is relatively small.

Hence a large company will focus on these crops, and the problems that occur in them will be their primary targets. Of course having discovered a new compound and confirmed its potential in a major crop, a company will seek additional income by establishing a use in a smaller "bolt-on" crop.

2 ESTABLISHMENT OF SCREENS

Having determined the crops that will be targeted, it is important to identify the organisms that occur in these crops against which it is essential to have biological efficacy. Typical examples of these are shown in Table 1.1. While the table indicates typical pathogens, weeds and insects, it is clearly not an exhaustive list but serves to illustrate specific species from each key crop area that should be considered when establishing a new screen.

A further consideration which often colours the selection of key species is the marketing/distribution strengths of the company. For example, if an organisation has specific strengths in Europe and has a significant share of the sugar beet herbicide and insecticide market, it may be more cost-effective to concentrate company resources in supporting and complementing that market. Thus, seeking a new product in, say, the rice market, a crop that has been dominated in crop protection usage by Japan, may not be a good idea.

Table 1.1 Typical pests of target crops

Crop diseases	*Insects*	*Weeds*
Cotton		
Seedling diseases	*Heliothis* spp.	*Echinochloa crus-galli*
	Pectinophora spp.	*Digitaria sanguinalis*
	Spodoptera spp.	*Ipomoea* spp.
	Aphis gossypii	*Datura stramonium*
	Bemisia tabaci	*Abutilon theophrasti*
	Tetranychus spp.	*Amaranthus* spp.
		Xanthium spp.
Corn		
Seedling diseases	*Diabrotica* spp.	All the above plus
	Ostrinia nubialis	*Sorghum halepense*
	Agrotis spp.	*Cassia* spp.
Rice		
Pyricularia oryzae	*Nilaparvata lugens*	*Echinochloa* spp.
Rhizoctonia solani	*Nephotettix cinciteps*	*Sagitaria* spp.
Gibberella fujikuroi	*Chilo* spp.	*Cyperus* spp.
		Monochoria vaginalis
Small-Grain Cereals		
Erysiphe graminis	Aphids	*Avena fatua*
Puccinia spp.		*Alopecurus myosuroides*
Pseudocercosporella		*Galium aparine*
Herpotrichoides		*Matricaria* spp.
Septoria spp.		*Viola* spp.
Fusarium spp.		*Veronica* spp.
Soybean		
Seedling diseases		As for cotton and corn
Vines		
Plasmopara viticola	Berry moth	Many weeds
Uncinula necator		
Botrytis cinerea		

For the sake of an example it will be assumed that the organisation has broad-crop interests and a "typical" broad-based screen will be described. In an ideal world all major pest species would be tested *in vivo*. Unfortunately some organisms are extremely hard to cultivate, have extremely long infection periods/life cycles or are restricted in certain locations by agricultural hygiene laws. Some examples are shown in Table 1.2.

Table 1.2 Difficult or restricted pest species

Pest	*Crop*	*Problem*
Pseudocercosporella herpotrichoides	Wheat/barley/rye	Very slow to infect; 3 month test. Resistant strains.
Botrytis cinerea	Vines plus many other crops	Main problem is as a pathogen of developing fruit. Difficult to simulate in a glasshouse. Resistant strains.
Diabrotica spp.	Corn	Larvae attack roots of corn plants. Main pest species show diapause. Prohibited in California.
Panonychus ulmi	Fruit trees	Very hard to maintain in culture in captivity.

Screens are divided into a number of stages, typically three. The primary screen is designed to ensure that all inactive compounds are identified as rapidly as possible and that only those showing biological efficacy are tested further. The secondary screen is designed to show how active a compound is in comparison to standards and related chemistries.

Secondary screening data should be sufficiently quantitative to allow more structure activity correlation work to be undertaken. Tertiary screening is a process in which compounds are tested at rates, spray volumes and timings that will coincide with those to be used in the field. It is data from these screens that are used to identify field candidates. Because different questions may arise at each stage, the three screening stages are often assessed in different ways, but in all cases it is essential that no potentially useful biological activity is missed. Typical screens are shown in Table 1.3. Screening results are either positive (active) or negative (inactive) at the primary level

Again Table 1.3 should be considered an example of the types of screening systems that might be employed by a large agrochemical company seeking new products for agriculture in industrialized countries. If the target markets were different the screens would

Table 1.3 Screening systems

Pest	*Indicating*	*Rate*	*Volume*
Insecticide Primary			
Megoura viciae	Aphid	High	High
Tetranychus urticae	Mites	High	High
Spodoptera littoralis	Noctuid	High	High
Heliothis virescens	Noctuid	High	Contact
Diabrotica undecimpunctata	Beetle	High	In soil
Aedes aegypti	Mosquito	High	Contact/feeding
Boophilus microplus	Cattle tick	High	Larvae in filter paper
Blattella germanica	Cockroach	High	Contact
Lucilia sericata	Blowfly	High	Larvae in diet
Insecticide Secondary			
Retest on species controlled at primary level. Reduce rate of application. Record percent mortality. Compare with standard.			
Insecticide Tertiary			
Reduce volume of application. Examine persistence. Introduce additional related pest species. Look for mobility in plant. Introduce resistant species.			
Fungicide Primary			
Erysiphe graminis	Powdery mildew	High	High
Puccinia recondita	Rust	High	High
Plasmopara viticola	Downy mildew	High	High
Pyricularia oryzae	Rice diseases	High	High
Septoria nodorum	Cereal diseases	High	High
Botrytis cinerea	*Sclerotia*-forming pathogens	High	High
Cercospora arachidicola	Leaf spots	High	High
Fusarium spp.	Imperfect fungi	*In vitro*	
Pseudocercosporella herpotrichoides	Eyespot	*In vitro*	
Fungicide Secondary and Tertiary	(as per insecticides).		
Herbicide Primary			
Avena fatua	Large-seeded grass	High	High
Alopecurus myosuroides	Small-seeded grass	High	High
Galium aparine	Important weed	High	High
Stellaria media	Small-seeded broad leaf	High	High
Chenopodium album	Important weed	High	High
Tripleurospermum maritimum	Composite	High	High

Table 1.3 (*Contd.*)

Pest	*Indicating*	*Rate*	*Volume*
Echinochloa crus-galli	Small-seeded grass	High	High
Digitaria sanguinalis	Small-seeded grass	High	High
Ipomoea purpurea	Important weed	High	High
Abutilon theophrasti	Important weed	High	High
Amaranthus retroflexus	Important weed	High	High
Xanthium pennsylvanaicum	Important weed	High	High

Herbicide Secondary

Run down tests on individual weeds against primary species affected plus close botanical relatives. Assessed as percent control. Introduce crops and commercial standards.

Herbicide Tertiary

Detailed studies of biological efficacy in comparison to crop selectivity. Persistence and mobility in plant and soil.

reflect this, but the principles would always remain the same.

It must also be remembered that it is important to ensure that screens are accurate and that they will accommodate rapid compound throughput. For this reason the formulation of new experimental compounds is always as simple as possible at the early stage of screening. Most compounds are applied in the primary and secondary stage as acetone/surfactant concentrates dispersed in water. Such a system ensures the even application of the compound to the target. Clearly, compounds which are insoluble in acetone and water cannot be applied in this way and in these cases the compounds should be finely suspended by grinding or sonication prior to application. The question of formulation should be addressed at the tertiary level, when a variety of formulation types should be compared with the primary screening system and the preferred composition selected for field evaluations.

3 SYNTHESIS STRATEGIES

Any organization involved in agrochemical discovery should have a strategy to introduce new compounds in the market. For basic survival, a company in this highly competitive field should aim to introduce, on average, two or three crop protection agents a year to meet small to medium size markets and aim for a big winner on an average once in 7-10 years. Added to this, it should introduce many newer, safer and more effective formulation and application technologies to stay in business. A major research and development department should have a big synthetic organic chemistry team providing new compounds for screening. On an average, 40,000 new compounds must be screened for every new agrochemical that goes into the market.

A number of approaches are open to a synthetic chemist when embarking on a synthesis programme:

(a) Empirical approach - random synthesis;
(b) Analogue synthesis - 'me-too' or bandwagon approach;
(c) Natural products as pesticides and leads;
(d) Biorational design - interfering with metabolic pathways;
(e) Computer graphics/physicochemical approach;
(f) Synthesis of specific active isomers;
(g) Combinatorial chemistry.

During the last 50 years these various approaches, which are in many ways linked to each other, resulted in the development of crop protection agents as shown in Table 1.4.

3.1 Empirical Approach - Random Synthesis

For a chemist it is of vital importance to plan a synthesis programme around a molecular structure of a lead compound, or model compound, to optimize biological activity. In this approach, random synthesis has played a key role in identification of leads. However, the structures of compounds such as DDT, paraquat and glyphosate were known long before their biological activity was discovered. Random synthesis is now becoming less and less favoured and is being replaced by testing compounds prepared by universities, research laboratories and other chemical companies.

3.2 Analogue Synthesis - "Me Too" or Following the Bandwagon

Whenever a major invention is announced by a company it is always an invitation to competitors to find 'me too' compounds

Table 1.4 Development of crop protection chemicals

Year	*Product*	
Prewar - period	• Natural products: pyrethrum, derris, tobacco, sulfur	
1940s - 1950s	• Organochlorine • Organophosphorus • Carbamates • Inorganics • Phenoxys • Triazines	
1960s - 1980s	New Era in Pesticides	
Insecticides	• Pyrethroids • Avermectins	
Herbicides	• Anilides; ureas • Paraquat • Glyphosate • Aryloxyplenoxypropionates • Sulfonylureas • Imidazolinones	
Fungicides	• Benomyl • Acylanilides • Azoles	
Plant Growth Regulators	• Ethrel • Cycocel	
1990s		
Insecticides	• Imidacloprid • Bt incorporated transgenic corn	
Fungicides	• ICI A 5504 (Azoxystrobin) • BAS 490 F (Kresoxym-methyl)	Strobilurin A related
	• CGA 2245704 — (Benzothiadiazole derivative)	Systemic activated resistance
Herbicides	• Mainly herbicide tolerant - soya, maize, cotton etc.	

that are equal if not better in biological activity than the original invention and outside the original patent. Classic examples are the invention of systemic fungicides, especially triadimefon (Table 1.5) by Bayer, and the sulfonylurea herbicides (Table 1.6) by DuPont. Following the invention of triadimefon in the early 1970s, more than 15 companies joined the bandwagon and today there are

Table 1.5 Azole fungicides

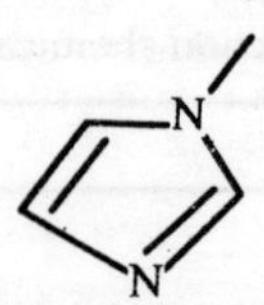 = Im 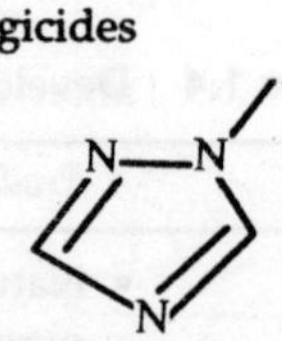 = Tr

Structure	*Name (company)*	*Inventor*
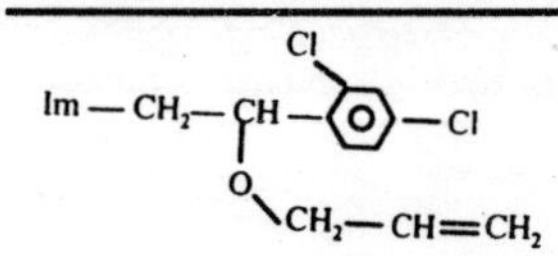	Imazalil (Janssen)	Godfroi et al.
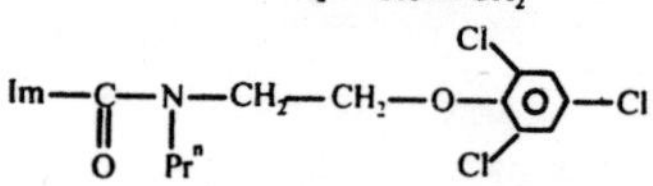	Prochloraz (Schering)	Brooks et al.
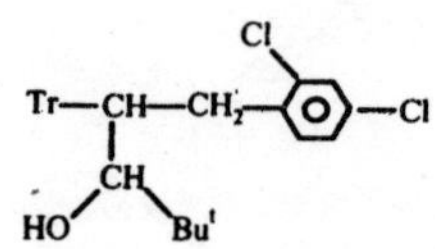	Triadimefon (Bayer)	Meiser et al.
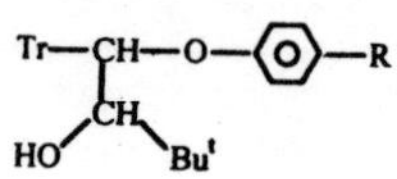	Triadimenol (R = Cl) Bitertanol (R = C_6H_5) (Bayer)	Kramer et al.
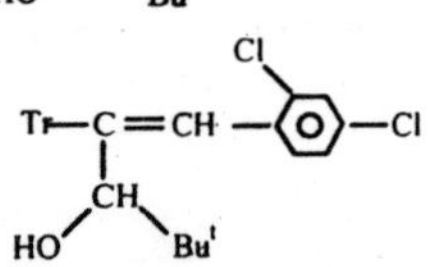	Dichlobutrazol (Zeneca)	Sugavanam et al.
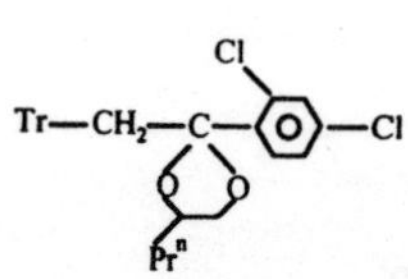	Propiconazole (Ciba)	Reet et al.
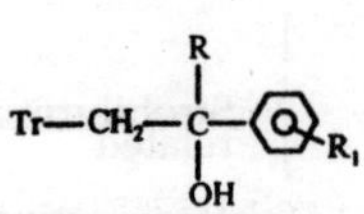	Flutriafol (R = oF.C_6H_4) (R_1 = p.F) Hexaconazole (R=Bu^n) (Zeneca) (R_1=2, 4 Cl_2)	Worthington et al.
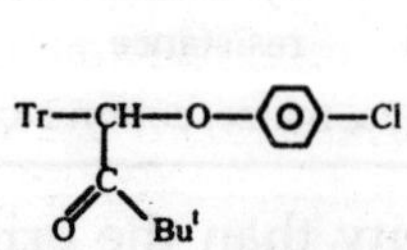	Diniconazole (Sumitomo)	Funaki et al.
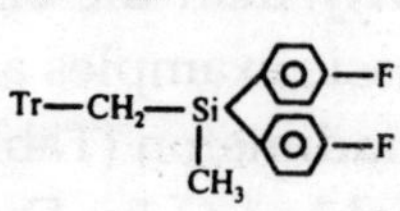	Flusilazole (DuPont)	—
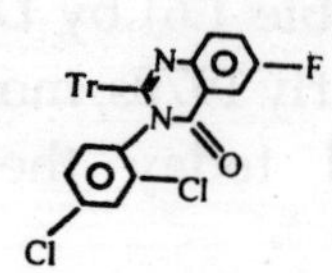	Fluquinconazole (AgrEvo)	Wilde et al.

Table 1.6 Sulfonylurea herbicides

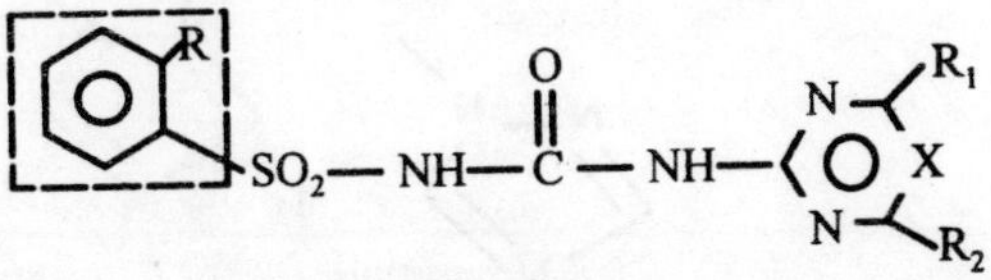

Name (trade name)	*Aromatic group*	*Heterocycle* X	R_1	R_2
Chlorsulfuron (Glean)	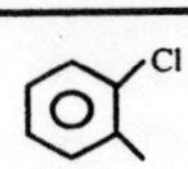	N	OCH_3	CH_3
Metsulfuron methyl (Ally)	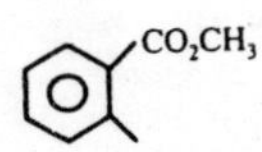	N	OCH_3	CH_3
Thifensulfuron methyl; also, thiameturon methyl (Pinnacle)	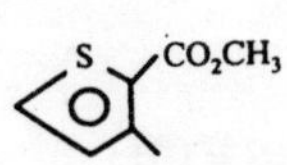	N	OCH_3	CH_3
Bensulfuron methyl (Londax)	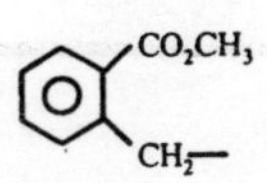	CH	OCH_3	OCH_3
Chlorimuron ethyl (Classic)	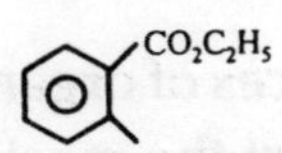	CH	OCH_3	Cl
Triasulfuron (Amber)	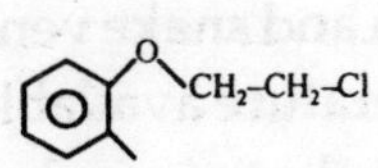	N	OCH_3	CH_3

than 25 azole compounds on the market (Table 1.5). It should be borne in mind that compounds based on imidazole, such as imazalil (Table 1.5), were invented long before triadimefon. In the case of sulfonylurea, changing the aromatic part of the molecule, there are many sulfonylureas on the market (Table 1.6), with only DuPont, AgrEvo, Novartis and Nissan having been successful. Another invention is the triazole plant growth regulator paclobutrazol, and only a few compounds in this area have come to the market (Table 1.7), mainly because of complexities involved in developing plant growth regulators and the small markets.

Table 1.7 Azole plant growth regulators

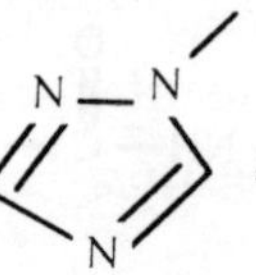
= Tr

Structure	*Name (company)*	*Inventor*
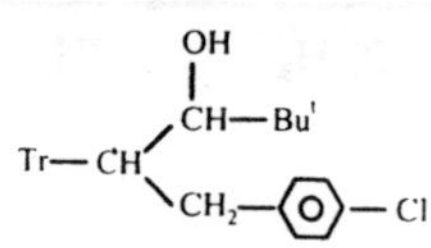	Paclobutrazol (2RS, 3RS) (Zeneca)	Sugavanam et al.
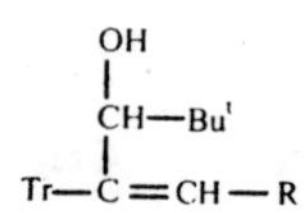	Uniconazole (E, S) R = p-Cl.C_6H_4 (Sumitomo) Triapenthenol (E) R = Cyclohexyl (Bayer)	Funaki et al.

3.3 Natural Products as Leads

Plants, fungi and bacteria are rich sources of organic molecules both simple and complex in structure. In fact the most toxic compounds such as curare, strychnine, aflatoxin and snake venoms are of natural origin. Despite a vast amount of literature available on the pesticidal properties of compounds of botanical origin, only three compounds have played an important role as pesticides. These are nicotine from tobacco, rotenone from derris roots and pyrethrins from pyrethrum flowers. Due to many drawbacks and severe competition from synthetic products these natural products have taken only a fraction of the pesticide market. Azadirachtin from the neem tree (*Azadirachta indica*) is being studied as a source of insecticidal active product.

No other natural product has given a more challenging and successful chemical lead than the pyrethrins. Following the extensive work at Rothamsted Experimental Station in England by Elliot and coworkers, today the synthetic pyrethroids are worth more than

US $2 billion at the user end. It is beyond the scope of this book to go into the details of the chemical synthesis but the broad approach of Elliot in changing the molecular structure to optimize the biological activity is shown in Figure 1.1.

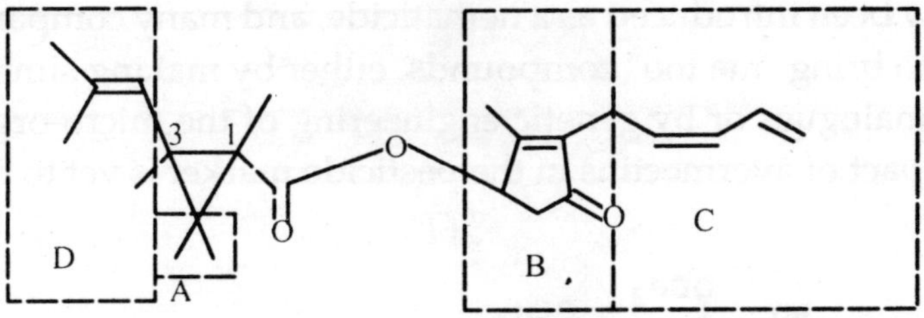

A. Two methyl groups on 2 carbon atom necessary
B. The ring can be replaced by similar groups
C. Side chain is necessary and should have unsaturation
D. Isobutenyl can be changed to dilhalovinyl

Figure 1.1 Synthetic modifications to pyrethrins; stereoisomerism around C_1 and C_3 should be either 1R, 3S or 1R, 3R (Ingold nomenclature)

Today there are more than 20 synthetic pyrethroids sharing the big insecticide market. Synthetic chemists clearly took the challenges of (i) solving the stereochemistry, especially in the case of deltamethrin (Figure 1.7(a)), (ii) of broadening the biological activity (aphicidal and miticidal) and (iii) of reducing the aquatic toxicology. One major problem is going to be the onset of resistance to pyrethroids unless a global approach for insecticide resistance management (IRM) strategy is adopted.

Another small success is based on the marine worm poison nereistoxin, which led to commercialization of cartap, Figure 1.2.

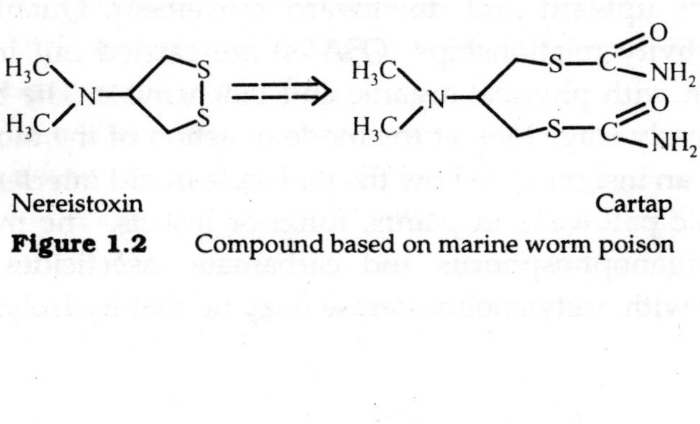

Nereistoxin Cartap

Figure 1.2 Compound based on marine worm poison

In addition, sex attractant pheromones and juvenile hormones have also provided leads for chemists to design synthetic programmes. A spin-off from the veterinary field is one of the metabolites belonging to avermectins (Figure 1.3), a complex macrocyclic lactone produced by *Streptomyces avermitilis*. This compound has now been introduced as a nematicide, and many companies are trying to bring "me too" compounds, either by making simple synthetic analogues or by genetic engineering of the micro-organism. The impact of avermectins in the pesticide market is yet to be seen.

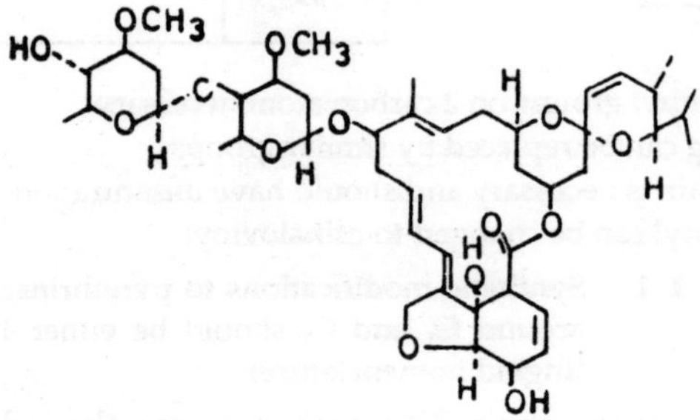

Figure 1.3 Avermectin B_{1a}

For some information on natural products, see Copping, 1996.

3.4 Biorational Design

Optimization of biological activity by synthetic chemists demands close collaboration with physical-organic chemists, biochemists and biologists. The partition coefficient (octanol-water) of a molecule plays a key role in taking the compound to the site of action by penetration through the cuticle of insects or the waxy surface of leaves. Once inside the plant, the physical properties of the molecule determine its upward and downward movement. Quantitative structure activity relationships (QSARs) are carried out in close collaboration with physical-organic and biochemists. The biochemists at an early stage look at the mode of action of the molecule, which gives an insight as to how the molecule might interfere with the metabolic pathways in plants, fungi or insects. The mode of action of organophosphorus and carbamate insecticides is by interference with acetylcholinesterase enzyme that hydrolyzes the acetylcholine generated in myoneural junctions during transmission of motor commands. The irreversible binding of organophosphorus insecticides and the reversible binding of carbamate insecticides to cholinesterase enzyme is well documented. All synthetic compounds in these two groups follow the phenomenon for suitable binding with the cholinesterase enzyme.

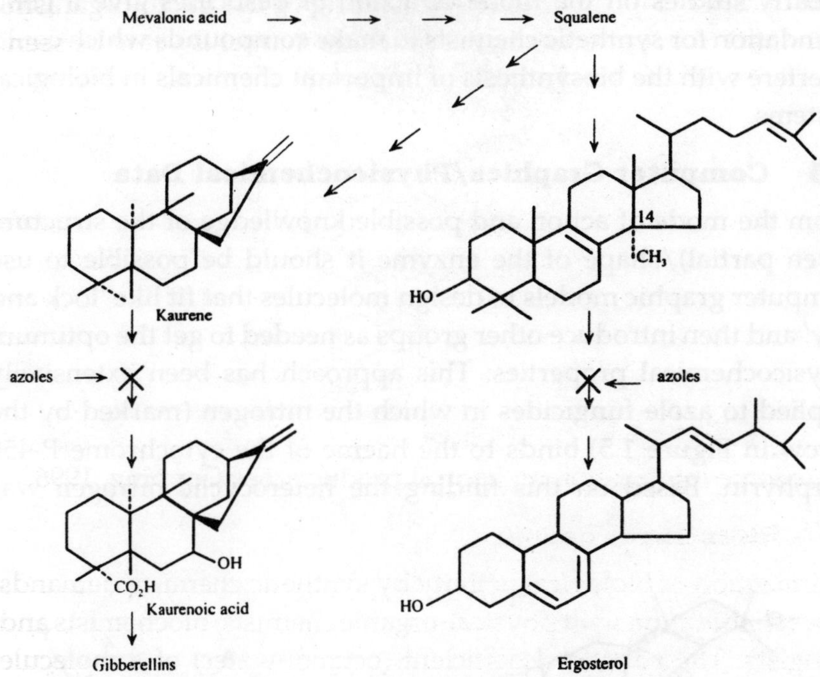

Figure 1.4 Mode of action of azoles as fungicide and plant growth regulator

Study of the mode of action of the systemic azole fungicides revealed that it is due to inhibition of ergosterol biosynthesis in fungi by inhibition of the C14 demethylation step (Figure 1.4). It is also established that morpholine-based fungicides such as tri-demorph interfere with the same pathway but at a different step(s). In the case of the plant growth regulator paclobutrazol, it has been established that it interferes with the biosynthesis of gibberellin, possibly by interfering with the conversion of kaurene to kaurenal (Figure 1.4). It is now well established that these azoles inhibit cytochrome P-450 enzymes in fungi and plants.

Recently introduced herbicides such as sulfonylureas and imidazolinones are known to inhibit biosynthesis of branched chain amino acids by interfering with acetolactate synthase enzyme. There are more than 20 herbicides having this type of action. Photosynthesis and plant pigment biosynthesis are well-established targets for a wide range of selective herbicides.

Early studies on the mode of action of pesticides give a good foundation for synthetic chemists to make compounds which could interfere with the biosynthesis of important chemicals in biological systems.

3.5 Computer Graphics/Physicochemical Data

From the mode of action and possible knowledge of the structure (even partial)/shape of the enzyme it should be possible to use computer graphic models to design molecules that fit like 'lock and key' and then introduce other groups as needed to get the optimum physicochemical properties. This approach has been extensively applied to azole fungicides in which the nitrogen (marked by the arrow in Figure 1.5) binds to the haeme of the cytochrome P-450 porphyrin. Based on this finding the heterocyclic nitrogen was

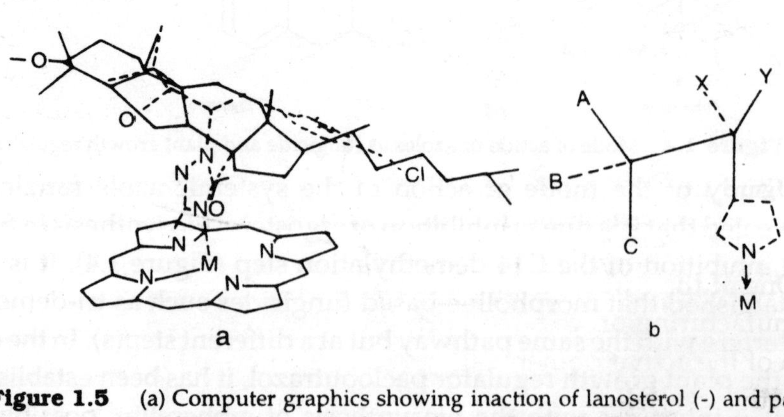

Figure 1.5 (a) Computer graphics showing inaction of lanosterol (-) and paclo butrazol (----) with cytochrome P-450 porphyrin(-------) and (b) basic structure for computer-aided design synthesis

retained in the basic structure (Figure 1.5) so that synthetic chemists could make new compounds for testing. In fact many compounds were made to achieve the best fit for getting a high level of activity. To date most of the success of computer graphics in agrochemicals has been in explaining structures after the invention of the highly active molecule.

3.6 Synthesis of Specific Active Isomer

Organic compounds, depending on their structure, can exist in different isomeric forms, generally called geometrical and optical-isomers. While nature makes specific isomers through the action of enzymes involved in biosynthesis, man-made chemicals are generally a mixture of isomers unless specific methods are employed during synthesis using stereospecific or resolution techniques. Since the shape of the active molecule is of vital importance for compounds with a specific mode of action, Figure 1.6 gives a simple illustration of a 'lock and key' approach necessary for biological activity for geometrical and optical isomers.

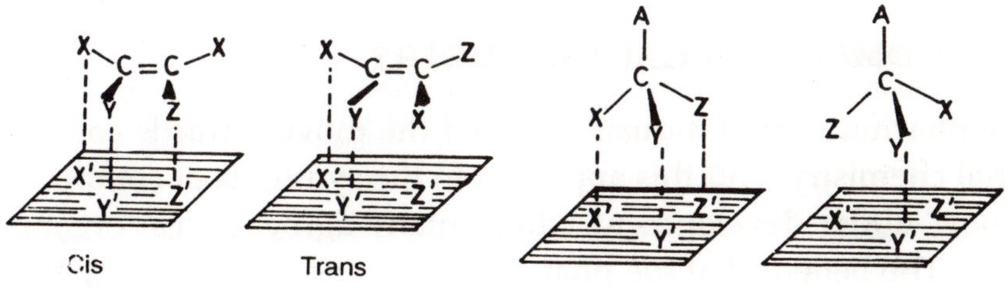

Figure 1.6 Simple illustration of 'lock and key' interaction of geometrical and optical isomers. Source: *Pesticide Formulations* by Van Valkenburg, 1973

One of the earliest examples is benzene hexachloride, where the manufacturing process has been modified to give a higher proportion of the active isomer, gamma-benzene hexachloride. By further selective crystallization the pure isomer could be obtained.

Pyrethroids have been a great challenge to organic chemists and the manufacturing process for the optically active single isomer deltamethrin (among 8 possible isomers) (Figure 1.7) has been successfully achieved by Martel and coworkers at Roussel Uclaf. Similarly, in the case cyhalothrin, in which one diastereoisomer (lR, αS, 1S, αR) (Figure 1.7(a)) has been selectively obtained for use in the field.

In the case of plant growth regulators, paclobutrazol (RS, RS), uniconazole (E,S isomer) and triapenthenol (E isomer) (Table 1.7) are specific isomers. Also the graminicide, fluazipop-p-butyl is a single R isomer (Figure 1.7(b)).

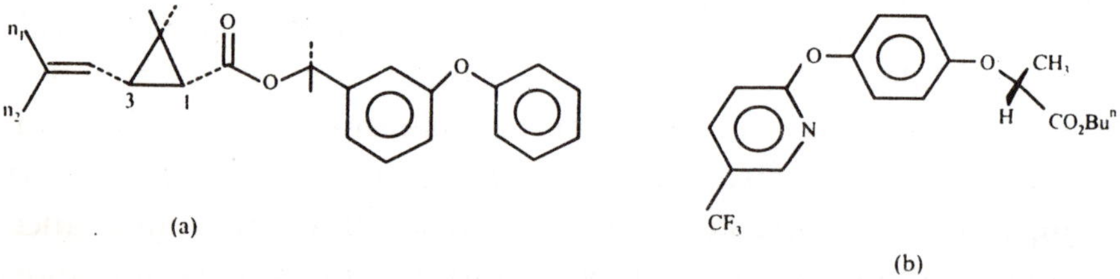

Figure 1.7 (a) $R_1 = R_2$ = Br, deltamethrin (lR, 3R, αR), $R_1 = CF_3$, R_2 = Cl, cyhalothrin diastereoisomer (Z-1RS, 3RS, αSR) and (b) fluazipop-p-butyl (R)

In the future, companies will be going more and more for marketing specific isomers so as to reduce unnecessary environmental contamination by chemicals that are not active.

3.7 COMBINATORIAL CHEMISTRY

The pharmaceutical industry has led the move towards combinatorial chemistry, and this approach to the synthesis of compounds is now under development within many agrochemical organizations. The benefit that the pharmaceutical industry has is that it is targeting specific biochemical reactions that have been identified as being effective and that have been accepted by the regulatory authorities as being of value in the treatment of a particular disease or illness. The introduction of a new compound that is more effective against a proven enzyme and/or less damaging to the patient will be completed easily and will bring the company an immediate return for the investment. This means that new compounds can be screened against *in vitro* enzyme targets quickly and evidence of a useful biochemical effect can be generated very rapidly. The problem is to provide the several hundred thousand compounds a year that can be accommodated by these screens. This has been solved through the application of combinatorial chemistry.

The successful application of combinatorial chemistry is based on the screening technique of taking the screen to the compound(s) rather than testing each compound separately in the screen To be an effective and a selective inhibitor of a target enzyme, the chemical must bind specifically to that target enzyme. It may bind to a co-factor in the biochemical reaction, and such compounds are generally less desirable as pharmaceuticals because co-factors are rarely enzyme-specific and the inactivation of a co-factor will lead to unwanted side effects. This may not be so undesirable as a mode of action of a new agrochemical.

Chemists make combinatorial collections, or libraries, in a rather simple way. Standard chemical reactions are used to assemble selected sets of building blocks into a huge variety of larger structures. To simplify the principles, imagine that there are four molecules, A1, A2, B1 and B2. A1 and A2 are related chemically as are B1 and B2, and those two different classes of compounds can react to form new molecules. Combinatorial chemistry allows the rapid synthesis of all possible analogues A1-B1, A1-B2, A2-B1 and A2-B2. The early work was undertaken with the construction of polypeptides and involved much larger numbers of compounds. If 30 compounds containing an amino group are selected and these are to be reacted with 30 compounds containing a carboxylic acid residue then there are 30 x 30, or 900 possible combinations. The addition of a third set of 30 building blocks would lead to 30 x 30 x 30, or 27,000 different possible combinations.

Two different techniques can be used to make these combinations. Parallel synthesis was invented in the 1980s by Mario Geysen, now at Glaxo-Wellcome. Reactions are usually carried out in 96-well microtitre plates. If the idea is to react 8 amines with 12 carboxlic acids the first amine would be placed in the first row of wells the second in the second and so on. The first carboxylic acid would then be added to the wells sequentially, allowing the synthesis of 96 compounds from only 20 different starting materials. It is common for the reagents to be attached to a solid support, such as a polystyrene bead, as this allows any unreacted material to be removed by washing, leaving the desired products attached to the beads. Such a procedure requires techniques to attach the starting material to the bead and then remove the product at the end of the reaction sequence, but the advantages of purification outweigh these problems. Much of this work can be done by robots, and some companies have developed methods for the preparation of over 1,000 compounds a day. When making this number of compounds it has to be remembered that doubling the number of products requires nearly twice as much time. This restraint limits the number of compounds that can be produced by parallel synthesis to tens of thousands of compounds rather than many more.

The other technique for combinatorial chemistry was developed by Arpad Furka at Advanced ChemTech and is known as split and mix. This technique can be addressed simply in the following way. If chemists attach three related compounds, A1, A2 and A3, to polystyrene beads separately and mix them together thoroughly and separate them into three vessels once again, then each of the new vessels will contain an equal amount of A1, A2 and A3. If the second reagents, B1, B2 and B3, are added to these vessels, then each will contain A1-B1, A2-B1, A3-B1 or A1-B2, A2-B2, A3-B2 or A1-B3, A2-B3, A3-B3. A further mixing separating and reacting with C1, C2 or C3 will produce the 27 products A1-B1-C1, A1-B1-C2, A1-B1-C3, A2-B1-C1, A2-B1-C2, A2-B1-C3, A3-B1-C1, A3-B1-C2, A3-B1-C3, A1-B2-C1, A1-B2-C2, A1-B2-C3, A2-B2-C1, A2-B2-C2, A2-B2-C3, A3-B2-C1, A3-B2-C2, A3-B2-C3, A1-B3-C1, A1-B3-C2, A1-B3-C3, A2-B3-C1, A2-B3-C2, A2-B3-C3, A3-B3-C1, A3-B3-C2 and A3-B3-C3. These molecules can be tested in enzyme assays for biological activity and the problem is then identifying which of the products is the most active. This can cause problems but these may be overcome by putting an indentifiable label on the bead that holds the molecule and checking this after the assay has been run or by the direct identification of the molecule using the advanced analytical techniques now available to the research chemist. The application of robots to this technique has allowed the synthesis of millions of molecules in a few weeks. The robots deliver the chemicals and perform the mixing and partitioning of the solid support.

These techniques allow the synthesis of large numbers of related compounds, which can be screened with relative ease against specific biochemical targets. Whole organism assays are less amenable to this type of screening, but it should be remembered that if a family of related compounds is under evaluation then it is likely that the biochemical mode of action of the compounds will be known. The most active against the target enzyme can be re-prepared in a quantity sufficient for a whole organism test and, in addition, data on a large number of compounds will have been generated, allowing a broadly based patent claim to be filed with supporting data.

4 FIELD SCREENING

All laboratory and glasshouse testing is designed to be rapid, accurate and unaffected by the time of the year it is run. To achieve this all laboratory-based screens are conducted under strict environmental control.

The real life situation is very different from this:

(a) Weed seeds germinate at different times and at different depths in the soil;

(b) Plant pathogen invasions may be rapid, reaching epidemic proportions, or they may be slow to develop;

(c) Insect attack is very dependent on weather conditions and the presence or absence of predatory species.

The first real test of a compound is how it performs in the field. However, as with laboratory-based tests it is important to ensure (a) that as much variability as possible is removed from the field test and (b) that compounds can be selected for development based on their biological efficacy rather than the good fortune of a preferred test site.

Most agrochemical companies have field research stations in all key crop areas. Typically these will be located in the United States of America (the Corn Belt, the mid-South or California), South America, the United Kingdom, France, Italy, Japan, the Philippines and Australia. Field screening tests are laid down under carefully controlled conditions using varieties of crop that are susceptible to the pest or disease under study and often with artificial infestation. Often field misting or irrigation is used to produce conditions that are ideal for pest attack. Weed seeds are often precision drilled to guarantee a stand.

Persistence of a compound can be determined by re-infestation until efficacy is lost.

It is important to include in any field screen the leading member of each chemical series such that improvements in efficacy can be assessed in the same trial. Leading products must also be included to confirm the validity of the trial and to compare a new compound's performance to a market leader.

Most important is the collection and interpretation of results. All too often some data are ignored because they do not meet the expectations of the experimenter or others are emphasized because they do. All data must be reported and assessed and a decision taken on the overall reliability of each product candidate. The worst decision that can be taken on any new field trial compound is that another season's trial is needed. It is important that the field screen permits the recommendation that the compound be dropped or elevated to development status.

5 FIELD DEVELOPMENT

Up until the time that a compound is recommended for developmental status it has been the responsibility of the research group. Once the decision is made to upgrade a compound, continued development, although organized by research, becomes the responsibility of each company region. This is clearly the correct procedure. Each country region has the responsibility for marketing a new product and it is the technical support officers of each region who are aware of the market needs for their particular area.

These trials represent the first time a compound is tested in a real "on-farm" situation and are frequently conducted on a crop sown by a collaborating farmer. Because it is not possible to predict the occurrence or intensity of pest attack, a large number of replicated trials must be conducted. It is important to determine the effective rate of use, the preferred formulation, the optimal timing of application, the strengths of the new compound in comparison to the standard and the weaknesses. It is only when these data are available that a decision can be made as to whether to proceed to commercialization and, if so, how to market the new product.

Research and field personnel should assure themselves that necessary staff functions have been performed. Patents should have been filed before a compound was selected for off-station testing. Decisions also should have been made as to whether or not mixture patents are of commercial value. A very important exercise is determining the probable manufacturing cost of the new compound. If a compound costs $100/kg and the market will stand a product cost of $25/ha, it is clear that a rate of use above 200g/ha would generate no profit while a rate of 100g/ha, or less would be acceptable. It is important to build into the manufacturing cost the price of formulation and packaging. In addition it is of fundamental importance that the toxicity of a compound be fully evaluated prior to commercialization.

6 REGULATORY REQUIREMENTS

As soon as a compound shows sufficient biological efficacy to justify field screening, acute oral and dermal toxicological data are generated. The field trial represents the first preparation of the compound in relatively large quantities and it is important to alert personnel from manufacturing, formulations and field evaluation of the potential toxicological hazards.

As the compound progresses through biological evaluation to a position of being a potential commercial compound, a full programme commences to generate basic data to support the development of the compound and to satisfy the regulatory authorities where commercialization is expected. These data are wide ranging and cover basic physicochemical properties, toxicity testing, environmental fate, ecotoxicity and crop residues.

Before a compound can be sold in any country, adequate data must be provided to show that the product is safe with regard to:

(a) Operator handling, mixing and spraying;
(b) Consumption of the treated crop;
(c) Environmental effects, long and short term;
(d) Crop effects (usually part of biological efficacy testing).

The data produced to satisfy national registration authorities are also used to assess the safety of the compound to research workers, manufacturing personnel and any other people who may come into contact with it during development, manufacture and, ultimately, commercialization. As emphasised earlier, the trend today is to concentrate development on compounds showing potential in major crops or against major pest species. This means that safety data are developed on a global basis, ensuring that the requirements of all the world's major regulatory authorities are considered. Countries such as Australia, Canada, Germany, Japan, the United Kingdom and the United States produce detailed guidelines on data required for the registration of a pesticide. These are supported by those generated by international organizations such as the Council of Europe, the Global Crop Protection Federation formerly known as the International Group of Associations of Manufacturers of Agrochemical Products (GIFAP), the Food and Agriculture Organization of the United Nations (FAO) and the World Health Organization (WHO). A typical timescale for the generation of regulatory data is shown in Table 1.8.

7 CONCLUSION

The discovery and commercialization of a new crop protection agent is a long-term, multidisciplinary project. It has been estimated that it now requires the synthesis of 40,000 new compounds to produce one new product. The cost of screening out the 39,999 compounds which were not quite good enough is now put at over $60 million. In addition to this is the cost of building a new manufacturing facility or modifying an existing plant. Typical timescales and expenditures are shown in Table 1.9.

It is of prime importance that any project that is initiated is based on chemistry that can be patented and with a biological effect that

Table 1.8 Typical regulatory data schedule

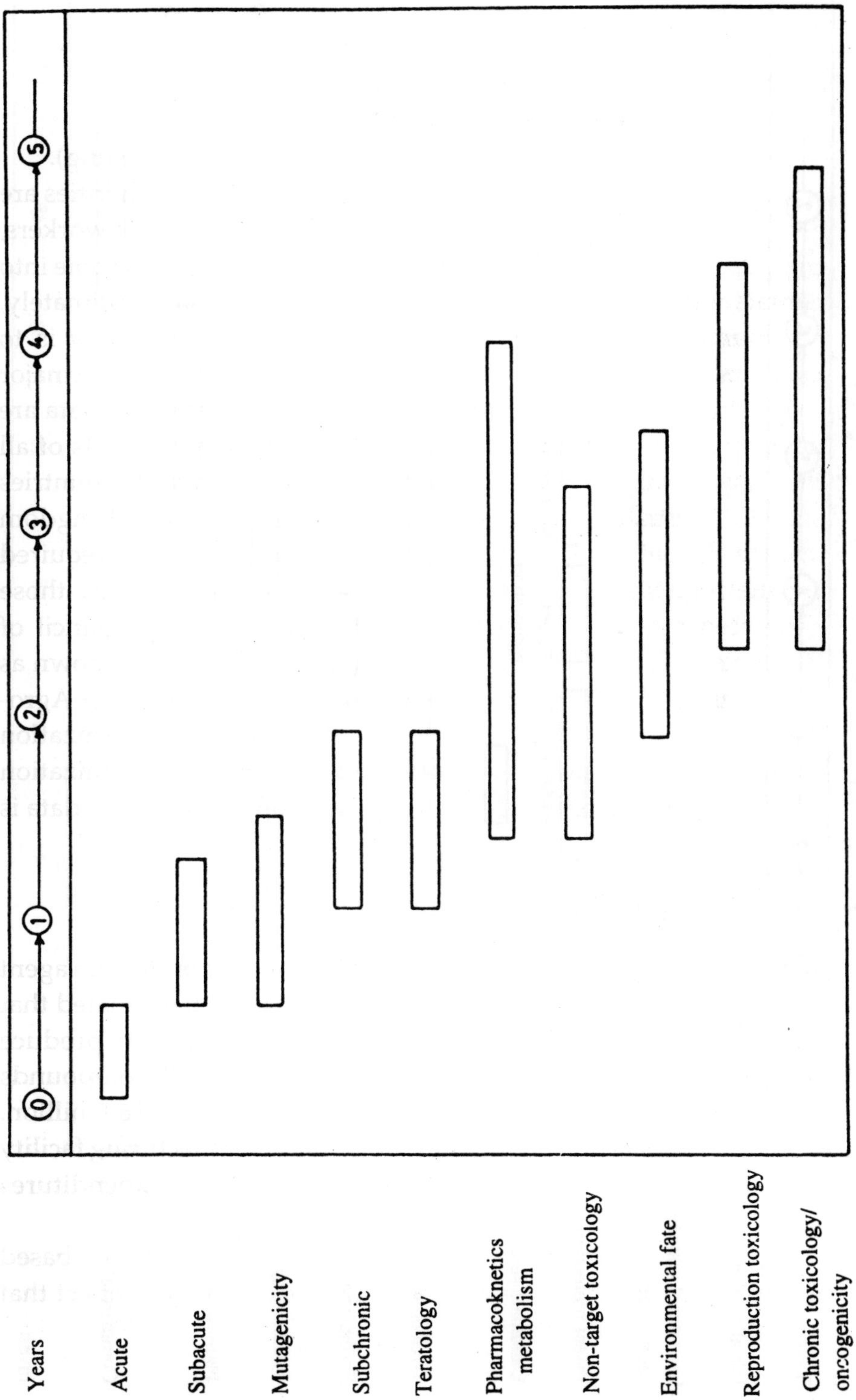

Table 1.9 Typical schedule for discovery to commercialization

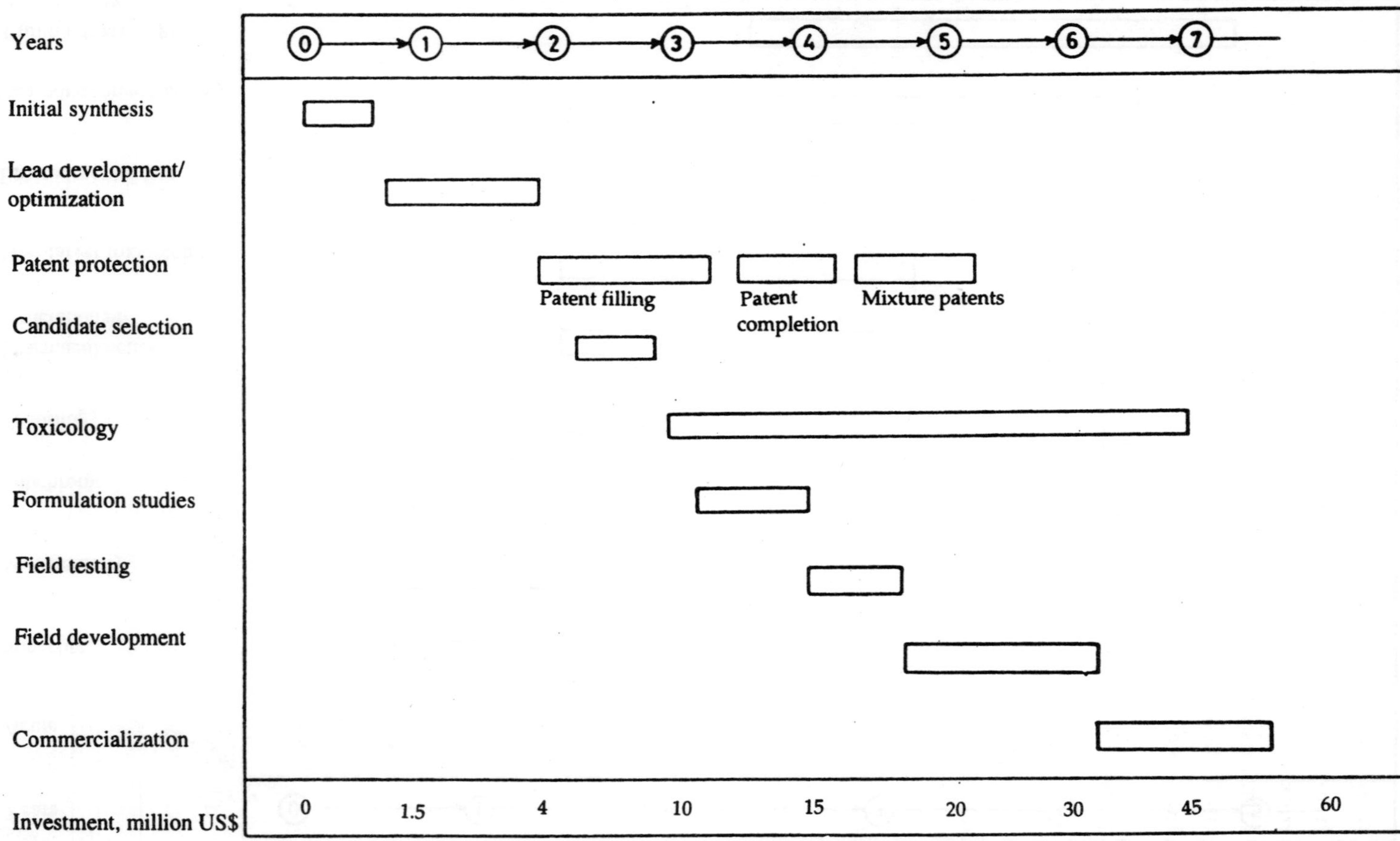

fulfils farmers' needs. Each decision taken in the 8-10 year timescale from discovery to commercialization calls upon the dedication of more resources to the project. Clearly, at the end of the day, the farmer buys cost-effective, reliable pest control. If any project is undertaken without the awareness of the final customer and the factors that determine his product selection, it is unlikely to succeed.

BIBLIOGRAPHY

(1) Buechel, K.H., ed. *Chemistry of Pesticides*. Wiley, 1983.
(2) Copping, L.E., ed. *Crop protection agents from nature: Natural products and analogues*. Royal Society of Chemistry, 1996.
(3) Corbett, J.R., K. Wright and A.C. Baillie. *The biochemical mode of action of pesticides*. Academic Press, 1984.
(4) Eliel, E. Stereochemistry of carbon compounds. McGraw-Hill, 1962.
(5) Pesticide science. Special issue, 29, November 1990.
(6) Selected publications, 1978-1987. Jealott's Hill Research, ICI Agrochemicals, 1988.
(7) Sugavanam, B. Recent developments in the chemistry of pesticides. UNIDO workshop report, 1989.
(8) *Synthesis and chemistry of agrochemicals*. D.R. Baker, J.G. Fenyes, W.K. Moberg and B. Cross. American Chemical Society, 1987.
(9) Van Valkenburg, W., ed. Pesticide formulations. Marcel Dekker, 1973.

CHAPTER

2

Pesticide Mixtures

Stefan Mosinski

Pesticide Formulation Laboratory
Instytut Przemyslu Organicznego
03-236 Warszawa-Zeran, Poland

1 INTRODUCTION

The use of pesticide mixtures is common, and their application will be growing in the near future for a number of reasons, which include improved biological performance and decreased costs of application. But there are other economic and environmental factors that are of critical importance.

By simultaneously applying more than one pesticide in a mixture, the applicator saves time, fuel costs and labour. At the same time he also reduces equipment wear and losses caused by mechanical damage to the crops and soil.

Using pesticide mixtures, one can combat a broad spectrum of pests. Different species of insects, pathogens of plant diseases and weeds can be controlled in a single application. Frequently the economic and effective control of an undesirable species can be achieved by using a mixture prepared from one relatively inexpensive component and one more expensive one as opposed to exclusive use of only the latter component.

Another important aspect of the usage of pesticide mixtures is synergism of two or more components in a mixture. In such a case the observed joint action of a multicomponent mixture is greater than the expected response calculated by adding the effects induced by individual components of the mixture.

An additional reason for using pesticide mixtures is the necessity to minimize the selection pressure, which may cause an increase in the population of resistant strains of pests and to delay the dominance of resistant strains as a result of the long-term use of some particular pesticides.

2 PESTICIDE INTERACTION TERMS

2.1 Compatibility and Incompatibility of Mixture Components

The basic prerequisite for high effectiveness of pesticide mixtures is the compatibility of their components. The term "compatibility" means that all components of the mixture are in harmony with one another (1). There is no disadvantageous interaction between them which could decrease the expected efficacy of the mixture or cause phytotoxicity and/or toxicity to warm-blooded organisms, including man.

Reduced mixture performance can result from chemical or physical interactions as well as from biological incompatibility. Chemical incompatibility results from hydrolytic and photochemical degradations and interaction with pesticide components including solvents, surfactants, fillers and other types of mixture ingredients. The physical properties such as melting and boiling points, volatility, solubility and other less clearly defined physical parameters may adversely affect the handling properties and performance of mixtures if the ingredients are not compatible.

Physical incompatibility is evidenced by formation of agglomerates, crystals, phase separation of emulsions or suspensions and thickening of spray liquids such that the spray application cannot be performed accurately or effectively.

The chemical and physical compatibility of the components of a mixture will be inadequate for high performance of the mixture if, at the same time, the components are not biologically compatible. Such incompatibility can happen if one component of a mixture is able to block the biological activity of the other. In such a case, a mixture can be less effective or even completely inactive.

Two other kinds of incompatibility are phytotoxic incompatibility and toxicological incompatibility. The former is used to describe the case when desirable plants are injured by pesticide mixtures and the latter, when the toxicity to warm-blooded organisms is higher than had been expected from the toxicity of individual components of the mixtures. Phytotoxic and toxicological incompatibilities may be induced by the chemical or physical incompatibility of the components as well as by biological incompatibility.

2.2 Synergism, Antagonism and Additivity

The efficacy of a pesticide can be increased by a totally inactive component or by another pesticide. Such a phenomenon is termed the "synergistic" action of the second component. Synergism is more precisely defined by the Weed Science Society of America (2) as an interaction of two or more pesticide components of a mixture such that the combined effect is greater than the predicted one based on the response to each component separately. This synergism is generally interpreted as a biological summation and not as an arithmetic addition of the biological effects of individual components.

A phenomenon opposite to synergism is antagonism, which is defined as a response to the joint application of two or more pesticides that is less than the response anticipated.

The third term, additivity, is defined as the combined action of two biologically active components, that turns out to be equal to the response predicted by adding the responses of each pesticide component if applied separately (3).

The above three definitions can be pictured graphically by the isobol method, which is illustrated for herbicides in Figure 2.1 (4). The term isobol designates a line passing through points of equal action or injury, e.g. a series of ID_{50} values resulting from administering two compounds in different ratios. The solid straight line is the isobol of a precisely additive performance at all dosage ratios. The line above this straight solid one represents some degree of antagonism and the line below the straight line represents some degree of potentiation.

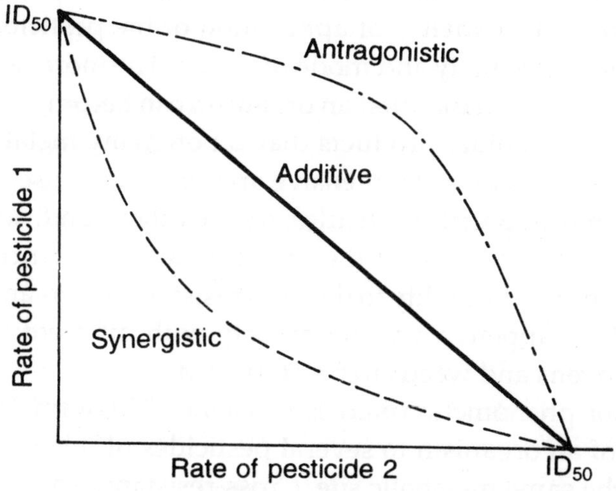

Figure 2.1 Types of plant responses to mixtures of pesticides expressed graphically

From the practical standpoint, synergism is the most desirable response because it enhances the activity of pesticides. By analogy, one can say that antagonism is the only response which is undesirable because it would reduce the efficacy of pesticides.

2.3 Pesticide Resistance

Pesticide resistance is the phenomenon of the loss of potency of a pesticide within a few years, sometimes even sooner, due to the development of a resistance gene in the pest (5). Resistance, in this context, refers to the loss of or decreased control over a pest population resulting from the use of chemicals for its control. Resistance of insects and mites to insecticides and acaricides, resistance of pathogens to fungicides and resistance of weeds to herbicides constitute the most serious and widespread aspects of the resistance phenomenon (6). It is the same sort of natural response shown by any life form in the face of an applied selection pressure. The pesticide kills the susceptible strains of pests or pathogens and if resistant individuals exist in the population, they survive. If the pesticide is used too intensively, the resistant strain becomes predominant, leaving the grower with a problem because he is unable to protect crops effectively by using a particular class of products. The rate at which resistant strains are likely to develop depends on the intensity and frequency of application of the pesticide and on the pesticide's specificity and mode of action. The more nonspecific a pesticide, the less chance that an organism can become resistant to it. However, even older products that act on many metabolic sites may induce resistance after intensive and prolonged use.

Modern, more specific pesticides, on the other hand, act only on one or two metabolic sites of the organism and resistance appears at an earlier time. The ability to develop resistant strains and the rate at which this happens vary enormously with different species of pests, pathogens and weeds to be destroyed.

A common phenomenon is cross-resistance. This term denotes the resistance of an organism to several pesticides of related structure acting on the same metabolic site. Cross-resistance can be induced after prolonged usage of only one of the related pesticides.

The relative abundance of insecticides in the 1950s and 1960s encouraged the wasteful, sequential usage of different insecticides to control resistance. This induced a multiplicity of resistance mechanisms within a given insect species, resulting in broad cross-resistance covering insecticides of many chemical groups (7).

From what has already been said, it follows that the key method of avoiding or delaying resistance is to minimize the selection pressure exerted by pesticides, thereby prolonging their useful life. The most effective way to do this is to establish an acceptable level of pest control (perhaps less than 100% control) and then apply the required amount of pesticides in a most efficient manner. If, however, resistance is detected, another pesticide (not related to the previous one) or an effective mixture of pesticides should be used.

The desirability of using pesticide mixtures to delay resistance is generally accepted in the case of fungicides and herbicides. However, in the case of insecticides the opinions of entomologists vary. There is a body of opinion that if a pest is controlled strongly with a combination of two insecticides, it is apt to develop resistance simultaneously against both insecticides. Persons who support this opinion do not recommend usage of mixed formulations for insect control.

3 KINDS OF PESTICIDE MIXTURES AND SOME OF THEIR PROBLEMS

Pesticide mixtures may be prepared by growers in the spray tank in the form of so-called tank mixtures or, alternatively, prepared by producers in the form of so-called prepacked mixtures or multi-pesticide formulations (MPFs).

The tank mixture is an aqueous dilution, brought to a specific use concentration suitable for application, utilizing two formulations of the individual compounds designed to be diluted with water. However, individual formulations may not be compatible. Each of them should be accurately adjusted to match physical and chemical properties. This requires, with respect to each active component, a special selection of the solvents and the systems of auxiliaries such as emulsifiers, dispersing agents, wetting agents and thickeners. The incompatibility of individual formulations to be employed may cause several difficulties, including separation, flocculation, agglomerate formation (resulting in the spray nozzles becoming blocked) or coagulation.

All of the mentioned drawbacks of tank mixtures can be eliminated by the use of MPFs comprising two or more individually active compounds and use of compatible auxiliaries and formulation ingredients.

The advantages of MPFs are the following:

(a) All components are compatible and there is no anxiety of bad behaviour of spray liquids prepared from MPFs;

(b) The proportion of active components is precisely chosen to gain the best biological result;

(c) The preparation of liquid sprays from MPFs is much easier and less troublesome than the preparation of tank mixtures;

(d) The producer of MPFs is better able to prepare different mixtures of good standards than are farmers, who prepare tank mixtures.

It is relatively simple to prepare MPFs of such formulations as wettable powders (WP), emulsifiable concentrates (EC) or water-dispersible granules (WG); more complicated to prepare are suspension concentrates (SC) and emulsion concentrates (EW). The most difficult to prepare are MPFs of the suspo-emulsion type. (Abbreviations are in accordance with *The Pesticide Manual*, 8th edition, 1987.)

Examples of some of MPFs are as follows:

Solubility characteristics of pesticide component		Possible MPFs	
(i)	Solid pesticide (SP) insoluble in water (P_1)	ia)	ECs of P_1 and P_3
	SP insoluble in organic solvents (P_2)	ib)	SC of P_2 in EC of P_3
	SP soluble in organic solvents (P_3)	ic)	SC of P_1 in EW of P_3
(ii)	SP insoluble in water (P_1)	iia)	SC of P_1 in SL of P_2
	SP soluble in water (P_2)	iib)	P_1 solubilized in SL of P_2
(iii)	Liquid pesticide (LP) insoluble in water (P_1)	iiia)	Emulsion of P_1 in SL of P_2
	SP soluble in water (P_2)	iiib)	SC of P_2 in EC of P_1
(iv)	SP insoluble in water (P_1)	iva)	SC of P_1 and EW of P_3 in SL of P_2
	SP soluble in water (P_2)		
	LP insoluble in water (P_3)		

It is apparent from the above list that some MPFs are multiphase systems in which at least one phase is a liquid. In such systems critical attention must be given to the method of manufacture, formulation approach, component materials selection and the effect of environmental factors (temperature and holding-time). These factors profoundly affect the variability in the product's performance, stability characteristics and a host of other properties.

There are many theories and experimental works dealing with disperse systems. Unfortunately most of them do not have a direct relation to the practical problems met in the production or usage of such systems. But, there is a large body of knowledge that, on proper use, enables the formulator to prepare better products more efficiently and evaluate them in a more systematic and realistic manner.

Problems associated with dispersed liquid pesticide formulations are primarily related to their nonequilibrium state. Each phase of such a system continuously seeks to reach thermodynamic equilibrium. In the case of an emulsion, equilibrium is represented by a completely coalesced emulsion. In the case of a suspension, equilibrium is synonymous with a settled, non-redispersible suspension. If both systems attain thermodynamic equilibrium, the products, for the most part, are of no value. Since, however, the final state of equilibrium is inevitable, the task of pesticide formulator is merely to delay its occurrence. Basically he must attempt to make the product as independent of processing variables as possible and to introduce the necessary energy barriers to delay thermodynamic equilibrium to achieve the desired shelf-life of the product. The concept of thermodynamic equilibrium, coupled with an understanding of free energy, constitutes a framework for design of the disperse systems of interest.

4 MPFs ANALYSIS

It is obvious that chemical analysis of MPFs is more complicated than the analysis of a formulation containing a single pesticide. The additional difficulties appear when the MPFs are multiphase formulations, e.g. in the case of SCs, EWs and suspo-emulsions. The most critical chemical analysis concerns the quantitative determination of the active ingredients of a composition of MPF to see if they are in accordance with the nominal composition given by the producer.

In order to determine what pesticide components are mixed together, their separation from the mixture is most often required. This can be performed using conventional techniques, especially in the case of multiphase MPFs.

Depending on the physical and chemical properties of the pesticide components, several methods of analysis must be considered. For example, in the case of three-component MPFs, the analytical methods should involve titration, an ultraviolet (UV) spectrophotometric method and a chromatographic method (either gas liquid chromatography (GLC) or high-pressure liquid chromatography (HPLC)). For the GLC method the components must be volatile and stable at the relatively high temperature of the determination. Nonvolatile compounds may be transformed into volatile derivatives, and the reaction must be quantitative, relatively fast, repeatable and reproducible. The modern on-column derivatization using N-methyl-N-trimethylsilyl trifluoroacetamide or trimethyl anilinium hydroxide may be carried out, as well as classical silylation or methylation. For the low-volatility pesticides (such as pyrethroids) or nonvolatile (such as phosphonomethyl glycine) and temperature-labile pesticides (majority of substituted ureas), HPLC is a more commonly used technique (8).

The lack of modern analytical equipment in local analytical laboratories, particularly in proximity to manufacturing operations in developing countries, is the main obstacle to performing quantitative chemical analyses of MPFs. For that reason, such formulations are seldom used in these countries.

5 TOXICOLOGICAL ASPECTS OF MULTICOMPONENT PESTICIDES

When combining pesticides, one must be concerned about a possible potentiation of the toxicity of the pesticides to warm-blooded organisms. The toxic effects of the pesticides may be additive, synergistic or antagonistic. For that reason, apart from the hygienic-toxicological investigations which are carried out on individual pesticides, several additional experiments should be conducted.

The fundamental test is the determination of an LD_{50} from measurements of the acute oral toxicity. From this value, the coergistic coefficient (J) is calculated for both individual pesticides and mixtures using the following formula:

$$J = \frac{LD_{50} \text{ predicted}}{LD_{50} \text{ experimental determination}}$$

where LD_{50} predicted is

$$\frac{\sum_{n=1}^{n} LD_{50}}{n}$$

and n is the number of pesticide components. In the same way one can determine the acute inhalation toxicity LD_{50} and also its coergistic coefficient J.

When coefficient J indicates an additive or antagonistic action of a mixture, further investigations are unnecessary. In the case of a synergistic effect, however, further tests are necessary. Measurements should be made of the subacute or subchronic toxicity of the mixture paying special attention to the toxicological profile and the mechanism of action. The results obtained on the mixture are compared with the summarized effects of the individual components.

The objective of the above tests is not only to discover the possible coergistic effect but also to determine the ineffective dose (NOEL = no effect level) and the minimum dose value corres-ponding to a visible effect (MEL = minimum effect level).

The toxicological results obtained from the mixture of half NOEL doses of pesticide A and B, as well as from the mixture of full MEL doses of these pesticides, are next compared with the corresponding results for the individual components.

If no enhancement of toxicological effects is found, it should be concluded that no extra additive action exists. This concentration can be accepted as a joint NOEL. If toxic effects are found, it should be concluded that this is the outcome of the mutual influence of the components on toxic effects.

If no NOEL is obtained for a mixture in this test, further investigations should be conducted to explain the coergistic effect (9).

6 EXAMPLES OF PESTICIDE MIXTURES

The numbers of individual formulations and MPFs available on the European market in 1989 are given in Table 2.1. The data show that the use of pesticide mixtures is often more advantageous than the use of single ones.

Table 2.1 Individual and MPFs available on the European market in 1989 (10)

Kind of pesticides	*Number of all formulations*	*Number of formulations containing the following number of pesticides* 1	2	3	4	5	6
Insecticides and acaricides	584	183	270	101	24	3	3
Fungicides	730	163	358	172	30	3	4
Herbicides	728	193	313	166	47	6	3

Mixtures of pesticides can lower the required combined doses if the mode of action of the pesticide components is additive. Furthermore, when a chemical mixture exhibits synergism, an even lower dose of the chemicals may be considered.

In spite of the apparent advantages of pesticide mixtures, their ultimate value must be measured in terms of a derived economic advantage to their user.

Examples of recommended MPFs of particular kinds of pesticides are given below.

6.1 Insecticides

Most of the organophosphates and carbamate insecticides that have been commonly used are ineffective against some insects, e.g. against green rice leafhopper. However, it has been observed that some organophosphates like malathion and diazinon have synergistic insecticidal activity in combination with carbamate insecticides such as metolcarb (MTMC), fenbucarb (BPMC), xylylcarb (MPMC) and isoprocarb (MIPC) resistant green rice leafhopper (11). Some of these mixed insecticides have been put on the market in Japan and are widely used for controlling resistant strains (12). The synergism found is believed to be due to inhibition by the carbamate of enzymes that decompose the organophosphate insecticide.

The other MPFs recommended by various producers against green leafhopper are chlorpyrifos methyl + dimethoate, chlorpyrifos ethyl + cypermethrin and malathion + fenitrothion + cypermethrin.

6.2 Fungicides

Fungicide mixtures are used to broaden the spectrum of activity of a product or to minimize the selection pressure inducing the formation of resistant strains. Examples for the enlargement of the spectrum are (a) triadimefon + carbendazim for controlling powdery mildews, rust and eyespot in cereals and (b) metalaxyl + mancozeb for activity against *Phytophthora* and *Alternaria* in potatoes. To delay the build-up of resistance, mixtures of fungicides with different modes of action are used, for example (a) benlate + captan against apple scab, (b) dicarboximides + folpet against *Botrytis* in grapes and (c) mixtures of phenylamide fungicides with suitable partners against potato late blight (13).

Demethylation inhibitors (DMIs) in combination with non-DMI fungicides are recommended for the control of apple scab (*Venturia inaequalis*), vine powder mildew (*Uncinula necator*) and cucurbit powdery mildew (*Sphaerotheca fuliginea*) (14).

Seed dressing against pathogens may be based on mixtures of thiram + carboxin for wheat, barley, oats, rice, cotton, soybeans and dry beans.

6.3 Insecticide/fungicide

There are numerous seed dressing formulations containing a combination of insecticides and fungicides. One example of such a combination used against wheat bulb fly and *Anthracnose* on beans is based on thiram + dichlofenthion.

6.4 Herbicides

The simultaneous application of more than one herbicide in a mixture is a standard practice in most weed control strategies. Historically the first cases of synergism were with hormone herbicides. Examples include combinations such as (a) 2,4-D or 2,4-DB and atrazine, which have been reported to be synergistic on yellow nutsedge and on bean and (b) chlorpham, which is active against grass weeds, and picloram, which is active against field bindweed etc. (15).

When the action of components is additive then the dosage level of each herbicide component can be reduced. This is particularly true in the commonly used mixture of 2,4-D + dicamba commonly used for weed control on maize. Reduction in the rate of use of the individual components results in reduced costs of application as well as an increase in crop safety (15).

Mixtures of a large number of highly active grass herbicides with selective broadleaf herbicides offer the potential for total post-emergence weed control in certain crops. A mixture of two sulfonylureas, chlorosulfuron and metsulfuron methyl, is a recent commercial mixture used on wheat. It utilizes the complementary strategy of components singly as well as their additive action on the more difficult-to-control broadleaf weeds.

A problem associated with a number of post-emergence herbicides, especially the wild oat herbicides, is that they must be applied at a particular growth stage. Mixtures of herbicides that are effective at different growth stages offer a way to reduce this growth stage specificity. A combination of barban and benzylprop or the methyl ester of flamprop were shown to have this effect (15). Fast-acting herbicides like paraquat are often used with long residual soil-sterilant herbicides to provide both rapid and long-term weed control (15).

REFERENCES

(1) H.F. Johnson and H.E. Kaldon, "Compatibility of pesticide tank mixtures," *Proceedings of the Second International IUPAC Congress of Pesticide Chemistry,* vol. 5, 1972, p. 485.

(2) Weed Science Society of America Terminology Committee, *Weed Sci. Suppl.* 33, 1985, pp. 1-23.

(3) J.L. Flint, Paul L. Cornelius and Michael Barret, "Analysis of herbicide interactions: a statistical treatment of Colby's method", *Weed Technology,* 2, No 3, 1988, p. 304.

(4) P.M.L. Tammes, "Isobols, a graphic representation of synergism in pesticides", *Neth. J.Plant Path.,* 70, p. 73.

(5) A.W.A. Brown, "The challenge of insecticides", *Bull. Ent. Soc. Am.,* 7, p. 6.

(6) F.P.W. Winteringham, "FAO international collaborative programme for the development of standarized tests for resistance of agricultural pests to pesticides", *FAO Plant Protection Bulletin* 17, No 4, 1969, p 73.

(7) M. Sawicki, "Strategies for countering insecticide resistance", *Shell Agriculture,* 1, 1988, p. 19.

(8) L. Ogierman, "On-column esterification and N-methylation carbetamide with N,N,N-trimethylanilinium hydroxide", *Pestic. Sci.,* 25, 1989, p. 155.

(9) R. Knapek, "Toxicological aspects of multicomponent pesticides", Warsaw, *Pestycydy, Special Issue UNIDO Workshop,* 18 September 1989, No 2-3, 1990, p. 181.

(10) *European Directory of Agrochemical Products,* 3rd ed., Hamish Kidd and others, (Royal Society of Chemistry, 1989).

(11) Iwata Toshikazu, "Effect of pesticide combinations on the development of resistance in green rice leafhopper", *Japan Pesticide Information,* 39, 1981, p. 3.

(12) Yoji Takanashi, "Present status of insecticides for controlling the resistant green rice leafhopper", *Japan Pesticide Information,* 36, 1979, p. 22.

(13) U. Gisi, H.Binder and E.Rimbach, "Synergistic interactions of fungicides with different modes of action", *Trans. Br. Mycol. Soc.,* 85, No.2, 1985, p. 299.

(14) Mark, Wade, "Fungicide resistance: putting the problem in perspective", *Shell Agriculture,* 2, 1988, p. 11.

(15) Jerome M. Green and Steven P. Bailey, "Herbicide interactions with herbicides and other agricultural chemicals", *Methods of Applying Herbicides,* C.G. McWhorter and N.R. Gebhardt, eds., WSSA Monograph 4, Weed Science Society of America, 1987, p. 37.

CHAPTER 3

Integrated Pest Management

Chalerm Sindhusake

Rice Entomology Group, Entomology & Zoology Division
Department of Agriculture, Bangkok, Thailand

1 INTRODUCTION

The philosophy and principles of integrated pest management (IPM) have been known for over 40 years and the procedure has been operational for many crops in developed countries (1). IPM was developed as a response to the total reliance on chemicals for insect control, which, as can be seen from chapter 2, causes resistance. The concept originated with the discipline of entomology, evolving from the earlier term, "integrated control," which originally meant using biological and chemical means for controlling insects (2). Although the IPM concept infers all disciplines of pest control and the associated resistant varieties of all pests, insect pests are the major target of most current IPM programmes. In practice, IPM implies a need-based application and the judicious use of pesticides in the context of a best-mix of control methods for a prolonged period. In the IPM system, the decision to use a pesticide is made after (a) determination of the magnitude of the pest problem, (b) application of sound technology for pest intensity assessment and sampling, (c) determination of yield losses caused by the pest and (d) determination as to whether these losses exceed certain threshold levels. In anticipation of implementing an IPM system, surveillance programmes have been implemented in several South-East Asian countries to provide an early warning of the spread of pests. This surveillance programme provides a safety period for taking corrective actions.

Yield losses due to pests and diseases are estimated at about 9%

in developed countries, while around 38% of the crops are lost in developing countries (3). These differences in yield losses are caused by the difference in infestation pressures associated with specific climatic conditions, pests and plant growth stage and other factors. In this review we shall describe an analysis of losses and a methodology to establish an economic threshold or action threshold for an IPM programme. Economic losses are of particular importance to developing countries.

IPM advocates that reasonable estimates of pest and pathogen levels can be made only through field observation. This has been a challenging task because there is a lack of knowledge of sampling methods for common pests and diseases that cause losses to crop production. Pest management programmes have contributed significantly to developing such sampling methods and to the training of the field observers who have responsibility for making reliable assessments of pest and disease severity.

2 DEVELOPMENT OF INTEGRATED PEST MANAGEMENT AND ITS IMPLEMENTATION

2.1 Crop Loss Assessment and Threshold

A pest management system can be developed to meet immediate requirements for a practical control programme without detailed ecological studies. However, to improve control strategies a thorough knowledge of a given pest population level and population dynamics is needed (4). Crop loss assessment is one of the important components of IPM. Then, a complete knowledge of the efficacy of pesticides is required to determine the necessity for chemical control in the light of economic damage thresholds and control thresholds. The effect of a pest or disease attack on yield is to be quantified and to be described by a mathematical model.

Walker (5) describes six main relationships between yield and pest infestation. The most common is the generalized model of the damage function worked out by Tammes (6). Crop losses in rice may be determined experimentally by field trials. However, precise measurements are difficult because of the dynamic nature of the crop, the pests, natural enemies and environmental interactions.

The occurrence of multiple pest species further complicates field trial measurements.

Many experimental approaches are available for the determination of crop loss data (7). Artificial infestations often induce different levels of pests in the greenhouse or fields. Egg masses or larvae may be placed on crops. Caging is necessary with highly mobile species or stages. The effect of caging on yield should be determined by comparing yield obtained in the absence of induced pest infestation with and without cages.

The effect of pesticide treatments on yield loss should also be studied. Crop yields may vary with the compounds used and the rate and frequency of application. A given pesticide regimen may result in different degrees of pest attack resulting from reduced pest populations or increased populations due to resurgence effects. Only those pesticides should be selected which have little or no direct effect on crop yield.

The problem of interplot interference due to the effect of a given pesticide on pest populations in adjacent plots is mentioned by Walker (8). Bias arising from these effects can be reduced by using large plots but assessing pest intensities and yield only in the central portion of the plot. One may also use guard rows as a barrier between plots. However, this is not a significant problem. After all, the objective of the experiment is not an accurate measurement of the effect of control measures but an assessment of the degree of yield reduction caused by various pest population levels. If pesticide treatments result in changes in pest population or cause injury in other plots, these effects will be perceived in the pest assessments that are carried out repeatedly in short time intervals.

Two or more of the above procedures may be needed to define workable thresholds. These thresholds must then be tested in the field over a range of pest densities and environments.

2.1.1 Natural infestation

Single plant or single hill trial: Here crop loss studies require variable pest intensities and corresponding variability in plant/crop yields. The objective is to determine yield as a function of different levels of pest infestation. In natural infestations, pest intensities are never distributed uniformly over a given area. On the other hand, there are many situations where single hills or plants may be used to provide data for characterizing a wide range of pest-loss relationships (9). Using this technique, chemical layout designed experiments and manipulation are not required. The single hill/plant technique uses tags on individual plants, where observations are taken on pest intensities. At maturity, all marked plants are harvested separately and yield per plant is determined. Pest-loss relationships are observed at intervals of time. If there are no damaged plants in the field a desirable yield will result. The analysis of data may be based either on the results of individual plants or on the mean values for data grouped into pest or disease intensity classes. The latter leads to better results in cases of high variability in yield resulting from factors other than pest and disease infestations. However, to reduce the effect of extraneous variables the sampling size has to be large so that a sufficient number of classes as well as entries per class are obtained.

The large amount of data generated in pest-loss relationships is difficult to interpret and use unless it can be synthesized into a quantitative relationship, commonly a model. A useful technique is a least-squares regression fit of statistical models based on minimal variance between the model and data. Regression models may be univariate or multivariate. The independent variables are also called predictors. Commonly, the dependent variable, y, is yield or yield loss while the independent pest variable is x. Common crop loss models are shown below:

$$y = a + bx$$

where y = crop loss and x = pest intensity at one time, and

$$y = a + b_1x_1 + b_2x_2 + \dots + b_nx_n$$

where y = crop loss and x_1, x_2, ... x_n = pest intensity of one species at different plant growth stages. This method of measuring plant responses has the distinct advantage that measurements are made without the effect of chemicals, artificial infestation or damage. Disadvantages of the method result from natural variations due to growing conditions, pest complexes and uncertainty of infestation levels. Partitioning and stratifying the material may help to account for and avoid such variations.

2.1.2 Simulated damage

The cause of plant damage may be misinterpreted as the effect of a pest attack because of the disappearance of plants in a specific pattern. The relationships between amount of damage and yield are easily found but it is difficult to relate pest densities and plant damage. This is due to the fact that under natural infestations other factors may be involved, such as growth stages of the pest, time and length of the pest attack, temperature and rainfall. However, simulation techniques require time at a specific location on plants to mimic exactly the action of an insect attack. Time and location variables will assist in the formation of a pest-loss response curve. Simulated damage was done on wheat cereals, grain legumes, sorghum, rice and mung bean to determine the factors that affect yield (10-12). Than et al. (13) simulated deadhearts and whiteheads in rice, as caused by stem borers, and Chiang (14) made artificial tunnels in maize stems.

Under natural conditions, the complex relationships among various factors in a biological system has inevitably led to simulation as a tool to describe how various factors affect one another in the infection process. Interpretation of a biological process through simulation is now linked to yield loss regression models to provide prediction capabilities. As environmental events change, computer simulations can prove the consequences of that change and interpret the results in relation to crop yield.

In addition to pest intensity, the magnitude of crop loss also depends on the growth stage of the plant at the time of pest attack and on environmental factors. These effects are to be quantified as well. Therefore, loss assessment studies have to be replicated over locations and time of growing seasons and the effect of pest attack has to be determined in relation to the growth of the plants.

2.1.3 Economic threshold

The concepts of economic thresholds and economic injury levels are the backbone of any sound pest management programme. In general, economic injury levels are not fixed in time or space for a given pest or a certain crop. It is not enough to establish one level for a pest on a given crop. Usually the level varies from region to region and from year to year, with variations in the value of the crop, the cost of the treatment and the stage of development of the plant. Certainly, economic variables are a primary consideration.

To compute economic thresholds, one needs a series of paired points which allow regression of yield reduction or population levels. Yield samples must be taken to detect some measurable losses in the quality and/or quantity of the product in plots with various levels of infestations.

2.2 Sampling Technique

The development of sampling plans to determine pest populations has two objectives: research and implementation. Research objectives must determine the degree of data reliability that is needed for implementation. Elements of a research sampling programme include the following: sampling universe; sampling method; sampling unit; sampling size; sampling pattern and sampling timing (15). In most pest management sampling programmes, sampling should be conducted at at least weekly intervals. Sampling devices such as an aerial net traps, are used to monitor immigrant brown planthopper populations in the Malaysian surveillance programme before starting on a full sampling programme (16).

The development of a sampling plan for a particular species also requires knowledge of the dispersion or spatial pattern of the species. There are three types of dispersion patterns which have been identified: random, aggregated and regular. Most of the major pests are found in an aggregated type (17). The dispersion pattern is calculated on the basis of a mean and a variance from the mean; if variance is equal to the mean, greater than the mean or smaller than the mean the dispersion types are random, aggregated or regular, respectively (15).

A sampling plan for practical pest management is critical in the decision-making process. The sampling methods reduce the sheer tedium of sampling and reduce the number of samples that must be taken. Several approaches have been used to replace the tedious counting methods with presence-absence sampling.

Presence/absence sequential sampling of *Heliothis armigera* was developed to replace a conventional samplirg method utilizing an economic threshold determination associated with 20 larvae/100 cotton plants in 0.8 ha in Thailand (18) (Table 3.1). Although sequential sampling systems have not been well worked out in many pest management systems, they will be useful in instances where pest populations are either very high or very low because they save time and labour costs. Furthermore, sequential sampling

Table 3.1 Presence/absence sequential sampling of *Heliothis armigera* (Hubner) in cotton (18)

Number of cotton plants in 0.8 ha	*Cumulative no. of larvae*	
	No treat	*Treat*
5	0	3
6	0	3
7	0	3
8	0	4
9	0	4
10	0	4
11	0	5
12	0	5
13	0	5
14	0	6
15	0	6
16	0	6
17	0	6
18	0	7
19	0	7
20	0	7
21	0	7
22	0	8
23	1	8
24	1	8
25	1	9
26	1	9
27	1	9
28	1	9
29	1	10
30	1	10

systems utilize a knowledge of the economic threshold to determine if further sampling is required before a pest management decision should be made. For instance, in Table 3.2, the sequential sampling of rice stem borers from economic threshold data corresponds to a proportion of 0.68 infested hills. The highest infested level that can be tolerated without control measures was set at 0.50 (50%) of the infested hills. The maximum number of sample units was estimated to be approximately 23.

Table 3.2 Sequential sampling of rice stem borers in transplanted rice, Supan Burl Rice Experiment Station, Thailand (19) (a)

Number of hill samples of 0.16 ha	*Cumulative number of damaged hills*				
		≤		≥	
1		–		–	
2		–		–	
3		–		–	
4	N	0	C	4	
5		0	O	5	
6	O	1	N	6	
7		1	T	6	
8		2	I	7	T
9		3	N	7	
10	T	3	U	8	R
11		4	E	9	
12	R	4		9	E
13		5	S	10	
14	E	6	A	10	A
15		1	M	11	
16	A	7	P	12	T
17		7	L	12	
18	T	8	I	13	
19		9	N	13	
20		9	G	14	
21		10		14	
22		10		15	
23		11		16	

(a) Economic threshold established at 50% infested hills.

Circumstances under natural conditions are more complex. For example, a sampling method has been developed which incorporates major predator species as well as the brown planthopper (20). In this case, predators were considered when the population of the brown planthopper was near the threshold. Utilizing this model allowed correct decisions to be made with 80-95% savings in time as compared to procedures that required a fixed number of samples (Figure 3.1) (20).

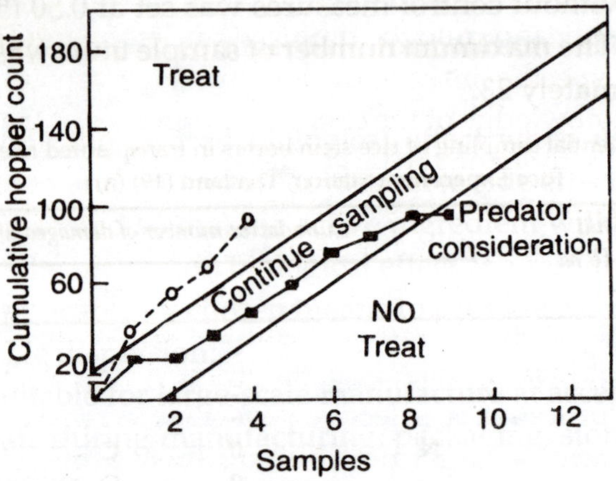

Figure 3.1 Sequel sampling model showing point of decision to treat with insecticide without considering predators or by considering predators (20)

2.3 Pesticides and Their Side Effect on Natural Enemies

Pesticides are an essential part of any integrated pest management programme because they contribute to increasing and maintaining high crop productivity and to saving human lives. Therefore, the use of pesticides has increased rapidly during the last 25 years in Southeast Asia. In Thailand, imports of insecticides rose by approximately 50% during the last 10 years. Half of the insecticides sold were used in rice insect control in the Philippines and in Thailand (21) and 20% were used in Vietnam (22).

A chemical pesticide will continue to be one of the most dependable weapons of the entomologist. However, there are many problems for which the use of chemicals has been ecologically unsound. Disadvantages of pesticides include insect pest resistance, outbreaks of secondary pests, adverse effects on non-target organisms, objectionable pesticide residues and direct hazards to the user (23). The misuse, overuse and unnecessary use of insecticides have been underlying factors in the growth of interest in pest management. Indeed, the pest management concept seeks to maximize the advantages and minimize the disadvantages of pesticides.

The criteria for using insecticides in pest management are complex. Ultimate success depends upon the insecticide selected, the method of application and the timing of the treatment. The informed choice of a specific insecticide for any programme is complex, involving a knowledge of the chemical properties of the compound, its biological activity on the target insect pest, its toxicity to humans and domestic animals, its effects on non-target organisms (crops, parasites, predators, pollinators and wildlife) and its environmental fate in air, water, soil and food.

Most plant protection groups conduct insecticide screening and thereby are knowledgeable about the efficacy of a pesticide. In sound pest control procedures, selective pesticides are preferable. Chemicals that control a target pest but are less harmful to beneficial organisms will be recommended. Thus, it is recommended that additional tests be conducted to investigate the effect of pesticides on the benevolent parasites and predators present in a certain crop. Testing the side effects of pesticides on beneficials involves two important aspects: (a) mortality rates of parasites and predators and (b) reduction in beneficial capacity induced by the chemical, changes in the parasitization rate and by prey consumption and fertility.

The use of an insecticide is not a good tactic for pest management if such use results in pest resurgence or pesticide resistance or if it disturbs the ecosystem balance due to deaths of natural enemies of insects and deaths of other non-target animals.

The main methods for testing and evaluating the side effects of pesticides on beneficial arthropods are summarized by Hassan as well as Heinrichs et al. in the *Proceedings of the FAO/IRRI Workshop on Judicicious and Efficient Use of Insecticides on Rice (24)*. These tests include laboratory, semi-field and field tests. Standardized assessment procedures have been developed by the International Organization for Biological Control of Noxious Animals and Plants (IOBC) to assess the initial toxicity of pesticides to predators and parasites.

The most important aspect of such studies, apart from measurement of direct mortality, is whether those which survive sublethal doses are able to search for a host oviposit. Sublethal doses may also prevent feeding. Thus, all available information on the effects of each chemical on natural enemies needs to be collected when attempting to initiate integrated chemical and biological control.

The impact of a pesticide on the beneficial capacity of natural enemies needs to be considered more carefully and the safest and least persistent chemical should be chosen. Table 3.3 shows that cypermethrin is harmful in the laboratory to *Lycosa pseudoannulata*, a major predator of the brown planthopper. Laboratory tests frequently indicate that a pesticide is harmful to beneficial insects. However, these harmful effects may be decreased in the field by localizing the treatment or by careful timing. Taking these precautions the beneficial species are less exposed to the spray deposits. Obviously, no pesticide should be applied if natural enemies are being released and have not yet become established. Apart from their direct effect on mortality, pesticides can be harmful to natural enemies by interfering with fecundity or searching behaviour, thus reducing their ability to exert biological control.

Table 3.3 Percentage mortality of the spider *Lycosa pseudoannulata* after exposure to an insecticide for 24-72 hr. under a fresh dry film on glass plates (25)

Insecticide	*Mortality (%)*		
	24 hr	*48 hr*	*72 hr*
MIPC	12.5	25	25
Monocrotophos	12.5	25	25
Buprofezin	0	0	0
Cypermethrin	7.5	100	100
Untreated	0	0	0

Semi-field tests are recommended by IOBC for two purposes: persistence tests and initial toxicity tests (24). Depending upon the objectives of the test, the beneficial organisms are exposed either to pesticide residues or to fresh dry insecticide films on plants in cages kept under field or field-simulated conditions. In these trials laboratory-reared arthropods of uniform age are used.

Under field conditions, the effect of pesticides can be measured by the change in population density sampled at intervals before and after pesticide applications. Sampling is dependent upon confined species.

Recently, the insecticide buprofezin, a thiadiazin (26), has been shown to be effective against the nymphs of the brown planthopper *Nilaparvata lugens* but, fortunately, low in toxicity to its natural enemies (27). In the field it will reduce a brown planthopper population and for a quick knock-down can be mixed with a carbamate insecticide.

The list of useful insecticides in a pest management programme can be broadened if care is taken to restrict the area of application and dosage to minimize undesirable side effects.

2.4 Implementation of Surveillance

The concept of IPM has been broadly accepted in developed countries since 1950. However it is still more of an aspiration than a reality in developing countries. The gap between concept and practice is large. One of the major problems is related to the implementation of the technology. Practitioners in developing countries are not knowledgeable about the technology of IPM.

Implementation may be defined as the process occurring after the decision to adopt an innovation has been made. The process includes orientation, initiation and decision, which are then integrated into a routine practice (28).

The majority of rice-growing countries in Asia have pest surveillance programmes. In the Philippines and Thailand, surveillance and early warning systems (SEWS) were implemented for rice during 1978-1988 under the aegis of a German aid programme. A pest surveillance system was implemented in 1979 in Malaysia following an outbreak of brown planthopper (16).

In Thailand, the concept of pest surveillance in rice as a tool in IPM has been practiced on a systematic basis since 1982. The first step emphasized data collection through a pest surveillance system. The surveillance of crop losses in the farmer's field and sampling methods were set up for surveillance technicians, as shown in Table 3.4. Following this the programme shifted to work with a target group of rice farmers.

Education of the user has often been cited as the key to successful implementation. Wearing (29) has confirmed that verbal communication is the most effective way of transferring IPM technology to the user. This may be accomplished by increasing the employment of consultants or in-house field personnel by large farmers or cooperatives. Extension services have been implementing IPM utilizing a training and visit system. These extension services set up demonstration plots at selected pilot areas in Malaysia and Thailand (30, 31).

Communicating IPM implementation is essential for motivating farmers. The main objective has been the technical aspects of surveillance, teaching the recognition of major pests and natural enemies and the judicious and efficient use of pesticides. The use of mass media, such as radio and television, has been inadequate for communicating extension information. Communication has been more successful through farmer centres by means of surveillance demonstration fields, exhibitions, group meetings and the use of selected farmers as information sources in the village. Through the systematic use of communication media, attractive posters and audio cassette dramas, one can successfully motivate farmers to demand IPM (32).

Incentives for adopting IPM rather than conventional means of control include the relative costs, pesticide resistance, environmental issues and grower hazards. The cost advantage is obvious. However, since the decision to use IPM will be made by the farmer (30), the cost advantage alone will be insufficient to promote IPM because the farmer may perceive that the financial risks are too great (33).

Table 3.4 Economic or action threshold and sampling method for rice insect pests in irrigated rice of the Thai surveillance programme

Pest	*Sampling method*[a]	*Economic threshold*[b]	*Growth stage*
Brown plant hopper	Count number of adults/ SU	15(10) adults/SU	Tillering to reproductive
		1.5(1.0) adult/SU in case of virus disease	Seedling to reproductive
Stemborer	Count number of deadhearts/ SU	1.5(1.0) deadhearts/SU 0.8(0.5) deadhearts/SU	Vegetative Reproductive
Rice gallmidge	Count number of onion shoots (galls)/ SU	0.5 onion shoot/SU 0.8(0.5) onion shoot/SU	Seedbed/ seedling
Rice leaffolder	Count number of folded leaves/SU	15(10) folded leaves/SU 9(6) folded leaves/SU	Vegetative Reproductive
Rice thrips	Count number of rolled leaves/SU	10 rolled leaves/SU	Seedling
Rice bug	Count number of bug/SU	0.8(0.5) insect/SU	Milky

(a) SU is sampling unit equals 1 hill in transplanted rice and 10 tillers (stems) in broadcast rice.
(b) First number is for transplanted rice; number in parentheses is for broadcast rice.

2.5 Obstacles to IPM Implementation

The complexity of IPM is accepted as a major obstacle to its implementation. Usually, information generated from research in IPM has not been fully disseminated to the farmers in developing countries. The process needs to be simplified as much as possible, particularly with reference to monitoring method or action and economic thresholds. These technical obstacles may be the greatest in the early stages of IPM implementation and will decline over time (34). In Thailand, IPM will be implemented by farmers if the new technology is simple, compatible, observable and available for trial.

Several authors have recognized that the complexity of IPM technology transfer and the poor educational standards of growers were also obstacles to implementation (23, 24). However, IPM programmes need not be complex and require technologically advanced delivery systems. It is unfortunate that in Thailand, the information supplied to farmers is still too inadequate in quantity and quality to facilitate rapid adoption of IPM (30). Hamelink and Kaewjantuk suggest the production of good quality radio and/or television programmes by public relations organizations. If Thai farmers have an education higher than level 4, IPM implementation is easier for them to adopt (35). However, variation in the willingness of farmers to adopt IPM is well recognized. Because farmers tend to look at the short-range aspects of income, the long-range cost advantages of IPM are an obstacle.

3 FUTURE PEST MANAGEMENT PRACTICES

Research, training and technology transfer are key components in developing effective IPM practices. Rice-growing countries in South-East Asia prefer resistant varieties as the first component in any IPM programme. Because of the simplicity of this new technology, farmers easily accept it. Broad genetic varieties with horizontal resistance will not only be accepted but will also minimize ecological changes.

Pest forecasting based on pest surveillance systems will be utilized in conjunction with advances in microcomputer and communication systems. Basic information on bioecology has already become well developed and farmers now need to be trained to make pest observations.

Chemicals are still in common use in most developing countries. With some of the newer pesticides there are trends towards fewer side effects on natural enemies and less environmental disruption. Proper timing in the application of control measures requires some information from surveillance and forecasting systems. Research on forecasting systems will develop in the future. However, the results of research work on IPM cannot be applied in different geographical regions because the biotic and abiotic interactions affecting pests and crops are often location-specific.

REFERENCES

(1) P.L. Adkisson, "Integrated pest management," *Bull. Entomol. Soc. Am.* 32 (4), 1986, pp. 136 - 141.

(2) K.M. Stern, R.F. Smith, van den Bosch and K.S. Hagen, "The integrated pest control concept", *Hilgaria,* 29, 1959, p. 81.

(3) PANS, *Crop Damage Estimates,* World Food Conference, Rome, 1974.

(4) M.J. Way, "Objectives, methods and scope of control", *Insects: Studies in Population Management,* P.W. Geier and others, eds., 1973.

(5) P.T. Walker, "Crop losses: some relationships between yield and infestation," *Med. Fac. Landbouww. Rijksuniv. Gent* 42 (2), 1977, p 919.

(6) P.M.L. Tammes, "Studies of yield losses: II, Injury as a limiting factor of yield," *T. Pl.- ziekten,* 67, 1961, p. 257.

(7) P.S. Teng, ed., *Crop Loss Assessment and Pest Management,* (APS Press, 1987) p. 27.

(8) P.T. Walker, "Methods for studying the pest/yield relationship and for appraising losses due to insects and mites," *Crop Loss Assessment Methods,* L. Chiarappa, ed., Supplement 3. FAO/Commonwealth Agric. Bur. Farmham Royal (UK), 1981, pp. 73-78.

(9) M.J. Richardson, M. Jacks and S. Smith, "Assessment of loss caused by barley mildew using single tillers," *Plant Pathol.,* 24, 1975, pp. 21-26.

(10) S.D. Wratten and P.C. Redhead, "Effects of cereal aphids on the growth of wheat," *Ann. Appl. Biol.* 84, 1976, pp. 437-455.

(11) E.G. Rubia and B.M. Shepard, "Simulation of insect damage on rice." IRRI Training Workshop on Systems Analysis and Simulation for Rice Production, Los Banos, Philippines, January. 1987.

(12) P. Sepsawat, "Crop loss assessment on major mungbean pests in rice-based cropping systems, *Proceedings of the Sixth Entomology and Zoology Division Conference,* Bangkok, Thailand, 1988, pp. 503-513 (in Thai).

(13) H. Than, G. Arida and V.A. Dyck, "Population Dynamics of the yellow rice borer, *Tryporyza incertulas,* and its damage to the rice plant," IRRI, Los Banos, Philippines, 1976, (mimeo).

(14) H.C. Chiang, "The effects of feeding site on the interaction of the European corn borer, *Ostrinia nubilalis,* and its host, the field corn," *Entomolo. Exp. Appl.* 7, 1964, pp. 144-148.

(15) T.R.E. Southwood, *Ecological Methods, with Particular Reference to the Study of Insect Population* (London, Chapman Hall, 1978), p. 524.

(16) A.C. Ooi, "A surveillance system for rice planthoppers in Malaysia," *Proceedings of the International Conference on Plant Protection in the Tropics,* K.L. Heong and others, eds., (Kuala Lumpur, Malaysian Plant Protection Society, 1982), pp. 551-565.

(17) P.M. Ives and R.D. Moon, "Sampling theory and protocol for insects," *Crop Loss Assessment and Pest Management* (St. Paul, Minnesota, APS Press, 1987), pp. 49-75.

(18) N. Nachapong, D.E. Legg., S. Kittibunya and S. Wangboonkong, "Presence/absence sequential sampling plant for cotton bollworm (*Heliothis armigera* (Hubner)) in cotton, Petchaburi, Thailand," *Proceeding of 1st IPC Workshop,* 1-3 March 1988, p. 413 (in Thai).

(19) B. Heuel-Rolf and P. Vungsilabutr, "Development of simple sampling methods for pest surveillance in rice," *Pesticide Management and Integrated Pest Management in Southeast Asia,* P.S. Teng and K.L. Heong, eds., 1988, pp. 113-122.

(20) B.M. Shepard, E.R. Ferrer and P.E. Kenmore, "Sequential sampling planthoppers and predators for insecticide treatment decision in rice," Second International Conference on Plant Protection in the Tropics, Genting Highlands, Malaysia, 1986, pp. 29-31.

(21) H. Ishikura, "Projected trends in the use of insecticides in rice pest control," *Proc. FAO/IRRI Workshop,* on *Judicious and Efficient Use of Insecticides on Rice,* Los Banos, Philippines 1984, pp. 57-66.

(22) R.P. Bateman and M.J. Watt, "Pesticide use in Vietnam," Fourteenth Session of the Asia Pacific Plant Protection Commission, Jakarta, Indonesia, 5-10 August 1985, *Provincial Agenda,* Stem 8, 1985, p. 14.

(23) E.H. Smith, "Integrated pest management (IPM): Specific needs of developing countries," *Insect Sci. Appl.* 4, 1983, pp. 173-177.

(24) S.A. Hassan, "Testing the side effects of pesticides on beneficial arthropods," *Proceedings of the FAO/IRRI Workshop on Judicious and Efficient Use of Insecticides on Rice,* Los Banos, Philippines, 1984.

(25) R. Pattrasudhi and T. Bhudhasamai, "Side effect of insecticides applied as foliar sprays to natural enemies of major rice insect pests", 1988 (unpublished).

(26) H. Kanno, K. Ikeda, T. Asai and S. Mackawa, "2- Tert-butylimino-3-isopropyl-5-phenylperhydro-1,3,5-thiadiazin-4-one (NNI-750), a new insecticide," *Proc. 1981 British Crop Protection Conference,* Pest and Disease, 1981, pp. 59-66.

(27) O. Kajihara, T. Asai, K. Ikada and S.S. Lim, "Buprofezin, a new insecticide for control of brown planthopper, *Nilaparvata lugens,*" Kuala Lumpur, *Proc. Int. Conf. Plant Prot. in Tropics,* (1982), p. 711 (abstract).

(28) C. Robert-Gray and T. Gray, "Implementing innovations — a model to bridge the gap between diffusion and utilization," *Knowledge: Creation Diffus. Util.* 5, 1983, p. 213.

(29) C.H. Wearing, "Evaluation of the IPM implementation process," *Ann. Rev. Ent.* 33, 1988, p. 17.

(30) J. Hamelink and S. Kaewjantuk, "Factors affecting the success of extension communication strategies used in a rice surveillance program in Thailand," *Pesticide Management and Integrated Pest Management in Southeast Asia,* P.S. Teng and K.L. Heong, eds., 1988, p. 229-234.

(31) A.H. Nik, "Overview of IPM infrastructure and implementation on food crops in Malaysia," *Pesticide Management and Integrated Pest Management in Southeast Asia,* P.S. Teng and K.L. Heong, eds., 1988, pp. 27-30.

(32) M.M. Escalada and P.E. Denmore, "Communicating integrated pest control to rice farmers at the village level," *Pesticide Management and Integrated Pest Management in Southeast Asia,* P.S. Teng and K.L. Heong, eds., 1988, pp. 221-228.

(33) P.G. Fenemore and G.A. Norton, "Problems of implementing in pest control: a case study of apples in the U.K.," *Crop Prot.* 4, 1985, pp. 51-70.

(34) M.E. Whalon and B.A. Crofl, "Apple IPM implementation in North America," *Ann. Rev. Entomol.* 29, 1984, p. 435.

(35) S. Arunchaichana, "Technology transfer, increasing of human capital and agricultural production: case study on rice IPM at Ladlumkaew district," *Pathum Thani*, National Institute of Development Administration, 1987, p. 32 (in Thai).

CHAPTER

4

Pesticide Formulation Design

Sushil K. Khetan

Hindustan Insecticides Limited
Scope Complex, Core 6, 2nd Floor
7 Lodi Road, New Delhi-110003, India

1 PROLOGUE

Formulations have been associated with a plethora of myths. They have been viewed as random mixtures of unrelated components or as product preparations with unusual properties, giving them an aura of exotic alchemy. Such perceptions may have prevailed partly due to the relatively scanty information available on the subject. Arthur C. Clarke, the eminent English science writer, in an empirical law named after him, propounded that any sufficiently advanced technology is indistinguishable from magic (1). Modern formulations often represent the fruits of high technology and could be construed by some to be akin to magic.

In the early stages, formulations evolved as a result of the general need to convert basic materials into products for practical use. The original emphasis was on the use of readily available ingredients and processing facilities. The initial empirical approaches have gradually given way to delineated scientific and rational methodology. In the process, well-thought-out formulation design principles have been established. Various advances in related scientific and engineering disciplines such as colloid and surface chemistry, reaction kinetics and size reduction technology have taken place. The fundamental formulation principles developed have diverse industrial and consumer applications in the fields of cosmetics, pharmaceuticals, industrial oils, rubber chemicals, paints, adhesives, pesticides etc.

1.1 Pesticide Formulations in Perspective

In the context of pesticides, the need for uniform distribution of a small quantity over a large area led to the evolution of some of the early formulations. The first application of pesticides was as dusts, which consist of a powdered blend of a pesticide and a mineral diluent. Thereafter, pesticides sprayed in water became popular and suitable application equipment was developed (2). For this kind of application, aqueous solutions, water-dispersible powders and emulsions evolved. The choice of aqueous solutions is limited to active ingredients that are water-soluble and hydrolytically stable. For water-insoluble solids, wettable powders (WP) became a preferred product. They consist of a micronized blend of an active ingredient, a dispersing agent, a wetting agent and a carrier/diluent. Emulsions were preferred for water-insoluble liquids so that these could be diluted with water. However the formulations suffered from thermodynamic instability; they were the forerunners of more recently developed emulsion concentrates (EW), which are more stable.

Emulsifiable concentrates (EC) which readily emulsified on dilution with water, were developed after the Second World War, when surfactants based on ethylene oxide became available. These formulations were formed by dissolution of an active ingredient in a water-immiscible solvent, usually an aromatic hydrocarbon, along with a blend of anionic and non-ionic emulsifiers. The result is a solvent-based solution, which on addition to water, gives a dispersion of fine oil droplets in water.

WPs, ECs and, to a limited extent, dust formulations continue to be widely used and form the bulk of production in most developing countries. These basic formulations are popular for their ease of manufacture, effectiveness, low cost and ease of application with conventional equipment (3). However, the handling and use of these formulations have given rise to concerns about user and environmental safety (4). These concerns include drift hazards associated with dust applications; dermal and respiratory toxic exposures during transfer and mixing operations of wettable powders under field conditions and inflammability and percutaneous toxicity due to the use of aromatic solvents in the emulsifiable concentrates.

1.2 Safe and High-Performance Formulations

Increasing awareness has been prompted by the following:

(a) Legislation seeking higher safety standards for the use of pesticide products (5,6);

(b) The ever-increasing expense of developing new compounds, necessitating the achievement of optimum biological activity per unit cost (7);

(c) Advances in related disciplines providing an improved understanding of mode of action and targeting have led to the development of safe and often high-performance formulation technologies (8) (Table 4.1).

Table 4.1 Safe and high-performance formulations

Basic formulation	*Corresponding safer, high-performance formulation*
Aqueous solutions/ soluble powders (SP)	Aqueous concentrates
Dusts (DP)	Dustless (DL) dusts, Microgranules (MG)
Emulsifiable concentrates (EC)	Emulsion concentrates (EW)
	Microemulsions (ME)
	Gels (GL)
Wettable powders (WP)	Suspension concentrates or Flowables (SC/OF)
	Water-dispersible granules (WG) or Dry Flowables (DF)
	Water-soluble package
Granules (GR)	Slow/controlled-release (CR) granules
	Encapsulated granules (CG)
Others	Suspo-emulsions (emulsion flowables) (SE)
	Capsule suspensions (CS)
	β-Cyclodextrin (CD) inclusion complexes
	Ultra low volumes (UL)

One such development is suspension concentrates (SCs) or flowable formulations. These are pre-formed suspensions of well-dispersed micronized active ingredients and can be viewed as a safe alternative to wettable powders (4). The vast majority of such formulations are aqueous-based, flow as a liquid and can be measured by volume instead of having to be weighed. For water-immiscible liquids, emulsion concentrates (EWs), (9) also being aqueous based, provide a safe alternative to EC's. On the other hand, microemulsions (MEs), the transparent and thermo-dynamically stable emulsions, present many useful features, including enhanced biological activity (9).

The wettable powders (WPs) have been improved to form relatively dust-free water-dispersible granules (WGs) (3). The product is in the form of small spheres, flows like a liquid and can be measured by volume. For this reason, this formulation is also known as a dry flowable (DF). DuPont, which helped pioneer DF technology, has introduced several of its sulfonylurea herbicides (Table 1.6) using this formulation technology. For improving their safety profile further, they are now increasingly being packaged in water-soluble polyvinyl alcohol bags in pre-measured quantities to avoid direct worker contact with pesticides. The introduction of WPs in water-soluble bags has also addressed dusting concerns.

In a related development, ECs have been improved to water-emulsifiable gels (GL). The gel technology developed by Rhone-Poulenc (10) transforms ECs into a product of gelatin-like consistency. The gel formulation possesses unique rheological properties that allow its packaging in water-soluble polyvinyl alcohol bags in pre-measured quantities. This eliminates worker exposure to chemical vapours or to dermal contacts in the process of transfer to the mixing tank. The container disposal problem, associated with EC formulations, is also eliminated here. Several gel products have been introduced by different companies, for example, bromoxynil gel under the name Buctril is being marketed by Rhône Poulenc.

An improved version of a dust formulation is a DL-dust, which is relatively non-drifting (also known as 'drift-less') and adheres adequately to foliage (11).

Basic granular formulations developed along with other first-generation formulations (dusts, ECs and WPs) have been primarily used for the application of insecticides and herbicides to soil. The application of new technology to granule formulations has improved their placement and persistence. For example, microgranules (size range 100-200 μm) are designed to adhere to foliage. Here persistence is affected by choosing the mode of formation (pre-formed, agglomerated or extruded) and by factors such as ground and climatic conditions and the method of application (12). Further improvements have been achieved by an encapsulation technique in which small particles of a pesticide are surrounded with a protective shell or coating for subsequent release at a controlled rate. Encapsulated products can be formed in different sizes (granular or microencapsulates). Microencapsulate suspensions are increasingly gaining currency by providing controlled release properties. They provide enhanced persistence as well as increased safety by reducing handling hazards and phytotoxicity.

The development of β-cyclodextrin inclusion complexes is yet another approach which is particularly suitable for poorly water-soluble, photolabile and volatile substances (13). Kwizda (14) has improved conazole fungicide formulations by complexation of flutriafol and triflumizole, respectively, with β-cyclodextrin.

A number of insecticide preparations have been developed expressly for public health and stored grain pests. These include ready-to-use kerosene-based solutions, aerosols, smoke generators, baits and fumigants. Other specialty pesticide uses relate to veterinary and wood treatment applications. However these formulations constitute a relatively small percentage of pesticide use, the bulk of which is for plant protection.

2 DEVELOPMENTS IN PEST CONTROL STRATEGIES

Modern formulations with high-performance characteristics are one of many concurrent developments in improving pest control strategies. The invention of highly potent and safer pesticidal chemicals has resulted in a substantial reduction in their use rates (15). A better understanding of the modes of action and translocation has provided leads for optimization of their biological activities(16, 17).

Enhanced concerns for user and environmental safety have led to more stringent regulations (5, 6, 18). And through a newer understanding of the physiology and biochemistry of resistance mechanisms, more rational approaches have been devised for the management of resistance to safer pesticides (19). These developments have had a profound influence on the design of pesticide formulations.

2.1 Increasingly Potent Pesticides

One of the most prominent general trends has been the progressive increase in the potency of all the classes of chemical pest control agents (15, 20). The activity has risen by several orders of magnitude in the last few decades. New pesticidal chemicals have, for the most part, high unit activity and low dose rates. For example, for insect control use, rates of dichlorodiphenyltrichloroethane (DDT) were around 1.5 kg active ingredient per hectare in the 1950s. With the introduction of pyrethroids in 1974, a level of 15-20 g/ha has been reached. Similarly, the introduction of sulfonylureas has brought herbicide rates down to 10-20 g/ha and ergosterol biosynthesis inhibitors (EBIs) as fungicides came down to a level of 200-100 g/ha. For the formulator, its implication is a greatly reduced margin of error, as a very small quantity of active compound now needs to be distributed efficiently over a large crop area.

2.1.1 Volatilization losses on application

Various chemical and physical processes cause loss of the pesticide during and after the application process. These losses are due to factors such as solubility, volatility, partition characteristics and stability. These properties are relatively important for achieving efficient distribution of the more potent compounds. For example, while application made at different dosages may give similar areas of deposit, the rates of volatilization are not dependent upon the active ingredient concentration but on the volume of solution delivered and the particle size of the spray solution. Thus, if a more active ingredient is sprayed at a lower dose and a lower volume, the volatility loss would represent a greater proportion of the toxicant applied at the lower dose. The evaporation loss may thus become a critical factor in determining the period over which pest control is achieved.

2.1.2 *Selective placement of toxicant*

The mechanics of distribution also play a significant role in the selective placement of toxicant within the crop environment. The importance of placement efficiency in the efficacy of the product becomes obvious from the observable differences in the laboratory and field toxicity of DDT, dimethoate and deltamethrin (Table 4.2) (21). While, obviously gaining an enormous increase in chemical efficacy, deltamethrin loses a significant amount of this efficacy advantage (over DDT) when placed in a field situation.

Table 4.2 Comparison of intrinsic activities and application rates of representative insecticides (after Graham-Bryce as reported in F.R. Hall)

Insecticide	*Typical application rate (g/ha)*	*Relative application rate*	*Relative lethal dosage**
DDT	1,000	50	1,600
Dimethoate	500	20	1,039
Deltamethrin	20	1	1

*Mean value calculated from data for four species: *Phaedon cochleariaea*[a], *Anopheles stephensi*[b], *Christoneura occidentalis*[a] *and Musca domestica* [b].
(a = Phytophagous pests; b = non-agricultural pests).

2.2 Optimization of Biological Activity

The effective activity of a pesticide is the product of its inherent activity and its concentration at the site of action (17). Biological activity can be improved by improving the quantity and timing of targeting a pesticide to its active biological site through physical, chemical or biological manipulation during its formulation (7, 8), for example, by optimization of particle size and/or by use of proper adjuvants.

2.2.1 *Effect of particle size*

From a biological standpoint the least complicated type of pesticide to formulate is that which exerts its effect on the surface of the plant (18). For solid fungicides, such as sulphur, zineb and dichloronaphthoquinone, a most critical parameter to optimize activity is small particle size and an even distribution over the surface. Sulphur, in a particle size range of 1-2 μm, was nearly 100 times more active on *Conidia* of barley mildew than a product whose particle size range was 15-16 μm (22). One of the reasons a smaller particle gave better fungicidal activity was its better deposition on foliage. The overall rain resistance has also been found to be better with smaller particles. Similar observations have been made with acaricides such as cyhexatin and cyclofentazine, which showed a good inverse correlation between activity and particle size (16).

In contrast to the above, in contact insecticides such as DDT and carbaryl, there is an optimum particle size, not necessarily a minimum, that gives the best results. The effectiveness of DDT on mean percentage kill of *A. aegypti* was optimal at a size range of 10-20 μm (Table 4.3) (16). Similarly, in the case of carbaryl, the first instar larvae of the cabbage white butterfly, *Pieris brassicae,* picked up and dropped the largest particles of carbaryl, while the smallest particles adhered too strongly to the plant surface. The optimum size was 15-20 μm.

Table 4.3 Effect of particle size on insecticidal activity

DDT crystal size (μm)	*Mean kill of* Aedes aegypti *on plaster of paris surface (%)* [a]		
	1 min.	*2 min.*	*4 min.*
0 - 10	18	45	80
10 - 20	70	98	..
20 - 40	5	33	60
40 - 60	..	0	0

(a) Application rate is 50 mg/ft^2

2.2.2 *Role of adjuvants*

The biological activity of **systemic** pesticides depends on their ability to penetrate into plants. This is more true of herbicides and plant growth regulators, which have no activity at all on the plant surface. A formulation that doubled uptake would double activity (17).

Formulation adjuvants have a significant effect on the uptake of pesticides. They may either change the permeability of cuticles to facilitate penetration of the active ingredient into the target organism or enable the pesticide to accumulate in plant tissues, where it is more accessible to a specific organism. Penetration in the plant is clearly essential for systemic insecticides and fungicides, which rely on translocation in the plant for controlling pests and diseases that occupy internal tissues. It is also important for leaf surface insects such as aphids that feed on internal tissues (phloem system), or fungal pathogens, such as powdery mildew, that send haustoria into epidermal cells. Similarly, any herbicide or plant growth regulator must penetrate at least as far as the photosynthetic tissues if it is to have any effect. Enhancement of herbicidal effectiveness through the addition of adjuvants is a common practice, particularly with the application of post-emergence herbicides (23).

A study jointly carried out at the State University of Ghent in Belgium and ICI (24) clearly demonstrated that oil/surfactant adjuvants greatly enhance the activity of many insecticides (Table 4.4). Similarly, atrazine herbicide, when used with a build-in adjuvant in its aqueous flowable formulation, controlled the weeds in a maize field with a reduced rate of application. Thus, use rate of 0.75 kg/ha of atrazine with 0.75 litre of an adjuvant (Atplus 230) was found equivalent in efficacy to 1 kg/ha of atrazine with 6 litre of emulsifiable spray oil (24).

2.3 Registration Considerations

Pesticides, by design, are biologically active. Since in most cases they have some degree of toxicity to man and animals, they pose some potential risks. Through registration procedures, pesticides are regulated in terms of their safety and efficacy. The data required for registration are often extensive and may vary from country to country.

To regulate the quality of products that are registered, international bodies such as WHO and FAO have published basic quality standards. Some countries, such as Indonesia, have adopted these

Table 4.4 Effect of spray-tank adjuvants on the bioefficacy of insecticide formulations

Insecticide	*Mortality of cotton leafworm (3rd instar larvae of* S. littoralis – *s. strain) (%)*			
	Insecticide alone		*Insecticide with 2% adjuvant* [a]	
	Formulation	*Formulation concentration (ppm)*	*Adjuvant concentration (ppm)*	*Mortality (%)*
Monocrotophos	LC_{25}	154	8.5	87.5
(O-P insecticide)	LC_{50}	275	14.0	97.5
Methomyl	LC_{25}	29	2.0	80.0
(carbamate)	LC_{50}	49	4.0	82.5
Diflubenzuron	LC_{25}	8	0.7	97.5
(Chitin synthesis inhibitor)	LC_{50}	12	1.0	92.5

(a) Adjuvant: Atplus 413, a blend of cottonseed oil (83%) + surfactants Atplus 298/299 (17%), from ICI Specialty Chemicals.

as such. However, many countries are devising their own local specifications to suit the general conditions and use patterns prevailing there.

The safe use of pesticides involves reducing toxic exposures to the farmer and minimizing environmental contamination. Thus a safe product would need to conform to optimized product efficacy and have a minimum impact on non-target organisms including non-target insects and plants, warm-blooded organisms and the environment (18). The risks during manufacture, transportation, storage and application would also need to be suitably covered. In this context packaging as well as application aspects would need to be considered as an integral part of any formulation design programme.

2.3.1 Reduction in toxic exposure

The vast majority of herbicides and fungicides being used have a relatively low acute and chronic toxicity, but some of the organophophorus (O-P) esters or carbamate insecticides have a high intrinsic acute toxicity to warm-blooded organisms (15). This feature creates some degree of risk at all stages of their use. For example, the worker in manufacturing, the farmer in the field and the consumer of treated commodities could be subjected to repetitive toxicological exposures. Whereas risk to the worker and to the consumer can be managed extensively by suitable legislation and monitoring, the risk to the farmer is more difficult to manage, as it depends upon education and understanding. Particularly in many developing countries, the handling of pesticides has been at a less than satisfactory level of safety. Farmers risk contamination from splashes of concentrated chemicals when they dilute or from spraying if they are not using proper handling precautions. Possible answers to these toxicity problems lie in providing better education for the user, selecting less hazardous formulation types (25), developing safer handling systems and improving application methods.

The use of a unit dose water-soluble sachet reduces farmer exposure during measuring and dosing operations. The entire package containing the product is discharged from an over-pack directly into the spray tank, whereupon the water-soluble film containing the product dissolves, allowing the product to dissolve or disperse (4). The Shell Chemical Company, for example, markets Mancozeb at 80% w/w concentration in water-soluble sachets (26). Another closed filling system for applicator safety is electrodynamic spraying employing a specially designed oil-based formulation packed in a ready-to-use 'bozzle', which is inserted into the sprayer (27) (see Chapter 18). One other way of protecting the farmer during spray preparation is through the use of dry-flowable formulations. These formulations have the major advantage of being essentially dust-free during handling.

2.3.2 Reduction in phytotoxicity

Controlling a pest on agronomic crops with minimal phytotoxicity to the target plants is an important requirement of the formulation (18). Encapsulated formulations have been successfully employed to meet such situations. For example, development of a controlled release formulation of an acaricide, propargite (Omite C.R.) (28) by Uniroyal, has enabled its use on crops susceptible to phytotoxicity such as citrus, young cotton plants and strawberries. With some herbicides one may wish to maintain activity against weeds but reduce phytoxicity to a crop. The use of safeners has been proposed for use with thiocarbamates (19, 29). These compounds stimulate defence metabolic reactions in the crop species but not in the weeds. For example, Eradicane, a combination of EPTC with an inert herbicide safener as a micro-encapsulated formulation, has been used to protect crops such as maize and sunflowers and kill their primary weeds. Other approaches to minimizing the phytotoxicity of herbicides include non-dispersion-type applications such as wipe-on application or highly directed sprays (3).

2.4 Resistance Management

An organism from a population initially controlled by a pesticide that loses its susceptibility such that it now resists control by that pesticide is considered to have developed resistance. The risk of resistance is a limitation common to all pesticides (30, 32). It occurs due to the continuous use of chemicals, setting in motion the selection process of insensitive individuals in the target populations, which eventually become dominant. Due to basic differences in biology, such as number of generations per season, size of population and mobility of individuals, the risk of the onset of resistance varies considerably between groups of organisms (19). For example, insects and pathogen populations are more susceptible to resistance, whereas in weeds it develops slowly. Resistance adversely affects the use of pesticides from the point of view of effectiveness, economy and safety.

2.4.1 Modes of resistance occurrence

Resistance to pesticides arises primarily through metabolism of the pesticides by increased enzymatic activity within the organism. Some pesticides are activated metabolically and thus it is theoretically possible to generate resistance through reduced activation.

However, it is much more common to observe increased detoxification in resistant strains. In some cases, decreased uptake or enhanced excretion also contribute (19).

New generation pesticides, which are selectively toxic by attacking a single enzyme or receptor peculiar to the pest group, are excellent candidates for the onset of resistance. A single mutation at this site may render a target insensitive and the pest highly resistant. On the other hand, compounds acting at multiple essential sites are less open to the development of target site resistance.

2.4.2 Pesticide combination formulations

Prolonging the effective life of valuable chemical control agents by slowing, preventing or reversing the onset of pesticide resistance is a desirable goal of resistance management. Two methods of delaying the onset of resistance include the use of pesticide combination formulations and the substitution of a dissimilar material with a distinctly different mode of action for every other application. The first approach is widely used in fungicidal treatment. A formulated fungicide mixture containing the at-risk fungicide is combined with a partner with a different mode of action (31).

A pesticide mixture should have synergistic action. This occurs, in broad terms, where the activity of a combination is much greater than that expected from their individual effects. For example, when a mixture of two herbicides controls more weed species than each herbicide separately, it would be attributed to synergistic action. Advantages of synergistic combinations might include the utilization of a lower dose rate due to increased efficiency, a broader spectrum of activity and reduced risk of developing of resistance.

In insect control, combination formulations have often used a relatively inexpensive, poorly active compound with more expensive and more active compounds. Salut, a combination formulation of chlorpyrifos and dimethoate by BASF, has an additive and synergistic action controlling a broad range of insect pests (32).

Similarly, fungitoxic combination products are used to broaden the spectrum of activity of a component or to achieve a higher level of activity by means of synergistic interactions. To alleviate the risk of fungal resistance due to single-site fungicides, non-systemic contact fungicides are often mixed with ones that have a multi-site mode of action. Thus, captan, captafol, maneb and mancozeb are all good mixture partners. For example, the resistance of a fungal pathogen to the fungicide benomyl is reported to be delayed by mixing it with maneb, which exhibits a different mode of action (33).

Herbicide mixtures fall into two categories (often overlapping). In the first, each herbicide is used at a full rate and the spectrum of weeds controlled by each is mutually exclusive. In the second, both control the same spectrum of weeds (34). The use of mixtures in the first case may increase the number of years over which a given herbicide is used separately. The lower dosage in the second case should lead to a much lower selection pressure for each herbicide, substantially delaying the appearance of resistance. The resistance to s-triazines is the most prevalent and best-studied example of herbicide resistance. To overcome the resistance problem, herbicide mixtures such as tridiphane and atrazine have been used that synergistically control grasses as well as weeds in maize.

Several considerations weigh in developing pesticide combination formulations. The physical properties of different pesticides may limit the type of formulations that can be developed. The need for the combination formulation of compounds of widely differing physical properties has led to the development of new formulation types such as Suspo-emulsions. These are formed by combining an emulsified organic phase with one or more active ingredients and a continuous aqueous phase also containing one or more active ingredients in the form of a solid dispersion (35). In this type of formulation, the chemical properties of the mixture partners need to be compatible, while the biological properties should lead to complimentary and higher activity. Some of these aspects are covered in detail in chapter 2.

3 FORMULATION DESIGN

Pesticide formulations must be physically and chemically stable pesticidal compositions that can be applied by the user and that result in effective and efficient control of pests (36). Effective pesticide products are also required to have a high safety profile without substantially increasing the cost to the consumer.

Most pesticides are not suitable for application in their original form. The difficulty arises in dispersing these materials, which exist in physical forms varying from volatile liquids through semiliquid or solids to crystalline materials and uniformly distributing them, in small quantities (hundreds of grams to just a few grams per hectare) over a large area (section 2.1) (4). Even if they could be applied as such, many pesticides would exhibit only a fraction of their biological activity (37). For example, many systematic insecticidal and fungicidal compounds and all the herbicides and plant growth regulators have to penetrate the outer lipid layers of foliage to reach the site(s) of action to exhibit their full potential biological activity (section 2.2). Only the pesticidal formulation compositions consist of agents that facilitate this process.

The correct design of a formulation is one of the critical steps in the development of a pesticidal product. If things go wrong, it could lead to various field problems adversely affecting their performance (4). The direct costs of a product failure can be very high, and the indirect costs (through bad product image, negative marketing potential etc.) can be devastating, particularly if a product recall and/or compensation becomes necessary.

In the early days of formulation technology the choice of formulation types was limited to a few standard formulations: emulsifiable concentrates for general spraying use, wettable powders for crops where phytotoxicity was a problem and granular formulations for soil applications. Today a technical pesticide can be formulated in a multiplicity of formulation types (Section 1.2). The selection of a suitable formulation type is dependent on factors like (a) market needs and competing products, (b) the physical limitations of the active ingredient and (c) the status of technology at the time the formulation is developed (18).

In designing a formulation for a specific end-use it is desirable to have access to a vast amount of information about the pesticide, including the pest and crops involved, the production facilities, the storage conditions, the methods of transport, application techniques and many related factors.

In addition, many countries have formed their own specifications for pesticide formulations taking into consideration local ground and climatic conditions and use patterns. Due to the high costs of shipping and time delays in receipt of imports from distant countries, it is desirable to use endogenously available carriers, solvents and surfactant systems. Thus, if one is exporting a product, one should develop versatile formulation design specifications.

3.1 Formulation Inputs

C.G.L. Furmidge (38) of Shell Research has compiled requisite feedbacks for formulation design and has classified these in a set of product design inputs, namely (a) compound inputs, (b) application inputs, (c) marketing inputs and (d) manufacturing inputs. These inputs lead to the establishment of a design specification that may ensure acceptability of the product throughout its life.

3.1.1 Compound inputs

The basic properties (physical, chemical and biological) of the technical pesticide constitute the compound inputs. The design of the final product would be subjected to the limitations of these properties. Physical properties that are important include physical state, solubility, melting and boiling points, vapour pressure and partition coefficient (octanol/water). Similarly, chemical properties that are important include hydrolysability (acid/base), thermal and photochemical stability and oxidation and reduction properties. Information on the reactivity of a given pesticide with solvents, carriers, surfactants and other possible formulation ingredients is also necessary. If the stability of the compound during formulation or use is likely to be affected, means of stabilizing the compounds also need to be established. Based on this information, a preliminary selection of formulation types can be made (Figure 4.1).

Biological properties to be considered include type of action (stomach, contact or systemic), spectrum, degree of activity, selectivity and mode of action. These properties would indicate

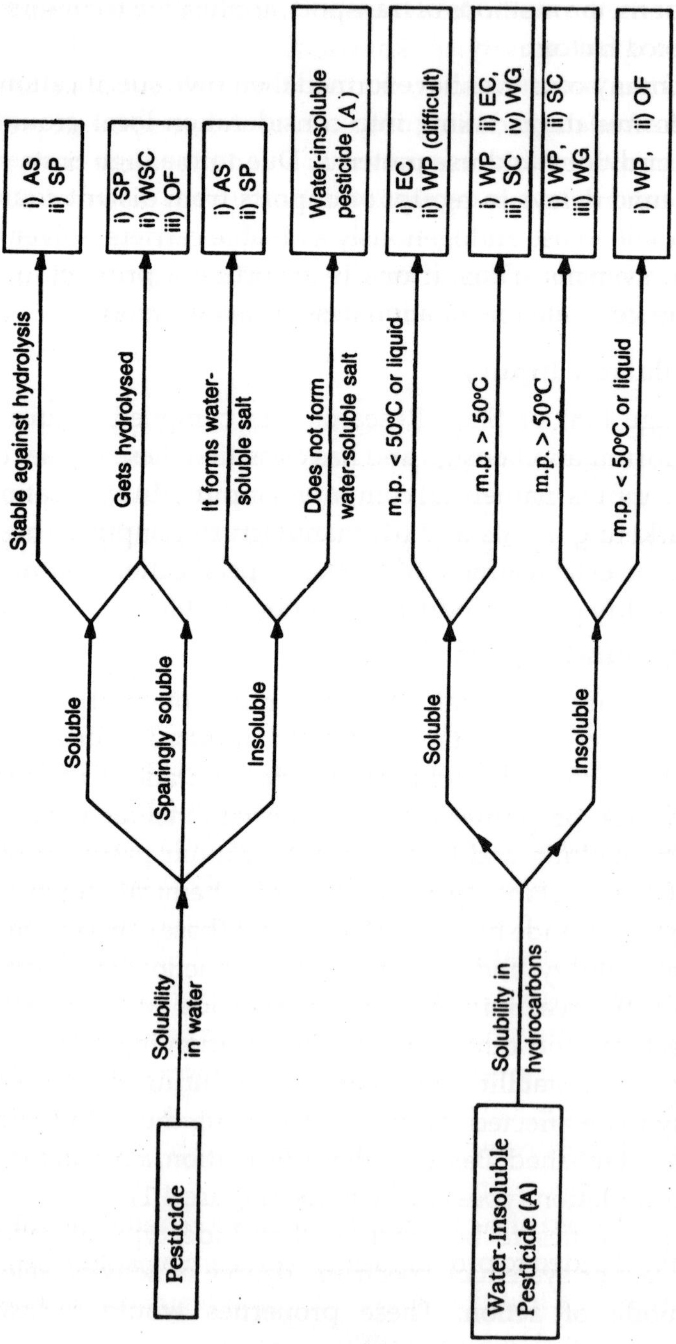

Legends : AS = aqueous solutions, SP = soluble powders, OF = oil-based flowable, SC = water based flowable, WSC = water-soluble concentrate, EC = emulsifiable concentrate, WP = wettable powder, WG = water-dispersible granules.

Figure 4.1 Influence of physicochemical properties of technical pesticides on the choice of formulations for aqueous spray

potential uses of the product, dose rates and possible approaches to maximzing biological activity. A knowledge of the toxicological properties, such as acute toxicity, irritation to eye and skin, the toxicities of possible degradation products of the compound and toxicity to fish, provides valuable information on the safety risks to non-target organisms and safeguards needed by the users of the formulation.

The physical, chemical and biological properties of a pesticide can profoundly influence the formulation developed for sale. For example, large differences in solubility characteristics of pyrethroid insecticides, cypermethrin and deltamethrin, as well as differences in their intrinsic activities, dictate vast differences in the concentrations of their respective EC formulations.

Phorate has a high acute dermal toxicity. Therefore its granular formulations are given a protective coating to safeguard the user (39).

3.1.2 Application inputs

The environments in which the product is to be used and information on the local constraints on its application are application inputs, which also include pests to be controlled, plants to be protected or, for public health requirements, a particular formulation type that is preferred. Also, information should be included on such likely factors as type of application equipment commonly used, application dose rate required, effect of expected climate following treatment (such as high temperature or rains) and effect of hardness of water used for dilution (for water-diluted sprayable formulations). These inputs describe the factors that affect the ways in which the compound is distributed over the target. The information is also useful in defining the physical properties that would have to be built into the product to ensure its most effective distribution. A few examples are cited to elaborate these points.

The liquid pyrethroid insecticides, formulated as emulsifiable concentrates, are intended to provide a good surface cover on plants. However, the parts of a leaf that are not covered may remain susceptible to pest attack. Thus pyrethroid insecticides sprayed by conventional equipment onto cotton give little protection against white fly. This happens because spray droplets land mainly on the upper surface of the leaves while the insects live on the lower surface. If the same pyrethroid insecticide is formulated for spraying by electrodyn equipment, each droplet receives an electrical charge, which attracts it to the leaf surface including its underside and stem, killing the white fly (40).

Some crops are rather sensitive to aromatic oil. However, leaf damage is not always due to the oil itself. Carbamate insecticides are known to be rather phytotoxic if used in an emulsified oil solution. For this reason carbamates are almost always used as wettable powders (2).

In the public health spraying of insecticides, the condition of wall surfaces should be taken into consideration. Wall surfaces that are porous tend to be catalytically active and rapidly deactivate insecticides. Thus an insecticide in the form of a wettable powder is the formulation of choice. Because of the particulate nature of a wettable powder, the insecticide remains on the surface of a treated area, where it is accessible to insects. When liquid solutions are used, the insecticide can be absorbed and become chemically deactivated (Figure 4.2) (41).

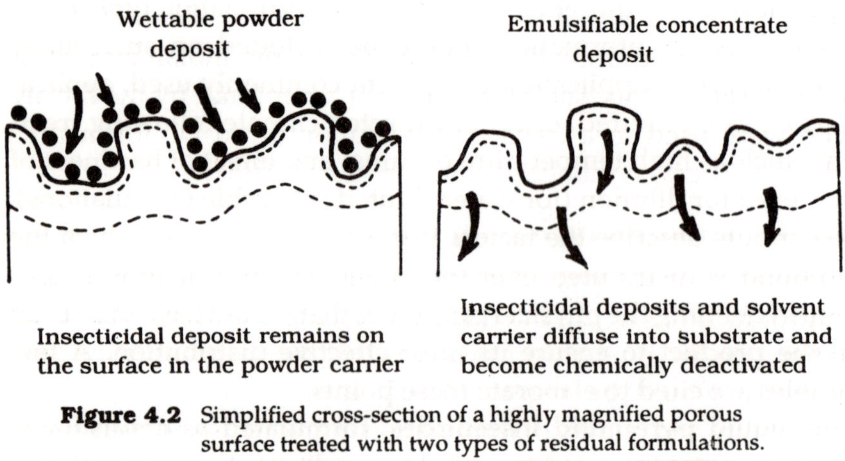

Figure 4.2 Simplified cross-section of a highly magnified porous surface treated with two types of residual formulations. (Source: Wellcome Foundation Brochure - COOPEX)

3.1.3 Marketing inputs

The properties of a product that are perceived by marketing personnel as being most desired by the user constitute the marketing inputs. These inputs relate to saleability of the product taking into account user convenience, attractiveness, safety, durability and economy. They should include user preference for a particular type of formulation, its presentation and packaging, local storage conditions and product shelf-life. It would be well to note product compatibility with locally preferred in-tank mixtures. The maximum cost a product can bear would be guided by the cost of competing products in the market.

Specific country registration requirements would obviously have to be met. In situations where the end-use of a given pesticide formulation is restricted because of toxicity considerations, registration might still be achieved by adjusting the level of active ingredient in the formulation. This approach might allow access to a desired market segment, such as a fairly toxic molecule to the home-owner market or a restricted-use product to a general-use market (18). The following examples will illustrate the concept of marketing inputs.

In the tropical climate of Indonesia, farmers are not inclined to use wettable powders, but readily accept water-dispersible granular formulations for their handling convenience and novelty (42).

Monsanto withdrew Butachlor EW from the Indian market due to its unsatisfactory shelf-life and poor acceptance by the farmers, who were used to a transparent solution of an EC formulation, which, to them, symbolized purity of the product.

3.1.4 Manufacturing inputs

Information on the availability of production equipment and quality control facilities at the intended processing plants, as well as limitations on the type of formulations and the manufacturing specifications, would be manufacturing inputs. Information on the availability and quality of formulation ingredients and their sources would also be necessary to provide a realistic design input for the product.

For example, the WHO specification for DDT 75% WP (43) provides for a relatively high suspensibility (minimum 70%). The formulation must be processed in a fluid energy mill to attain the requisite fine particle size. For overseas shipment, the specification also stipulates a heat stability test at 90°C for 24 hr. To meet this requirement, one must use a superior grade technical DDT (m.p. > 91°C). In addition use of a particle size analyser would be valuable for monitoring the change in the particle size on heating, to learn whether or not the product meets specifications.

3.2 Formulation Design Profile

The formulation design inputs contribute to the development of a formulation design profile by listing various requisite characteristics of the product. In turn, each characteristic would be specified by certain properties, forming the basis of formulation development.

Many times, specific information is not available. Marketing inputs, such as user preference for a new formulation type, may be lacking because the product is new to the market. This presents an element of risk for product design. Positive, negative and limiting information must be balanced to select acceptable properties of the formulation. Thus, the design profile, at best, represents a compromise between what is theoretically desirable and what is practically and economically possible.

The limitations imposed by the physical and chemical properties of the pesticide itself on the type of formulations that can be developed may be aggravated by other factors such as shelf-life, safety and cost considerations. The act of balancing various conflicting requirements to arrive at a particular design profile requires unique training, skill and experience on the part of the formulation designer. This is a common feature in many dynamic and innovative professions. For example, an architect employs well-founded principles of civil engineering and related disciplines. He uses readily available building materials to shape his specific-need-based creation, which should be functional as well as aesthetic. Similarly in the design of a pesticide formulation, an understanding of chemistry is a means of arriving at a viable and practical solution to its application. Close collaboration with personnel in many allied disciplines is necessary to devise optimal solutions.

3.3 Stages of Formulation Development

The development of a commercial formulation for use in crop protection proceeds through different stages: (a) preliminary studies, (b) investigational stage and (c) design of a commercial formulation (18, 44).

3.3.1 Preliminary studies

Initially, all the related information pertaining to compound inputs and application inputs is compiled and feasible types of formulations are suggested. The preliminary formulations are prepared and initial testing of their physical properties carried out. A preliminary biological test is also carried out.

3.3.2 Investigational stage

If the preliminary formulation(s) show promise, various factors that influence the biological activity and phytotoxicity of the compounds in the formulation are investigated. Similarly, factors that influence the application of the compounds are studied. At this stage, product analytical methods are developed and the combination of ingredients that will go into the formulation is fixed. Testing for long-term stability of the compound(s) in the formulation is also initiated. A small-scale field trial is carried out. If it becomes necessary, alternative formulations are prepared.

3.3.3 Design of a commercial formulation

Candidate formulations are subjected to comprehensive biological testing. The suitability of various types of application equipment is also checked. After successful completion of these tests, the process of manufacturing the product is finalized, including a check on the availability of all the ingredients and their conformity to national regulations. Determination is made of the formulation's physical compatibility with other pesticides likely to be used in tank mixtures.

The work on the formulation packaging development is carried out simultaneously and data required for registration purpose are accumulated.

A well-designed commercial formulation should do the following:

(a) Provide the desired biological effect when used as recommended and not induce undesirable side effects;
(b) Be presented so that the active ingredient will be able to produce the maximum biological effect at a minimum cost;
(c) Be applicable by chosen means to provide easy, effective and reliable dispersion;
(d) Be suitable for large-scale manufacture at acceptable costs;
(e) Be safe during manufacturing, packaging, storage and transportation;
(f) Have an adequate shelf-life, normally 2 years or more;
(g) Be acceptable to the registration authority and the consumer.

4 DESIGN OF TAILOR-MADE FORMULATIONS

Most formulations are designed for the broadest possible uses. Some situations nevertheless warrant a specialty formulation that provides optimum performance for a single site, pest or use pattern (45). The deliberate orientation of the formulation towards a narrow and well-defined set of conditions may make its application impractical or economically unacceptable for other uses. In some respects, the design approach to a specialty formulation can be less restrictive, as product performance is to be optimized for a specific outlet only. A case-study of a formulation designed to control surface-feeding aquatic pests is presented to illustrate the rationale of tailor-made formulations.

4.1 A Self-Spreading Formulation for Control of Anopheline Larvae Populations

Stagnant clean water pools serve as breeding sites for anopheline mosquitoes, the primary vectors of malaria. Fallen leaves, refuse and vegetation in and around the water body result in a protected breeding habitat. The pool water is drunk by stray cattle and wild animals and local inhabitants may also use it for fishing. The use of a broad-spectrum insecticide would not be suitable in this situation. Microbials such as *Bacillus thuringiensis var. israelensis* (sero type H-14) and *Bacillus sphaericus* (strains 1593, 2362 etc.) are highly specific in their activity against the mosquito larvae and are safe to other organisms. They produce proteinous protoxins that must be ingested by the larvae to be effective.

The anopheline larvae are adapted to collect particulates from the air-water interface, randomly ingesting the particulate matter. An effective microbial larvicidal preparation could make use of the feeding behaviour of the larvae by suspending the toxin in the feeding horizon, i.e. at or near the water surface. The other desirable features of the formulation could include an ability to penetrate barriers, such as natural films and vegetation, and remain floating for a long duration without degradation (46). The standard water-diluted sprayable formulations such as WP and SC formulations disperse in the whole water body and soon settle to the bottom. Thus significant quantities of the toxin particulates move out of the feeding zone soon after treatment. Treatment with granules, pellets and briquettes is also not appropriate, as they do not have the requisite physical properties (47).

Ramdas and Khetan (48) reported a self-spreading oil formulation, which is designed to incorporate all the above desired properties. The product is a stabilized suspension of the micronized toxicant in an oil phase containing lyophilic surfactants dispersed in an alcohol. It is suitable for ULV application. The droplets of the product, on contact with water, spread spontaneously with a great force (spreading pressure > 60 dynes/cm^2) into a micro-reticulum (Figure 4.3a), which soon breaks down into evenly distributed discrete microglobules (Figure 4.3b). These oil-solubilized toxicant globules are of optimal dimensions (~5 µm) for rapid tapping and ingestion by the mosquito larvae. As expected a substantially reduced dose of the toxicant in this formulation was found to suffice in controlling larvae populations as compared to corresponding aqueous flowable formulations, which spread out in three dimensions.

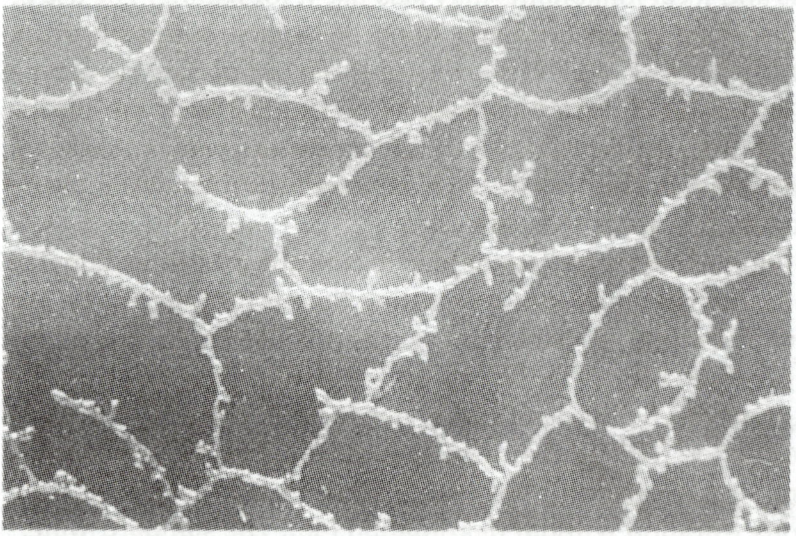

Figure 4.3(a) The self-spreading oil formulation forming a microreticulum on the surface of water

Figure 4.3(b) The microreticulum breaks down to discrete microglobules, which can be ingested by anopheline mosquito larvae

Thus, a *B. thuringiensis var. israelensis* (IU 12,000) preparation, at a dose rate of 14 µg (biomass by wt./ft^2) is reported to achieve >90% control of the 3rd instar larvae of *A. stephensi*. Due to its high spreading pressure the formulation penetrates into natural films and vegetation and remains floating. It breaks down to micro-globules, maintaining oxygen supply to non-target aqueous organisms. This kind of oil film breakdown does not occur in the case of mineral-oil-based ULV formulations, which form continuous films on the water surface.

5 CONCLUSION

Pesticide formulation design is influenced by the inherent properties of the active ingredient, the end-use of the product, regulations, competition, technological advances and cost-effectiveness. The advent of high-potency, low-dose-rate, highly selective pesticides has manifested the importance of optimal targeting with minimal impact on non-target sites. It has also given fillip to the market trend towards multi-component active-ingredient products employing complex formulation technologies.

With safety being the driving force, user-friendly and environmentally attractive formulation technologies have increasingly made inroads, with the possibility of commanding a significant portion of the pesticide market. A resultant shift from ECs to gel, EWs or microemulsion formulations can be foreseen, although EC formulations with safer solvents would remain. Similarly, WP formulations would be displaced by WGs or in some cases by SCs or would be packed in water-soluble bags for niche markets. Controlled-release formulations such as capsule suspensions would also play a greater role on account of their safety to users and the environment.

Formulation design can also modulate the toxicity profile of the end-use products. Thus, formulations could be designed to reduce hazard potential, increase initial or residual toxicity, improve the cost-performance ratio and expand the area approved by regulatory officials. An integration with packaging and application technologies in the system design also remains a desired goal of the formulator. In conclusion, the formulator has to take a holistic view when designing a pest control product that includes the pest, the plant, the worker, the user, the bystander and the environment.

REFERENCES

(1) A.C. Clarke, *The Clarke laws: Profiles of the future* (London, Gollencz, 1982).

(2) G.S. Hartley, Formulants of pesticides, *The Expanding Uses of Petroleum* (London, British Institute of Petroleum, 1982), p. 15.

(3) C. Corty, "Developments in formulations and application technology for safe use of agrichemicals," *Proceedings of the fifth International Congress on Pesticide Chemistry*, J. Miyamoto and P.C. Kearney, eds. (Oxford, Pergamon, 4, 1983), p. 281.

(4) B. Speight, "Effects of formulation upon the safe use of pesticides," *Field Worker Exposure during Pesticide Application*, W.F. Tardoir and E.A.H. Van Heemstra, eds. (Amsterdam, Oxford, New York, Elsevier, 1980), p. 233.

(5) B. Cross, "Trends in agrochemical formulations," *Regulation of Agrochemicals: A Driving Force in their Evolution*, G.J. Marco, R.M. Hollingworth and J.R. Plimmer, eds. (Washington, D.C., American Chemical Society, 1991), pp. 89–100.

(6) J.D. Bailey, "Safer pesticide packaging and formulations for agricultural and residential applications," *Reviews of Environmental Contamination and Toxicology, Vol. 129,* (New York, Springer-Verlag, 1992), pp. 17–27.

(7) J.M. Deming and R.W. Magin, "Optimisation of biological efficacy through formulation," *Proceedings of the Sixth International Congress on Pesticide Chemistry,* Roy Greenhalgh and T.R. Roberts, eds (Oxford, Blackwell, 1986), p. 231.

(8) G.B. Beestman, "Emerging technology: the bases for new generations of pesticide formulation," *Pesticide Formulation and Adjuvant Technology,* C.L. Foy and D.W. Pritchard, eds. (CRC Press, Inc., 1996), pp. 43–68.

(9) Th.F. Tadros, "Colloidal aspects of pesticidal and pharmaceutical formulations – an overview," *Pestic. Sci.* (U.K.), 26,1989, p. 51.

(10) L.E. Hodakowski and A.K. Taori, "Water emulsifiable gels in water soluble packaging," American Society of Agriculture Engineers/Canadian Society Agriculture Engineers, Spokane, Washington, International Summer Meeting, Paper No. 931119, June 20-23, 1993; L.E. Hodakowski, "Gel in water soluble bag technology," Rhone-Poulenc Internal Communication, July 1996.

(11) T. Uejima, "On new formulations of pesticides in Japan," *J. Pestic. Sci.* (Japan), 8, 1983. p. 125.

(12) C.J. Marrs and M.R. Middleton, "The formulation of pesticides for convenience and safety," *Outlook Agric.* (U.K.), 7, No. 5, 1973, p. 231.

(13) J. Szejtli, "Cyclodextrins in pesticides," *Starch* (Weinheim), 37, No. 11, 1985, p. 382.

(14) W. Zsifkovits, "Improvement of conazole fungicides by complexation with β-cyclodextrin," Kwizda GmbH, Leobendorf, Austria, 1995.

(15) F.J. Schwinn, "Importance, possibilities and limitations of chemical control now and in future – an industry view, *Ecolo. Bull.* (Copenhagan), 39, 1988, p. 82.

(16) Q.A. Geering, "A biological approach to pesticide formulation," *Proceedings of the Asia Pacific Regional Workshop on Pesticide Formulation Technology*, New Delhi, Pesticide Development Programme India, March 1987.

(17) C.E. Price, "Biological activity and formulation chemistry," Concept Programme on Pesticide Formulation, New Delhi, Pesticide Development Programme India, 3-4 March 1986.

(18) W.R. Landis, "Registration considerations for pesticide formulations," *Pesticide Formulations and Application Systems*, G.B. Beestaman and D.I.B. Vander Hooven, eds. (Philadelphia, American Society for Testing and Materials, 1987), p. 123.

(19) R.M. Hollingworth, "Vulnerability of pests," *Study and Exploitation for Safer Chemical Control in Pesticides, Minimizing the Risks*, N.N. Ragsdale and R.J. Kuhr, eds. (Washington, D.C., American Chemical Society, 1987), p. 54.

(20) R.A. Herrett, "Low-dosage pesticides: less is better," *Chemtech*, April 1988, p. 220.

(21) F.R. Hall, "Pesticide formulations and other parameters affecting dose transfer," *Pesticide Formulations: Innovations and Developments*, B.Cross and H.B. Scher, eds. (Washington, D.C., American Chemical Society, 1988), p. 260.

(22) W. Mass, "Influence of particle size on pesticidal activity," Advances in Pesticide Science, *Proceedings of the Fourth International Congress on Pesticide Chemistry*, H. Geissbuhler, ed. (Oxford, Pergamon, 1979), p. 772.

(23) G.D. Wills and C.G. McWorter, "The effects of adjuvants on biological activity of herbicides," *Proceedings of the fifth International Congress on Pesticide* Chemistry, J. Miyamoto and P.C. Kearney, ed. (Oxford, Pergamon, 1983), p. 289.

(24) M. Ibrahim, "Critical approach of the problem of insecticide resistance in *Spodoptera littoralis* and *Musca domestica*," Doctoral thesis, Faculty of Agricultural Science, University of Ghent, 1986; ICI Specialty Chemicals, "Agricultural chemicals – oil/surfactant adjuvants for insecticides and herbicides," Brochure 20-17 E.

(25) J.L. Hudson and O.R. Tarwater, "Reduction of pesticides toxicity by choices of formulation," *Pesticide Formulations: Innovations and Developments*, B. Cross and H.B. Scher, eds. (Washington, D.C., American Chemical Society, 1988), p. 124.

(26) Anonymous, *Agrow*, 115, No. 22, 13, July 1990.

(27) G. Matthews, "The attack of the charged bigade," *New Scientist* (U.K.), 14 July 1988, p. 55.

(28) R.F. Peterson, "Reduction of propargite phytotoxicity through spray drying," *Advances in Pesticide Formulation Technology*, H.B. Scher, ed. (Washington, D.C., American Chemical Society, 1984), p. 163.

(29) Anonymous, "How to use less to do more," *Chemical Week*, May 23, 1984, p. 32.

(30) L.B. Brattsten and others, "Insecticide resistance: Challenge to pest management and basic research," *Science*, 231, 1986, p. 1255.

(31) D.W. Hollomon and K.J. Brent, "Fungicide resistance – can we cope with it?" *Chem. and Indust.* 1989, p. 177.

(32) U. Neumann and V. Harris, "BAS 270 001 – A versatile broad spectrum dual component insecticide formulation for use on pomefruit, grapes and other crops," 1984 British Crop Prot. Conf., *Pests and Diseases*, 5 A-6, p. 437,

(33) H. Geissbuhler, "How safe are pesticides?" *Experimentia* (Basel) 38, 1982, p. 890.

(34) J. Gressel, "Strategies for prevention of herbicide resistance in weeds," *Rational Pesticide Use*, K.J. Brent and R.K. Atkin, eds. (Cambridge University Press, 1987), p. 183.

(35) S. Memula, P. Berger and C. Chellappa, "Suspoemulsions with improved stability and correlation of long term stability with the zeta potential" *Pesticide Formulations and Application Systems: Vol. 15*, H.M. Collins, F.R. Hall and M. Hopkinson, eds. (West Conshohocken, PA, American Society for Testing and Materials, 1996), pp. 132-144.

(36) H.B. Scher, "Advances in pesticide formulation technology: an overview," *Advances in Pesticide Formulation Technology*, H.B. Scher, ed. (Washington, D.C., American Chemical Society, 1984).

(37) J.W. Van Valkenburg, "Terminology, classification and chemistry," *Adjuvants for Herbicides*, (champaign, Il., Weed Sci. Soc. Am., 1982), p. 1.

(38) C.G.L. Furmidge, "Pesticides: trends in formulation design," *Span* (U.K.), 23 January 1980, p. 35.

(39) C.G.L. Furmidge and T. Shanton, "Formulation: some factors affecting pesticide performance," *Span* (U.K.), 16 February 1973, p. 82.

(40) C.E. Price, "Testing for biological activity," Seminar Series., New Delhi., Pesticide Development Programme India, March 1986.

(41) *Coopex, A Novel Approach to Insect Control*, Wellcome Foundation Berkhamsted, Herts, Brochure.

(42) S.K. Khetan, "National Pesticide Development Centre Indonesia: technical findings and recommendations," UNIDO Report, October 1990.

(43) DDT water-dispersible powder for overseas shipment: Specification; WHO/IS/2 bis 0016.1, " *Specifications for Pesticides Used in Public Health*, 7th (Geneva, World Health Organization, 1988), p. 16.

(44) "Formulation of agrochemical products," AF/WTCH/MPC, FBC Ltd. U.K., 21 March 1980, (hand out).

(45) M.E. Kotz, "Functional Formulation for prescription performance," *Pesticide Formulations and Application Systems*, L.D. Spicer and T.M. Kaneko, eds. American Society for Testing and Materials, 5, 1986, p. 3.

(46) L.A. Lacey, "Production and formulation of *Bacillus sphaericus*", *Mosquito News*, 44, 1984, p. 153; WHO Consultation on the Development of *Bacillus sphaericus* as a Microbial Larvicide, Geneva, 7 October 1985.

(47) S.K. Khetan, "Insecticide formulation technology," Pondicherry, India, WHO Consultation on Bacterial Formulations for Cost-effective Vector Control in Endemic Areas, 19 October 1988.

(48) P.K. Ramdas and S.K. Khetan, "A self-spreading formulation for the control of surface feeding/inhabiting aquatic pests, Abstracts, Vol. 2, Hamburg, Germany, *Seventh International Congress on Pesticide Chemistry*, H. Frehse, E. Kesseler-Schmitz and S. Conway, eds. 1990, p. 39.

LABORATORY PERSPECTIVES

CHAPTER

5

Pesticide Formulation Analysis

Warren R. Bontoyan[+]

Maryland Department of Agriculture
State Chemist Section
Annapolis, Maryland 21401, United States

1 OVERVIEW

Developing nations purchase pesticide products to protect plants and for the control or eradication of biological vectors for health protection. Not only must these products be efficacious but they must also present a minimum danger to human health and the environment. Purchases are, to a large degree, dependent on the advice and guidance of FAO and WHO. These United Nations specialized agencies develop specifications for pesticide products through the cooperation of industrialized nations that have highly developed pesticide industries regulated by scientists and officials who have a thorough understanding of pesticide manufacturing, storage and use. The specifications include many of the same requirements needed to satisfy the stringent registration and enforcement policies of these industrialized countries. To satisfy requirements pertaining to safety, a pesticide manufacturer must develop and submit toxicological and ecological data in support of product registration. The potential registrant must also furnish appropriate chemical methods that can be used by regulatory agencies to monitor the quality of pesticide formulations and any residues or metabolites that may be present in the environment or

[+]Former Director of the United States Environmental Protection Agency, Office of Pesticide Programs Laboratory, Beltsville, Maryland.

food as a result of agricultural or human health protection programmes that use pesticides. In order to ensure that reliable and accurate methods of analysis are used to assess the quality of pesticide products used by developing nations, these methods are part of the specifications.

2 PHYSICAL TESTING ASPECTS OF PESTICIDE FORMULATION

Health and environmental concerns generally have received the most attention by regulatory agencies. However, the attention given to the physical nature of pesticide formulations and technical material in some countries (e.g. the United States) is much less than that given to the chemical ingredients. Products with optimum physical qualities relevant to application and storage may be more efficacious than formulations with greater quantities of active ingredients. Use of these products reduces the quantity of pesticides necessary to obtain control.

Therefore, the importance of the physical qualities of technical and formulated products is at least as great as the necessity for a product to contain the guaranteed active ingredient(s) (AI). In a world of limited monetary resources, developing countries must use products that have maximum efficacy with a minimum of AI. The saving is not only measured by the AI cost but is also measured in terms of future health and environmental costs. Both FAO and WHO are cognizant of these considerations and have expended much effort, time and money in the preparation of specifications and a manual on the development and use of specifications for plant protection products. Both FAO and WHO, in close cooperation with the Collaborative International Pesticide Analytical Council (CIPAC) and the International Group of National Associations of Manufacturers of Agrochemical Products (GIFAP) have developed both physical and chemical specifications for pesticide products.

Pesticide formulation physical properties that should be evaluated are described in FAO Plant Production and Protection Paper 85 (1). Some of the more important properties are:

(a) *Particle size by dry and wet sieve tests.* The wet sieve determination is important for wettable powders and suspension concentrates to determine if a product could clog the nozzles of spray applicators. For granules and dustable powders, particle size is important as it directly relates to AI effectiveness from the perspective of release and surface area;

(b) *Flowability of powders test.* This test is important to determine that powder formulations will flow freely from the application machine to give a uniform application;

(c) *Suspensibility.* This test is important to determine if sufficient active ingredient remains in suspension to give a satisfactory and effective mixture during spray application. The test is applicable to suspension concentrates, wettable powders, capsule suspensions and water granules;

(d) *Wettability of the product.* This test determines if powders are rapidly wetted when mixed with water. The test will ensure the use of material that can be totally and evenly applied from spray applicators;

(e) *Persistent foam.* This test is to determine if excess foam will occur when filling spray tanks. It is applicable to wettable powders, emulsifiable concentrates, capsule suspension concentrates and water-dispersible granules;

(f) *Emulsion stability and re-emulsification.* This test is to determine the stability of the dilute emulsion on standing and its ability to be re-emulsified. The test is applicable to emulsifiable concentrates;

(g) *Wetting of crops.* This test is to give an indication if a dilute spray will satisfactorily wet the leaves of a crop in accordance with instructions;

(h) *Acidity, alkalinity or pH.* This test is to determine the potential for active ingredient degradation, deterioration of the physical properties of the formulation and corrosion of the container. The test is applicable to all products;

(i) *Flash point:* This test is to determine the flammability hazard of a product;

(j) *Stability (storage above 20°C/below 20°C)*: This test gives an indication of the physical stability of products below and above normal temperatures. It is conducted for short intervals. For above 20°C, the test is conducted on samples stored at 54 ± 2°C for 14 days. For temperatures below 20°C, the test is conducted on samples at 0 ± 1°C for seven days.

3 CHEMICAL ASPECTS OF FORMULATIONS

As previously mentioned, much attention has been focused on the chemical characteristics of compounds proposed for registration. In addition to examining the potential harmful effects of the AI, equal attention is given to the evaluation of degradation and/or metabolic products, as well as related chemicals present in technical material either as reaction products and/or impurities. Once a chemical impurity or reaction product is recognized as a potential danger, a chemical specification for it in either the technical or formulated product is established. Therefore, accurate and precise methods of analysis are a must in order to determine adherence to a chemical specification. For both FAO and WHO specifications, CIPAC and the Association of Official Analytical Chemists (AOAC) are the primary organizations that develop standard chemical methods of analysis for both technical and formulated materials. A method is accepted for use by FAO/WHO and most regulatory agencies throughout the world only after the successful completion of a well planned CIPAC/AOAC collaborative study. The methods are generally very sophisticated and are specific for the analyte of interest. The CIPAC and the AOAC (2) methods are available and published at regular intervals.

Today, almost all methods for formulation analysis are either GLC or HPLC. Most of the methods are reliable, less prone to interference than methods such as infra-red (IR) analysis, fairly rapid and quite specific. However, there are some who maintain that IR methods provide reliable and rapid analysis especially if the analytes are properly grouped for analysis. In terms of time, IR analysis compares favourably to GLC/HPLC and is perhaps even faster. However, as previously stated, IR analysis is subject to interference and is less specific than GLC/HPLC.

Because of the continued emphasis on GLC and HPLC analysis, IR methods are not being developed to any significant degree. Therefore, demand for GLC/HPLC instruments in developing nations will probably continue to increase. From a production point of view, the author believes thin layer chromatography (TLC) can also make a significant contribution to analysis of AIs and impurities. A brief discussion of TLC's potential role will be discussed later.

For years, there has been a question as to whether methods using sophisticated instruments are practical for use by developing nations. The concern is valid; however, it is most probable that scientists will continue to develop methods that require the use of GLC or HPLC with highly specific and sensitive detectors. The problems with such methods are obvious:

(a) Are parts and repair of these instruments readily available in developing nations ?

(b) Do electrical currents have significant fluctuations ?

(c) Will the instruments be used in labs with proper temperature and humidity control ?

Experience has indicated that in a significant number of developing nations, these questions will be answered, at least for now, in the negative. What then is the solution? For some, it is the establishment of central laboratories in different regions of developing nations, with sufficient funding for the training of repair technicians. Others have suggested the use of TLC, which can detect the presence of many AIs, which can then be quantified by a trained technician. Usually, visual estimation of the AI is acceptable to determine the potential efficacy of a product with regards to the analyte. A combination of TLC and elemental analysis is not an unreasonable approach for those countries which cannot, on a routine basis, meet the criteria for maintaining or performing analysis with sophisticated electronic instruments. Because FAO/WHO specifications require identity tests, a number of TLC systems have been incorporated. FAO/WHO specifications also address the presence and amount of toxic impurities in technical and formulated products. In addition to these toxic chemicals, levels of impurities associated with stability or product acceptance are also of concern. Although TLC may be applicable for the detection of many impurities, it is probably safe to assume that GLC/HPLC methods are preferred for impurities that are biologically significant at the ppb or low ppm levels.

In general, the methods preferred for technical and formulation analysis are GLC and HPLC. There are some who maintain that GLC methods using thermal conductivity detection (TCD) should be the system of choice because the detector is the least sensitive of various detectors, thereby negating errors from dilution. Although a number of early methods were developed for thermal detection, work in method development has focused on flame ionization detection (FID). FID methods require a dilution at least a magnitude greater than TCD; however, analytical errors from dilution have been overcome with the use of internal standards. Flame ionization also has the sensitivity required for detecting and quantifying low level impurities of toxicological or environmental significance. In addition to FID and TCD there are the nitrogen/phosphorous detectors (NPDs) and flame photometric detectors (FPDs). These have distinct advantages for formulation analysis in that they give added specificity (they are nitrogen- and phosphorous-specific). However, these detectors are very sensitive and present problems of accuracy and precision because of the need to dilute sample solutions by several magnitudes. Again, internal standards can, to some degree, compensate for this deficiency.

Besides the TCD, FID, FPD and NPD detectors, other GLC detectors have potential use for the analysis of pesticides. However, their use is limited due to high sensitivity and the absence of applicable methods: Electro conductivity detector (ElCD, Hall), Infra-red detector (IRD), Mass selective detector (MSD) and Atomic emission detector (AED).

ElCD is highly specific for halogenated and nitrogen- and sulphur-containing pesticides. Specificity is based on reduction or oxidation of the pesticide to form either HCl, NH_3 or SO_3. The detector does need a fair amount of maintenance and may not be cost-effective for analyzing a large number of formulations. These considerations, combined with the high degree of sensitivity (need for large dilution), do not at this time make ElCD a practical choice for formulation analysis.

IRD is based on the absorption of infra-red radiation. The resulting signal is enhanced by Fourier Transform. This detector is not used at this time to any significant extent for pesticide analysis.

MSD is used for confirmation of pesticide residue analysis performed by ECD, FID, NPD or FPD detectors. In addition to confirmation, the MSD can be used in the single-ion mode (SIM) to quantify pesticide residues at low ppb levels. It has little practical use for routine formulation analysis other than confirming the presence of low levels of toxic impurities or cross-contamination.

AED is based on the excitation of atoms to a higher energy level and measuring the specific wavelength of the light emitted as atoms return to the ground state. Wylie and Oguchi (3) analyzed 13 different herbicides for eight elements with a GLC-AE chromatograph and auxiliary columns.

Table 5.1 gives a description of the various detectors and their applicability to pesticide analysis (formulation and residues). Although ECD, NPD, and FPD detectors have applicability for formulation analysis, almost all GLC methods in AOAC and CIPAC use FID detection (see Tables 5.2 and 5.3).

There are disadvantages to GLC analysis. Many compounds are heat-labile and as such may degrade or undergo rearrangement in the injection port or column. If the decomposition temperature is in the range of 200°C or slightly greater, the problem can be overcome by slightly lowering the injection port temperature, which is generally operated at about 200°C. Column temperatures can also be lowered to prevent decomposition. However, attention must be given to the effect lower temperatures may have on retention times and peak shapes. One should also take precautions to ensure that detector temperatures are high enough to prevent distillation of analytes that emerge from the column to detectors. The use of metal

Table 5.1 Various detectors and their applicability to pesticide analysis

Detector	*Specificity*	*Formulation Analytical Range*	*Formulation*	*Residue*
TCD	No	0.01–100% AI in formulation	Excellent (external and internal Stds.)	No
FID	No	0.001–100% AI in formulation	Excellent with I.S.	Yes (0.01 ppm or above)
ECD	Specific compounds only	Requires several magnitudes dilution	Some with I.S.	Excellent (to below ppb level)
NPD	Yes (P, S)	"	"	Excellent (to 1 ppb)
FPD	Yes (P, S)	"	"	Yes (to 0.01 ppm)
ElCD	Yes (halogen, N and S)	"	"	Yes (to 1 ppb)
MSD	Yes	"	Confirmation	Yes

columns will also significantly increase the chances for decomposition or rearrangement of analytes. Only glass columns should be used for pesticide analysis. In addition, all transfer lines should be glass (e.g. glass sleeves for on-column injection and the transfer line from column to detector).

With the advent of HPLC, investigators focused their attention on the advantages it offered over GLC. Obviously, pesticide compounds that are heat-labile are not analyzed by GLC. There are no heat decomposition problems with HPLC. HPLC not only eliminates problems of degradation in high-temperature injection ports and columns, but it also allows the analyst to use numerous mobile solvent combinations for achieving separation of an analyte. In addition, the polarity (pH) of the mobile solvent can be adjusted with buffers. Ion suppression techniques can also be used to influence rates of analyte movement through a column. While GLC columns are based on the relative affinity of the analyte to the coated column stationary phase, HPLC gives the added advantages of analyte partitioning between column material and the mobile phase (reverse and normal phase chromatography), as well as separation by stearic exclusion, ion exchange and absorption.

Carbamate pesticide residues which may degrade in GLC analysis should be analyzed by HPLC with post column derivatization with analysis should be analyzed by HPLC with post column derivatization with a fluorescence detector. HPLC also eliminates the need for esterifying phenoxy acid herbicides and their corresponding salts which will not chromatograph on GLC columns.

HPLC can also be used to increase the specificity of analysis by first obtaining a complete UV scan to determine if several maximas are present. If there are, then the final extraction solution can be subjected to analysis at the several wavelengths of maximum absorption. Some instruments have the ability to stop flow and obtain complete UV scans. Others use a diode array system to determine purity of the peak.

Although HPLC usually has both refractive index and UV detectors, UV is the choice for pesticide analysis. UV detectors are available either as the single or variable wavelength instruments. There is some indication that variable wavelength detectors are somewhat less sensitive then fixed wavelength ones; however, the versatility of the variable instrument more than outweighs this questionable sensitivity difference. Although GLC-FID analyses are more sensitive, for formulations this has little significance. It is only when performing residue analysis that sensitivity has a significant impact. Even then, the sensitivity factor can in some instances be overcome by concentrating of the extracting solvent and/or injecting a larger aliquot of sample extract. HPLC columns can accommodate larger amounts of analytes (greater amount of extraction solvent) than GLC columns.

There are some differences of opinion as to methods for quantifying of analytes by HPLC, i.e. external standard or internal standard methods. Some are of the opinion that with HPLC's accurate loop injection systems, it doesn't matter whether the analyst uses internal or external standards for quantification. Internal standards may, however, serve as a marker if questions arise as to slight changes in retention times. Tables 5.2 and 5.3 give parameters for a significant number of GLC/HPLC pesticide formulation methods which are official (successfully completed collaborative study) and appear either in the AOAC or CIPAC books of methods. It is apparent that acceptable AOAC and CIPAC

Table 5.2 GLC methods

Pesticide	*CIPAC*	*AOAC 16th Ed.*	*Column*	*Detector*	*Internal standard*
Alachlor					
Technical	Handbook D	985.04	10% SP–2250 on	FID	Di-n-pentyl phthalate
EC	"	"	100–200 Supelcoport	"	
Granules	"	"	"	"	
Capsule suspension	Handbook E	983.03	"	"	
Allethrin {d-trans}					
Technical	Handbook 1A	973.12	5% OV-1 on 80–100	FID	Dibutyl phthalate
Formulation	"	"	Chromosorb W(HP)	"	
Ametryn {a}					
Formulation		971.08	3% Carbowax 20M on 80–100 Gas-Chrom Q	FID	Dieldrin
Atrazine					
Technical	Handbook 1A	971.08	3% Carbowax 20M on	FID	Dieldrin
Formulation			80–100 Gas-Chrom Q	"	
Benfluralin					
Technical	Handbook 1B	973.14	5% DC200 on 80–100	FID	Diisobutyl phthalate
Formulation	"	"	Chromosorb W(HP)		
Bromoxynil					
Technical	Handbook 1C		2.5% Apiezon L on 80–100 Chromosorb G-AW-DCMS	FID	Diazomethane methylation of bromoxynil + 2, 4 -dichlorobenzoic acid
Bromoxynil octanoate					
Technical	Handbook 1C	980.05	10% SP–2100 on 100–120	FID	Docosane
Formulation	"		Supelcoport	"	

(a) Applicable to atrazine, prometon, prometryn, propazine and terbutryn.

(Contd...)

Table 5.2 GLC methods *(Contd...)*

Pesticide	*CIPAC*	*AOAC 16th Ed.*	*Column*	*Detector*	*Internal standard*
Buprimate					
Technical	Handbook1C, F		10% SE-30 on 100–200 on	FID	n-Tetracosane
EC	"		Chromosorb W-HP	"	
Wettable powder	"		"	"	
Butachlor					
Technical	Handbook D	986.04	10% SP-2250 on 100–120	FID	Triphenyl phosphate
Formulations	"	"	Supelcoport	"	"
Butylate					
Technical	Handbook 1B	974.05	3% SE-30 or OV-1 on	FID	EPTC
Formulations	"	"	60–80 Gas-Chrom Q	"	
Captan					
Technical	Handbook 1C	971.05	3% XE-60 on 100–200	TC/FID	Tech. dieldrin
Wettable powder	"		Chromosorb G-AW	"	"
Dustable powder	"		"	"	"
Chlordimeform					
Technical	Handbook D	985.05	3% Carbowax-20M on	FID	Diethyl terephthalate
Formulation	"	"	80–100 Gas-Chrom Q	"	
Chlorobenzilate					
Technical	Handbook 1A	971.08	3% Carbowax 20M on	FID	Dibenzyl succinate
Dispersible powder	"	"	80–100 Gas-Chrom Q	"	"
EC	"	"	"		"

(Contd...)

Table 5.2 GLC methods *(Contd...)*

Pesticide	*CIPAC*	*AOAC 16th Ed.*	*Column*	*Detector*	*Internal standard*
Chloropropylate					
Technical	Handbook 1A	971.08	3% Carbowax 20M on 80–100 Gas-Chrom Q	FID	Dibenzyl succinate
EC	"	"		"	"
Chlorpropham					
Dustable powder	Handbook 1C		4% OV-1 on 100–120 Chromosorb W-HP	FID	n-Octadecane
				"	
Chlorthal-Methyl (Dacthal)					
Technical	Handbook 1C	970.05	10% silicone UC-98 on 80–100 Silanized Diatoport S	FID	No
Wettable powder	"	"			
Granule	"	"			
Chlorthiamid					
Technical	Handbook 1A		3% OV-225 on 100–120 Gas-Chrom Q	FID	No
Dispersible powder	"			"	
Granule	"			"	
Cycloate					
Technical	Handbook 1A	974.05	3% SE-30 or	FID	Pebulate
Formulation	"	"	3% OV-1 on 60–80 Gas-Chrom Q	"	
Cypermethrin					
Technical	Handbook 1C,	986.02	3% OV-101 on 100-120 Chromosorb W-HP	FID	Di-(2-ethylhexyl) phthalate
Wettable powder	"	"		"	
EC	"	"		"	

(Contd...)

Table 5.2 GLC methods *(Contd...)*

Pesticide	*CIPAC*	*AOAC 16th Ed.*	*Column*	*Detector*	*Internal standard*
Diazinon					
Technical	Handbook 1A	971.08	10% silicon DC 200 on 80-100 Gas-Chrom Q	FID	Aldrin
Dusts	"	"		"	"
Dispersible powder	"	"	"	"	"
EC	"	"	"	"	"
Microencapsulated		982.06	3% OV-17 on 80–100 Supelcoport		
Dichlobenil					
Technical	Handbook 1B	979.03	10% Carbowax 20M on 100–120 Chromosorb PAW	FID	Methyl myristate
Wettable powder	"	"		"	"
Granules	"	"	"	"	"
Disulfoton					
Technical	Handbook 1C	980.10	10% Silicone SE-30 on 80–100 Chromosorb W-HP	FID	Di-n-Butyl phthalate
EC	"	"			
Endosulfan					
Technical	Handbook 1C	983.08	10% OV-210 on 80–100 Chromosorb W-HP	TCD, FID	Di-(2-ethylhexyl) phthalate
Dustable powder	"	"			
Wettable powder	"	"			
EC	"	"			
EPTC					
Technical	Handbook 1B	974.05	3% SE-30 or OV-1 on 60–80 Gas-Chrom Q	FID	Butylate
EC	"	"			
Granules	"	"			
Fenitrothion					
Technical		989.02	7.5% OV-210 on 100–120 Chromosorb W-HP	FID	Dibutyl sebacate

(Contd...)

Table 5.2 GLC methods *(Contd...)*

Pesticide	*CIPAC*	*AOAC 16th Ed.*	*Column*	*Detector*	*Internal standard*
Wettable powder	"	989.02	7.5% OV-210 on 100–120 Chromosorb W-HP		Dibutyl sebacate
EC	"	"			
Fensulfothion					
Technical		986.07	5% OV-330 on 80–100 Chromosorb W-HP	FID	4-Chlorophenyl sulfoxide
Formulation					
Fenvalerate					
Technical	Handbook D,F		2% Apiezon L on 100–120 Chromosorb W-HP	FID	Diphenyl phthalate
EC	Handbook G				
Fluazifop-butyl					
Technical	Handbook D,G	984.08	3% OV-17 on 100–120 Chromosorb W-HP	FID	Dibenzyl succinate
EC	"	"			
Granules	"	"			
Fluometuron					
Technical	Handbook 1B	977.07	2% OV-3 on 80–100 Gas-Chrom Q	FID	Diethyl phthalate
Wettable powder	"	"			
Iprobenfos					
Technical	Handbook D		5% Silicone SE-30 on 60–80 Chromosorb W (AW-DMCS)	FID	Diallyl phthalate
EC	"				
Isofenfos					
Technical	Handbook D	987.01	10% SP-2100 on 80–100 Supelcoport	TCD, FID	Diisobutyl phthalate
Wettable powder	"	"			
EC	"	"			
Granules	"	"			

(Contd...)

Table 5.2 GLC methods *(Contd...)*

Pesticide	*CIPAC*	*AOAC 16th Ed.*	*Column*	*Detector*	*Internal standard*
Malathion					
Technical	Handbook 1B	979.05	5-7.5% OV-210 on 100–120 Chromosorb W-HP	FID	1,3-Diphenoxybenzene
Dustable powder	"	"			
Wettable powder	"	"			
EC	"	"			
Mecoprop					
Technical	Handbook 1C		2.5% Apiezon L on 80–100 Chromosorb G(AW,DMCS)	FID	Undec-10-enoic acid
Salts (Tech)					
Metolachlor					
Technical	Handbook D	985.06	3% OV-1 on 80–100 Gas-Chrom Q	FID	Dipentyl phthalate
Formulations	"	"			
Metribuzin					
Technical	Handbook D	984.11	3% OV-225 on 80–100 Gas-Chrom Q	FID	Di-n-butyl phthalate
Formulations	"	"			
Parathion					
Technical	Handbook 1B	978.06	1.5% SE-30 + 1.5% OV-210 on 80–100 Gas-Chrom Q	FID	Dipentyl phthalate
Formulation	"	"			
Parathion methyl					
Technical	Handbook 1B	977.04	Same as Parathion	FID	p, p′-DDE
Formulation	"	"			
Microencapsulated		980.11	3% OV-1 on 80–100 Supelcoport	FID	Bis(2-methoxyethyl) phthalate

(Contd...)

Table 5.2 GLC methods *(Contd...)*

Pesticide	*CIPAC*	*AOAC 16th Ed.*	*Column*	*Detector*	*Internal standard*
Permethrin					
Technical	Handbook 1C, F	986.03	3% OV-210 on 100–120 Chromosorb W-HP	FID	n-Octacosane
Wettable powder	"	"			
EC	"	"			
Dustable powder	"	"			
Phosalone					
Technical	Handbook D		10% SE-30 on 80–100 Chromosorb W-HP	TCD/FID	Dioctyl sebacate
Wettable powder	"				
EC	"				
Pirimicarb					
Technical	CIPAC	982.08	10% Silicone-SE-30 on 100–120 Chromosorb W-HP	FID	Nonadecane
Formulations	Proceedings 81				
Pirimiphos-ethyl					
Technical	Handbook D		10% SE-30 on 100–120 Chromosorb W-HP	FID	Nonadecane
EC	"				
Pirimiphos-methyl					
Technical	Handbook 1C, F	991.34	10% Silicone SE-30 on 100–200 Chromosorb W-HP	FID	n-Octadecane
Wettable powder	"	"			
EC	"	"			
Dustable powder	"	"			
ULV	"	"			
Aerosol	"	"			

(Contd...)

Table 5.2 GLC methods *(Contd...)*

Pesticide	*CIPAC*	*AOAC 16th Ed.*	*Column*	*Detector*	*Internal standard*
Prometone					
Technical	Handbook 1C	971.08	3% Carbowax 20M on 80–100 Gas-Chrom Q	FID	Tech. dieldrin
Formulations	"	"			
Prometryn					
Technical	Handbook 1B	971.08	3% Carbowax 20M on 80–100 Gas-Chrom Q	FID	Tech. dieldrin
Formulation	"	"			
Propachlor					
Technical	Handbook D	986.05	10% SP-2250 on 100–200 Supelcoport	FID	Di-isobutyl phthalate
Formulation	"	"			
Propazine					
Technical	Handbook 1A	971.08	3% Carbowax 20M on 80–100 Gas-Chrom Q	FID	Tech. dieldrin
Formulation	"	"			
Pyrethrin					
Formulation		982.02	5% OV-1 on 80–100 Chromosorb W-HP	FID	Dicyclohexyl phthalate
Terbutryn					
Technical	Handbook 1A	971.08	3% Carbowax 20M on 80–100 Gas-Chrom Q	FID	Tech. dieldrin
Formulation	"				
Tetradifon					
Technical	Handbook 1B	981.02	3% SE-52 on 100–200 Chromosorb W-HP	FID	Hexacosane
Formulation	"	"			
Trifluralin					
Technical	Handbook 1A	973.14	5% DC200 on 80–100 Chromosorb W-HP	FID	Diisobutyl phthalate
Formulation	"				

(Contd...)

Table 5.2 GLC methods (*Contd...*)

Pesticide	*CIPAC*	*AOAC 16th Ed.*	*Column*	*Detector*	*Internal standard*
Vernolate					
Technical	Handbook 1B	974.05	3% SE-30 or OV-1 on 60–80 Gas-Chrom Q	FID	Cycloate
Formulation	"	"			
Vinclozolin					
Technical	Handbook D		10% Silicone SE-30 on 60–80 Chromosorb G AW-DMCS	FID	Fluoranthene
Wettable powder	"				
Suspension concentrate	"				

Table 5.3 HPLC methods

Pesticide	*CIPAC*	*AOAC 16th Ed.*	*Column*	*Mobile solvent*	*UV*	*Wave-length*	*Internal standard*
Aminocarb							
Technical	Handbook D	985.02	C18 bonded Silica Gel (e.g) Particil-10 ODS-3, Micropak MCH-10, Ultrapack-ODS, Zorbax-Sil	Tetrahydrofuran buffer-(60+40) Buffer Sol.–1.36g KH_2PO_4 and 2.68g Na_2HPO_4 $7H_2O$ in one liter.	✓	246nm	n-Butyrophenone
Formulations	"	"					
Bendiocarb							
Technical	Handbook D	986.09	Partisil-10 ODS-2	CH_3CN+H_2O (800 ml CH_3CN diluted to 2L with water)	✓	254nm	Propiophenone
Wettable powder	"	"					
Benomyl							
Technical	Handbook D	984.09	C18 Reverse phase	CH_3CN+2% acetic acid soln. (80+20)		290/ 280nm	No
Wettable powder	"	"					
Bentazone							
Technical	Handbook 1C	993.02	μ- Bondapak-C18	MeOH/.075N Na-acetate buffer (pH 6) (40+60)	✓	340nm	No
Aqueous formulation		"					
Brodifacoum							
Technical	Handbook 1C	983.11	Zorbax ODS (Rev. Phase)	MeOH-H_2O-HOAc (94.2+5.0+0.8)	✓	254nm	1,3,5-Triphenyl benzene
Bait concentrate	"	"	"				
(powder and liquid)	"	"					

(Contd...)

Table 5.3 HPLC methods *(Contd...)*

Pesticide	*CIPAC*	*AOAC 16th Ed.*	*Column*	*Mobile solvent*	*UV*	*Wave-length*	*Internal standard*
Captan							
Technical	Handbook 1C	980.06	Partisil 10	CH_2Cl_2	✓	254nm	Diethyl phthalate
Formulation	"	"	"			"	
Carbofuran							
Techncial	Handbook D	986.10	C18 Reverse phase	Deionized H_2O + MeOH (50+50)	✓	280nm	Acetophenone
Wettable powder	"	"	"			"	"
Suspension con-centrate	"	"	"			"	"
Granule	"	"				"	"
Chloridazon							
Technical	Handbook D		LiChrosorb RP18-7μm	MeOH + H_2O (60 + 40)	✓	286nm	No
Wettable powder	"					"	No
Suspension con-centrate	"		"				
Chlorphoxim							
Techncial	Handbook D		LiChrosorb Si 60	Hexane + EtOAc (950+50)	✓	254nm	No
Wettable powder	"		"			"	No
EC	"		"	"		"	No

(Contd...)

Table 5.3 HPLC methods *(Contd...)*

Pesticide	*CIPAC*	*AOAC 16th Ed.*	*Column*	*Mobile solvent*	*UV*	*Wave-length*	*Internal standard*
Chinomethionat							
Technical	Handbook 1C		LiChrosorb Si60	n-Hexane + tetrahydro-furan (95+5)		250nm	No
Wettable powder	"		"				
Dustable powder	"		"				
Chlorpyrifos							
Technical	Handbook 1C	981.03	Zorbax ODS	CH_3CN+H_2O+HOAc (82+17.5+0.5)		300nm	1,4-Dibromo-naphthalene
Formulation	"	"					
Cyanazine							
Technical	Handbook 1C	991.32	Lichrosorb NH_2	CH_2Cl_2+2-propanol (99+1)	✓	254nm	o-Nitroaniline
Wettable powder	"	"	"			254nm	
Suspension con-centrate	"	"	"			254nm	
Cyhexatin							
Technical	Handbook D	988.02	Lichrosorb RP-18 (ODS Bonded Silica)	MeOH+H_2O+HCl +NaCl {93+7+.001M+.005M}			n-Decylbenzene
Wettable powder	"	"				214nm	
Water disp. granules	"	"	"				
Suspension con-centrate	"	"	"				
Ultra low volume	"	"	"				
Cypermethrin							
Technical	Handbook 1C		Spherisorb S5W	0.5% v/v EtOAc/ Iso-octane	✓	278nm	Methyl benzoate
Formulations	"						

(Contd...)

Table 5.3 HPLC methods *(Contd...)*

Pesticide	*CIPAC*	*AOAC 16th Ed.*	*Column*	*Mobile solvent*	*UV*	*Wave-length*	*Internal standard*
2, 4-D							
Formulations (acid, salts, esters)	Handbook 1C	978.05	Partisil 10 μm ODS	CH_3CN + deionized H_2O (18+82) pH 2.95	✓	280nm	p-Bromo-phenol
Dalapon							
Technical	Handbook D	984.06	Partisil 10-25 ODS-3	CH_3CN+n-octylamine + $(NH_4)_2HPO_4$ (200ml+1.6ml +2.4g) diluted to 1L with water		214nm	
Formulation	"	"					
Deltamethrin							
Technical	Handbook D		Lichrosorb Si 60-5	Iso-octane + dioxane (80 + 20)	✓	254nm	No
Wettable powder	"						
EC	"						
Dustable powder	"						
Ultra low volume	"						
Dicamba							
Amine salts		984.07	Partisil 10-25 ODS-3	H_2O + CH_3CN + 17.7 N · NaOH (610+330+16.9)	✓	280nm	Salicylic acid
Dichlofluanid							
Technical	Handbook 1C		Spherisorb ODS	CH_3CN+H_2O (48+52)	✓	254nm	No
Wettable powder	"						
Dichlorprop							
Technical	Handbook G		Nucleosil C18-reverse phase	CH_3CN+water+H_2SO_4 (350+650+5) v/v (racemat determination)	✓	220nm	

(Contd...)

Table 5.3 HPLC methods *(Contd...)*

Pesticide	*CIPAC*	*AOAC 16th Ed.*	*Column*	*Mobile solvent*	*UV*	*Wave-length*	*Internal standard*
Wettable powder	Handbook G		Nucleosil chiral-2	n-Heptane + n-pentanol + tri-fluoroacetic acid (1000 + 0.2 + 0.5) v/v (enantiomeric ratio determination)		229nm	
Soluble liquid	"	"					
Dicofol							
Technical	Handbook D	986.06	Zorbax C8	MeOH + H_2O + HOAc (75 + 25 + 0.2)	✓	254nm	No
EC	"	"					
Diflubenzuran							
Technical	CIPAC Proceeding 81	983.07	Zorbax BP-C8	CH_3CN + 1,4- dioxane + H_2O (45 + 45 + 10)	✓	254nm	Linuron
Formulations							
Ethion							
Technical	Handbook 1B	979.04	Waters µBondapak C18 or DuPont ODS Permaphase	MeOH + H_2O (90 + 10)	✓	254nm	PCNB
EC	"	"					
Wettable powder	"	"					
Oil formulation	"	"					
Fensulfothion							
Technical	Handbook 1C	983.09	Partisil PXS 10/25 ODS-2	MeOH + H_2O (80 + 20) buffered to 0.0025M with H_3PO_4		230nm	Benzophenone
EC	"						
Granule	"						
Fluazifop-p-butyl							
Technical	Handbook G		Chira-Chrom-1	2,2,4-Trimethyl pentane	✓	254nm	Benzil

(Contd...)

Table 5.3 HPLC methods *(Contd...)*

Pesticide	*CIPAC*	*AOAC 16th Ed.*	*Column*	*Mobile solvent*	*UV*	*Wave-length*	*Internal standard*
EC	Handbook G		Chira-Chrom-1	+Propane-2-ol+ trifluoroacetic acid (1000 + 7 + 2)		254nm	Benzil
WG	"						
Folpet							
Technical	Handbook 1B	977.03	10μm Silicagel particles (Waters Asso. Inc. No. 27477)	Degassed CH_2Cl_2	✓	254nm	Dibutyl phthalate
Wettable powder	"	"					
Dustable powder	"	"					
Isoprocarb							
Technical	Handbook D		LiChrosorb Si 60	n-Hexane + tetrahydrofuran (82 + 18)	✓	254nm	No
Wettable powder	"						
MCPA							
Technical	Handbook 1C	980.07	Partisil 10μm ODS	$CH_3CN + H_2O$ (15 + 85) pH 2.83	✓	280nm	Salicylic acid
Wettable powder	"	"					
Esters	"	"					
Mecoprop							
Technical	Handbook 1C		μBondapak C18	MeOH + NaOAc buffer (45 + 55) pH 4.17	✓	280nm	2-(2,4-Dibromo -phenoxy) propionic acid
Methiocarb							
Technical	Handbook D	984.10	Partisil - 10 ODS-3	$CH_3CN + H_2O$ (60 + 40)	✓	266nm	Acetophenone
Wettable powder	"	"					
Suspension concentrate	"	"					
Dustable powder	"	"					

(Contd...)

Table 5.3 HPLC Methods *(Contd...)*

Pesticide	*CIPAC*	*AOAC 16th Ed.*	*Column*	*Mobile solvent*	*UV*	*Wave-length*	*Internal standard*
Parathion							
Technical	Handbook 1B	978.07	10μm silicagel particles (Waters Assn. Inc. #27477)	$CHCl_3$	✓	254nm	Benzophenone
Formulation	"	"					
Parathion methyl							
Technical	Handbook 1B	977.05	10μm silicagel particles (Waters Assn. Inc. #27477)	H_2O saturated $CHCl_3$+ $CHCl_3$ (200 + 300)	✓	254nm	Acetophenone
Formulation	"	"					
Phenmedipham							
Technical	Handbook 1C		10 μm Bondapak C18	CH_3CN+H_2O+Dioxane (480 + 520 + 10)	✓	238nm	Butyl benzoate
EC	"						
Phoxim							
Technical	Handbook 1C		LiChrosorb Si 60	n-hexane+tetrahydro-furan (96 + 4)	✓	254nm	No
EC	"						
Granules	"						
Propoxur							
Technical	Handbook D	984.12	Partisil - 10 ODS 3	CH_3CN + H_2O (60 + 40)	✓	280nm	n-Butyro-phenone
Formulations	"	"					
2, 4, 5-T							
Technical	Handbook 1C	980.08	Partisil 10 μm ODS	CH_3CN + H_2O (18 + 82) pH 2.95	✓	280nm	p-Bromophenol
Formulation	"	"					

GLC methods use internal standards. However, the use of pesticide chemicals as internal standards should be discouraged for future methods. A number of those presently utilized as internal standards have been canceled or suspended by regulatory authorities.

In spite of the excellent specific instrumental analyses available to analytical chemists, there are instances when wet analysis is the only alternative. As previously stated, a combination of TLC and total elemental analysis may have some utility when instrumental methods are unavailable. At this writing, the ethylene bisdithiocarbamate compounds in formulations are analyzed by the AOAC method of CS_2 evolution to quantitatively form a xanthate in alcoholic KOH solution, which is then measured by iodine titration. In such cases, classical wet chemistry methods are acceptable. Many such methods are published in the AOAC Official Methods of Analysis. Tables 5.4 and 5.5 give a few of the official methods, using spectrophotometer and classical analytical methods, respectively.

4 QUALITY ASSURANCE ASPECTS

Although there is vast array of excellent methods available for pesticide analysis, analytical results are only as reliable as the standards and reagents used in the analytical scheme. The availability of reliable standards is critical to an analytical laboratory. The question as to whether developing nations have easy access to reliable standards and proper storage facilities is valid. Obtaining reliable standards is also a problem in technically developed nations. Assuming reliable standards are available, specific conditions of storage and use must be followed. In general, all organic pesticide standards and standard stock solutions should be stored in the dark. Moisture must be as close to zero as possible.

Standards that can withstand high temperatures must be thoroughly documented. Those requiring refrigeration should be stored at about 0°C; however, freezer storage of some standards and stock solutions may result in precipitation. Perhaps one of the greatest dangers of storing organophosphorous pesticides under cold conditions is condensation. All such standards, when removed

Table 5.4 IR and other spectrophotometric methods

Pesticide	*CIPAC*	*AOAC 16th Ed.*	*Wavelength*
Aldicarb			
Technical	Handbook 1A	974.04	IR(5.2-6.0 μm)
Granules	"	"	1,900-1,600 cm^{-1}
Aldrin			
Technical	Handbook 1B	961.05	IR(11.5-13.0 μm)
Dusts	"	"	870-769 cm^{-1}
Granules	"	"	"
Wettable Powder	"	"	"
EC	"		
Bromofenoxim	Handbook 1A	–	IR 820-800 cm^{-1}
Carbaryl			
Dusts		976.04	IR (5.2-6.0 μm)
Liquid suspension			1,900-1,600 cm^{-1}
Cartap			
Technical	Handbook D		Colorimetric (412 nm)
SP	"		"
Chloramben			
Technical	Handbook 1B	971.06	UV (297 nm)
Granules	"	"	"
Aqueous soluble	"	"	"
DDT			
Technical	Handbook 1	960.13	IR pp′ (834 cm^{-1})
Dry formulation	"	"	op′ (750 cm^{-1})
Dicamba			
Technical	Handbook 1A	969.07	IR (9.3 – 10.7 μm)
Amine salts	"	971.07	1100 – 930 cm^{-1}
Diquat			
Formulation (salt soluble concentrate)	Handbook E	969.08	UV (310 nm)

from cold storage, must be allowed to reach room temperature before removing the cap to weigh material. Even the smallest amount of moisture condensing in a standard organic phosphorous material may rapidly hasten hydrolysis (decomposition). If an instrumental method is used, the analyst should obtain a standard of performance for the reference material in order to compare future instrument response (e.g. retention times, number of peaks and instrument response to the same quantity of standard analyte). Without such minimum comparison, questionable results are

Table 5.5. Wet chemistry methods

Pesticide	*CIPAC*	*AOAC 16th Ed.*	*Method*
Amitrole			
Formulations	Handbook 1B	967.06	Titrimetric
Arsenic		922.03	Titrimetric (hydrazine sulfate distillation)
Total			
"		924.04	Titrimetric (iodometric)
"		963.06	Titrimetric (ion exchange)
Water soluble		925.02	Titrimetric
Atrazine			
Dispersible powder	Handbook 1A		Total chlorine (titrimetric)
Bromacil	Handbook 1C		Titration
Chlormequat chloride	Handbook D		Titration (organic chloride)
Chlorotoluron			
Technical	Handbook 1A	977.06	Titration of liberated dimethylamine
Dispersible powder	"	"	
Dicofol			
Technical		976.02	Titration of hydrolyzable organic chlorine
Formulations		"	
Diuron			
Tech, formulation	Handbook 1A		Titration of liberated amine
Ethylene bisdithio-carbamates		965.15	Titration of formed xanthate
Linuron			
Technical	Handbook 1A		Titration of N, O-dimethyl hydroxylamine
Dispersible powder	"		
Monuron			
Technical	Handbook 1A		Titration of liberated dimethylamine
Dispersible powder	"		

obtained. In addition, the general physical condition of new standards must be recorded and compared to the physical appearance of the material when used for future analysis.

When analysing pesticide products, one not only considers the methods, types and techniques used but also QA protocols or plans. In the United States, Federal regulatory agencies are insistent that all laboratories and companies under their jurisdiction have acceptable quality assurance (QA) guidelines as part of their routine activities. Such requirements are not unique to the United States. It behooves all legitimate analytical laboratories to consider basic QA protocol requirements. Some of these are:

(a) **Complete and accurate record-keeping pertaining to samples**

(i) Description of sample
(ii) Where sample was obtained
(iii How sample was and is stored
(iv) Do samples need refrigeration
(v) Are samples properly identified ?
(vi) Who obtained samples ?

(b) **Proposed method of analysis**

(i) Has method been collaborated/studied by CIPAC or AOAC ?

(c) **In-house or uncollaborated methods should meet the following criteria**

(i) Is method scientifically sound ?
(ii) Determine linear range of analysis
(iii) Determine coefficient of correlation
(iv) Determine recovery values
(v) Is method rugged ?
(vi) Repeatability

(d) **GLC considerations**

(i) Does injection port temperature degrade analyte or standard ?
(ii) Does chromatogram indicate overloading ?
(iii) Is gas pure and dry (molecular sieve) ?
(iv) Does column give excessive bleed ?
(v) Reproducible retention time for analyte
(vi) Is peak Gaussian or spiked ?
(vii) Have column characteristics changed as evidenced by different retention time of standard reference solution ?
(viii) Has detector response for standard solutions changed ?
(ix) Linearity ?
(x) Use of internal standard (constant response ratio should be obtained)

(xi) Interfering peaks (co-elution and shoulders on peaks)
(xii) Excessive baseline noise
(xiii) Solvent purity
(xiv) Acceptable/practical retention time
(xv) Allow for elution of material from previous injection before next injection

(e) **HPLC method considerations**

(i) Reproducible retention time
(ii) Interference from other eluants
(iii) Linearity ?
(iv) Change in column characteristics when chromatographing a standard solution
(v) Change in detector response for standard solution
(vi) Selection of wavelength to give maximum absorbance
(vii) Selection of other wavelength maxima to give additional selectivity
(viii) Purity of mobile solvents
(ix) Does chromatogram exhibit column overload ?
(x) Are peak shapes acceptable ?
(xi) Are results improved with internal standards ?
(xii) Is there excessive baseline noise ?

(f) **Balance maintenance**

(i) Are balances periodically checked for accuracy by comparing to standard weights over several ranges ?
(ii) Are calculations recorded in method ?
(iii) Are balances checked periodically for proper operation (e.g. inoperable pans or drifting)
(iv) Are balances serviced periodically ?
(v) Excess vibration
(vi) Drafts
(vii) Extreme temperature
(viii) Corrosion
(ix) Cleanliness

(g) **Calculation considerations**

(i) Triangulation for area determination
(ii) Electronic calculation
(iii) Determine acceptable standard deviation
(iv) Have calculations on deficient samples, overformulation, impurities etc. checked by another analyst, signed and dated

(h) **Data considerations**

(i) Has all work been peer-reviewed to determine if all QA protocol and standard operating procedures were followed ?
(ii) Have all weighings, observations etc. been recorded in a notebook ?

5 SAFETY ASPECTS OF PESTICIDE ANALYSIS LABORATORY

Modern chemical laboratories must give priority to safety considerations. This is particularly true in today's world of concern for the safety of the environment. It also makes sense to be concerned for the health of laboratory personnel who work with pure and concentrated materials of known chronic and acute toxicity. In addition to toxicity, equal concern must be given to fire, explosion, asphyxiation etc. Some of the significant safety concerns are:

(a) **Fire**

(i) All smoking is prohibited
(ii) Are all fire exits properly marked ?
(iii) Is there an emergency exit plan ?
(iv) Are all fire extinguishers properly placed throughout the facility ?
(v) Are fire extinguishers of the appropriate type ?
(vi) Are personnel trained in proper use of extinguishers?
(vii) Are there enough emergency alarms ?
(viii) Are there proper alarms for deaf personnel ?
(ix) Is there automatic fire suppression equipment ?

(x) Are there enough and properly placed emergency showers and eye wash stations ?

(b) **Explosion**

(i) Do all personnel wear eye protectors ?

(ii) Do personnel use proper explosion shields when necessary ?

(iii) Are all personnel familiar with the explosive nature of different reagents/chemicals ?

(c) **Storage**

(i) Are volatile solvents stored in proper storage cabinets, rooms, vaults etc.

(ii) Is there a larger quantity of flammable solvents in laboratory than necessary for a day's operation ?

(iii) Are chemicals segregated in storage according to reactivity ?

(iv) Are chemicals stored or placed on shelves higher than lab bench height ?

(v) Are explosion-resistant refrigerators available ?

(d) **Disposal**

(i) Are there specific proper containers for broken glass, disposable pipettes etc.?

(ii) Is there proper removal of laboratory waste ?

(iii) Is there a protocol for disposal of samples ?

(iv) Solvents should not be discarded to local sewage systems

(v) Solvents should not be evaporated in hoods

(e) **Toxicity**

(i) Are material data safety information sheets available for review by all employees ?

(ii) Are efforts made to not use chemicals that are known carcinogens, mutants, teratogens etc. ?

(iii) No food or drinking is permitted in laboratory

(iv) Is ventilation adequate to keep inhalation exposure within acceptable limits ?

(v) Are protective gloves, respirators, clothing etc. mandatory and available ?
(vi) Are personnel aware of acute toxicity dangers ?
(vii) Have personnel been instructed to not pipette anything by mouth ?
(viii) Are blood analyses of lab personnel done annually ?
(ix) No personnel should work in a laboratory without a colleague in the nearby vicinity

6 NEW ANALYTICAL TECHNIQUES

Before concluding a general presentation on pesticide analysis, it is important to be aware of four new techniques that are now in use. These techniques will undoubtedly become routine in the near future and will further increase the chemist's analytical capability.

Optical isomers of various pesticides have been analyzed by GLC and HPLC without separation. Isomers elute as one peak and calculations give a total isomer contact. Because only one isomer usually has biological activity, the need to analyze specific isomers is obvious. Enantiomers may be analyzed by GLC/HPLC stationary phases. Muller and Bosshardt (4) have successfully separated six propionic acid ester herbicides enantiomers by chiral liquid chromatography and achiral gas chromatography. Tandy and Husband (5) also demonstrated successful separation and analyses of six pesticides by HPLC chiral stationary phases. Chiral column analysis both by HPLC and GLC can be expected to increase with successful collaborative studies.

In addition to chiral column analysis, two recently developed extraction techniques are now used by a number of state and federal laboratories in the United States. The first is a supercritical fluid extraction (SFE). A number of studies have indicated the practicality of this CO_2 extraction procedure. Results obtained compare favorably to commonly accepted extraction techniques. Presentations by Tang et al. (6) and Avila et al. (7) have demonstrated the usefulness of SFE in the analysis of environmental samples for various organic compounds. SFE uses CO_2 as the extraction solvent. The advantages of SFE are (a) the elimination of the disposal problem, (b) increased extraction efficiency (less than half the time required for conventional extraction) and (c) no organic solvents are used except alcohol, which is a modifier (adjust SFE polarity). Although there are at best very few methods or studies applicable to formulations, it is logical to assume that SFE will have a future significant impact on separation and extraction of pesticides from solid formulations.

The other extraction procedure is a very recent development and makes use of an accelerated solvent extractor. This instrument has been successfully used for extracting all types of environmental samples and reduces extraction time to less than an hour and in many instances will perform highly efficient extraction in 10-15 minutes. Usually no more than 25-50 ml of extraction solvent is used per sample. This instrument is also being used on pesticide formulations and is particularly valuable in extraction of solid formulations. The extractions are accomplished in a very short time by use of a heated solvent passed through a sample in a stainless steel cell under very high pressure. Although both SFE and the accelerated solvent extractor are excellent additions to an analytical laboratory's repertoire, they are very expensive and may be beyond the budgets of many government-funded laboratories.

The third development is capillary electrophoresis, which has been successfully used by the United States Environment Protection Agency (EPA) and a number of other laboratories in the United States for the analysis of pesticides extracted from environmental samples. However, its use on formulated material may be a problem because a capillary electrophoresis column can accommodate only microgram quantities of analyte; this would result in very large dilutions and increase the error. In the area of environmental analysis studies, Dinelle et al. (8) and Garcia and Henio (9) indicated successful separation and quantitation of sulfonylureas. Studies by Carneiro et al. (10) and Kaniansky et al. (11) have detected paraquat and diquat at the ppb levels in water. An informational overview of the application of CE for environmental analysis by Brumley (12) shows its potential for the analysis of numerous pollutants.

REFERENCES

(1) *Manual on the Development and Use of FAO Specifications for Plant Protection Products*, 3rd revised ed. (Rome, Food and Agriculture Organization of the United Nations, 1987).

(2) *CIPAC Handbook, Analysis of Technical and Formulated Pesticides*, Collaborative International Pesticides Analytical Council (Cambridge, England, Black Bear Press, Vol. 1, 1970; 1A, 1980; 1B, 1983 and 1C, 1985).

CIPAC Handbook, Analysis of Technical and Formulated Pesticides, A. Martin and G. Dorbat, eds., Collaborative International Pesticides Analytical Council (Cambridge, England, Black Bear Press., Vol. D, 1988; E, 1993 and G, 1995).

CIPAC Handbook, Physico-chemical Methods for Technical and Formulated Pesticides, W. Dorbat and A. Martin, eds., Collaborative International Pesticides Analytical Council, (Cambridge, England, Black Bear Press., Vol. F, 1995).

Official Methods of Analysis, 16th ed., Patricia Cunniff, ed., (Arlington, Virginia, Association of Official Analytical Chemists, 1995).

(3) P. Wylie and R. Oguchi, *GC-Atomic Emission Spectroscopy Using Microwave Plasmas* (Avondale, Pennsylvania, Hewlett-Packard Avondale Division).

(4) M.D. Muller and H.P. Bosshardt, *J. Assoc. Official Anal. Chem.*, 71m 1988, p. 614.

(5) M.J. Tandy and R. Husband, 30th Annual CIPAC Symposium, Vienna, Austria, 1986.

(6) Hua-Tang Tang, James Ho and James Eichelberger, "The analysis of organic pollutants in water by solid phase extraction followed by a supercritical fluid extraction elution," 105th AOAC Annual International Meeting and Exposition, Phoenix, Arizona, 1991.

(7) Viorica Lopez Avil and Werner F. Beckert, "SFE methods in environmental analysis," 105th AOAC Annual International Meeting and Exposition, Phoenix, Arizona, 1991.

(8) G. Dinelli, A. Vicari and P. Catizone, *J. Agric. Food Chem.* 41, 1993, pp. 7712-7746.

(9) F. Garcia, and J. Henion, *J. Chromatogr*, 606, 1992, pp. 237-247.

(10) M.C. Carneiro, L. Puignou and M. T. Galceran, *J. Chromatogr*, A669, 1994, pp. 217-224.

(11) D. Kaniansky, F. Ivanyi and F.I. Onuska, *Anal. Chem.* 66, 1994, pp. 1817-1824.

(12) William C. Brumley, *LC-GC*, 13, No. 7, 1995, pp. 57-67.

CHAPTER

6

Modern Equipment for a Pesticide Formulation Laboratory and Pilot Plant

Sushil K. Khetan

Hindustan Insecticides Limited
Scope Complex, Core 6, 2nd Floor
7 Lodi Road, New Delhi-110003, India

1 INTRODUCTION

A large array of modern equipment is available for processing, testing and analysing of pesticide formulations. When establishing a formulation laboratory, the purpose and function of the laboratory should determine the choice of equipment. The laboratory can have any of the following three basic objectives:

(a) To support general production activities, including quality control;
(b) To carry out formulation development, including specification and development of the required ingredients;
(c) To undertake basic research in pesticide formulation and formulation ingredients.

In practice, each of these may be combined with any or all of the others (1). The role of a formulation laboratory can also evolve: from initially supporting production activities it may subsequently undertake formulation research and development. This can be achieved by gradually adding equipment to meet a required need and degree of sophistication. However, if a clear objective is known in the beginning, it would be better to plan the laboratory layout and provide the requisite support facilities and utilities.

Formulation laboratory equipment may be broadly classified as either processing equipment or analytical equipment. Processing equipment includes equipment that carries out various unit operations such as size reduction, blending, agglomeration and granulation. Analytical equipment can be further divided into two classes: measurement of physical characteristics and chemical assays. Methods for chemical assay are covered in detail in chapter 5. Equipment commonly employed for processing and formulating at an intermediate level as well as for physical testing will be discussed here.

2 FORMULATION LABORATORY TO SUPPORT GENERAL PRODUCTION AND QUALITY CONTROL ACTIVITIES

The functions of such a laboratory would normally be to provide support to production in the areas of physical testing, including (a) assessment of the suitability of alternative fillers or emulsifiers, (b) determination of the quality of incoming raw materials and (c) maintenance of the quality of day-to-day production. The laboratory functions as an integrated component of the production system, leading to fewer lost batches and hence a better output rate. It needs to be suitably equipped to carry out basic quality control tests and storage stability tests on formulations such as (a) dusts and wettable powders, (b) emulsifiable concentrates and emulsion concentrates and (c) suspension concentrates.

Tests to assess the performance of dusts, wettable powders and emulsifiable concentrates can all be carried out in simple glassware, using suitable balances, a stopwatch and a set of sieves. Tests such as dry sieve test, flowability, wettability, suspensibility and foaming do not need elaborate equipment. Similarly, the performance of both wettable powders and emulsifiable concentrates can be assessed using simple glassware.

A simple volumetric assessment of settling, using a graduated measuring cylinder to measure the volume of compacted sediment, comparing it with that of a good quality sample, would give a very good indication of whether a formulation will meet specifications. The effect of changing surfactants or fillers in a wettable powder can be assessed in a similar way. For emulsifiable concentrates and emulsion concentrates, a comparison of the volume of settled cream with that of a standard, using a graduated conical cylinder, will provide emulsion stability information.

It would be obvious that a new formulation or changes in adjuvants in formulations can be usefully assessed without carrying out a special chemical determination in expensive analytical equipment. In many cases, a chemical assay can also be carried out by a volumetric method. Here, an autotitrator (ex Mettler, Switzerland) is very handy and eliminates the possibility of manual errors in repeat analysis. A Brookfield viscometer is another handy device to check the stability of suspension concentrates by measuring the non-Newtonian flow characteristics of the composition at different spindle speeds.

It would be desirable to include in the processing equipment a small hammer mill, or equivalent. However, if this not available, a simple coffee mill or a kitchen grinder can be equally effective provided the material to be ground is not too hard.

3 FORMULATION DEVELOPMENT LABORATORY

For a laboratory designed to develop new formulations, the equipment requirements are basically the same except that more sophisticated and specialized equipment is needed. In particular, more sophisticated powder mills are required which can be used to yield variable particle sizes. Air mills are best when it is desired to obtain the smallest particles. Additionally, equipment for wet grinding and for granulation would also be necessary.

3.1 Processing Equipment

At its simplest, a small hammer mill with exchangeable screens is required, together with a pin mill and, if possible, an air jet mill. Pin and hammer mills produce materials in a particle size range from 5 to 30 μm in diameter, with the majority of the particles being between 10 and 20 μm. Air mills bring the particle size down to 2-5 μm. These three mills produce a range of particle sizes suitable for the study of wettable powders and their physical properties.

Suspension concentrates, on the other hand, are generally prepared by wet milling the pesticide down to 1-2 µm (2). Thus, a laboratory size wet process bead mill such as a Dynomill (ex W. Bachofen, Switzerland) or a Molinex PE 075 vertical mill (ex Gebr. Netzsch, Germany) is essential if any serious work is too be carried out on suspension concentrates. This is often complemented by a Silverson-type high-sheer homogenizer (ex Silverson Machines, England) to provide a lump-free slurry for feeding into the Dynomill. High speed homogenizers are also useful for the preparation of emulsifiable concentrates and emulsion concentrates.

The Dynomill is an agitated-type horizontal bead mill and produces a fine (below 5 µm in diameter) and narrow particle spectrum. It consists of a short horizontal shaft with an agitator disc assembly that turns the beads used as the grinding media. These beads are distributed along the shaft in a stationary chamber (Figure. 6.1).

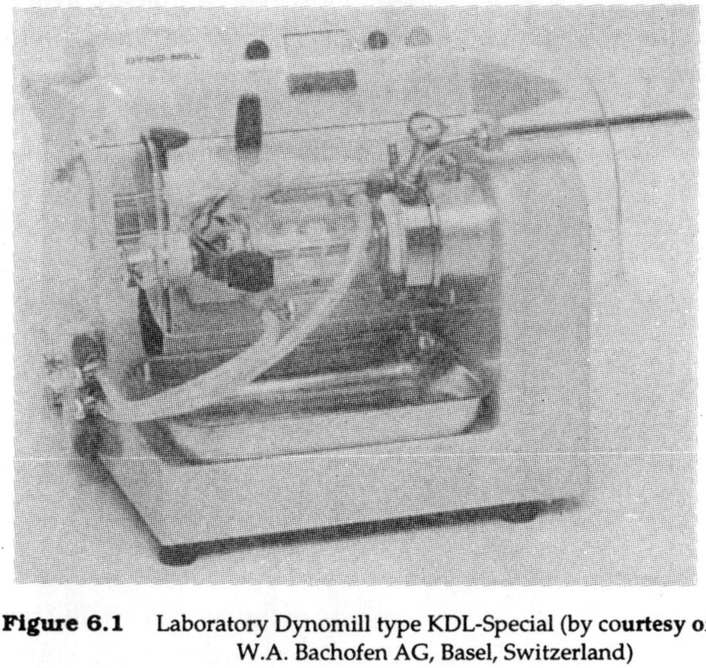

Figure 6.1 Laboratory Dynomill type KDL-Special (by courtesy of W.A. Bachofen AG, Basel, Switzerland)

The high sheer rate enhances media action for particle size distribution (impact action) and agglomerate separation (shear action) (3). This is a closed-type mill, which eliminates air entrapment and foaming during milling. Laboratory models with 0.8 litre and 0.6 litre capacities are available. However, for processing smaller quantities (as small as 75 ml), an Eiger mini motor mill (ex Eiger Engineering, England), which functions on a similar principle, can be used. One of the advantages of a Dynomill is that a laboratory process can be scaled up to production batches without going through intermediate pilot plant studies, because Dynomills and Silverson high sheer mixers (Figure 6.2) of varying processing capacities are available.

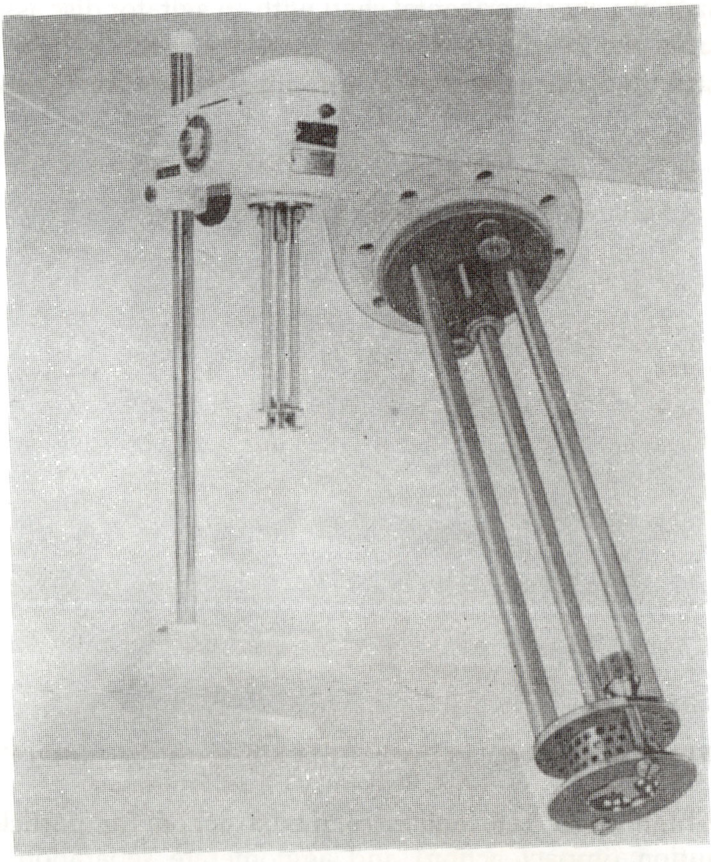

Figure 6.2 Silverson Laboratory Mixer (by courtesy of Silverson Machines Ltd., Chesom, Bucks, England)

For the development of granular formulations, including water-dispersible granules (WGs), compositions can be formed using tumbling, compacting (4) and fluid-bed granulating techniques. If the base material is a liquid, a spray-drying agglomeration technique is applicable.

Pan granulation is a well known technique and is currently used for many agrochemical WGs. The process works on a tumbling and granulating principle. In a broad description of the process, a micronized mixture of the pesticide and additives is placed in an inclined rotating pan. Liquid (acting as a binder solution) is sprayed upon the powder in such a quantity that the powder balls up and produces granules. This technique can be made to work as a continuous process relying on the fact that larger particles migrate to the top of the pan and overflow when additional powder is added to the pan.

Compacting granulation is exemplified by an extrusion process. The most sophisticated machines are twin-screw extruders in which the finely divided dry base is kneaded with water or a binder solution. The resulting dough is forced through a screen perforated with small holes. The spaghetti-type extrusions are dried, comminuted to the desired size and screened. This technique seems to have been most commonly used for granules that are totally or mainly water-soluble. A multiplicity of machines is available for the above-described operations. However for a pesticide formulation process laboratory, the most desirable equipment is an Erweka multi-purpose processor (ex Erweka Apparatebau, Germany). This apparatus is similar to a food processor, with attachments such as a coating pan, ribbon blender, extruder and a palettizer. (Figure. 6.3). It offers an ideal choice in terms of cost and space requirements.

Fluidized-bed granulation is a versatile technique to form agglomerated granules. Granulation is effected by spraying a binder solution on the pulverized material, which has been brought to a fluidized state. A laboratory-size fluidized-bed piece of equipment suitable for preparing water-dispersible granules is offered by Aeromatic, Switzerland. This apparatus has a transparent covering useful for monitoring the process during the development of a

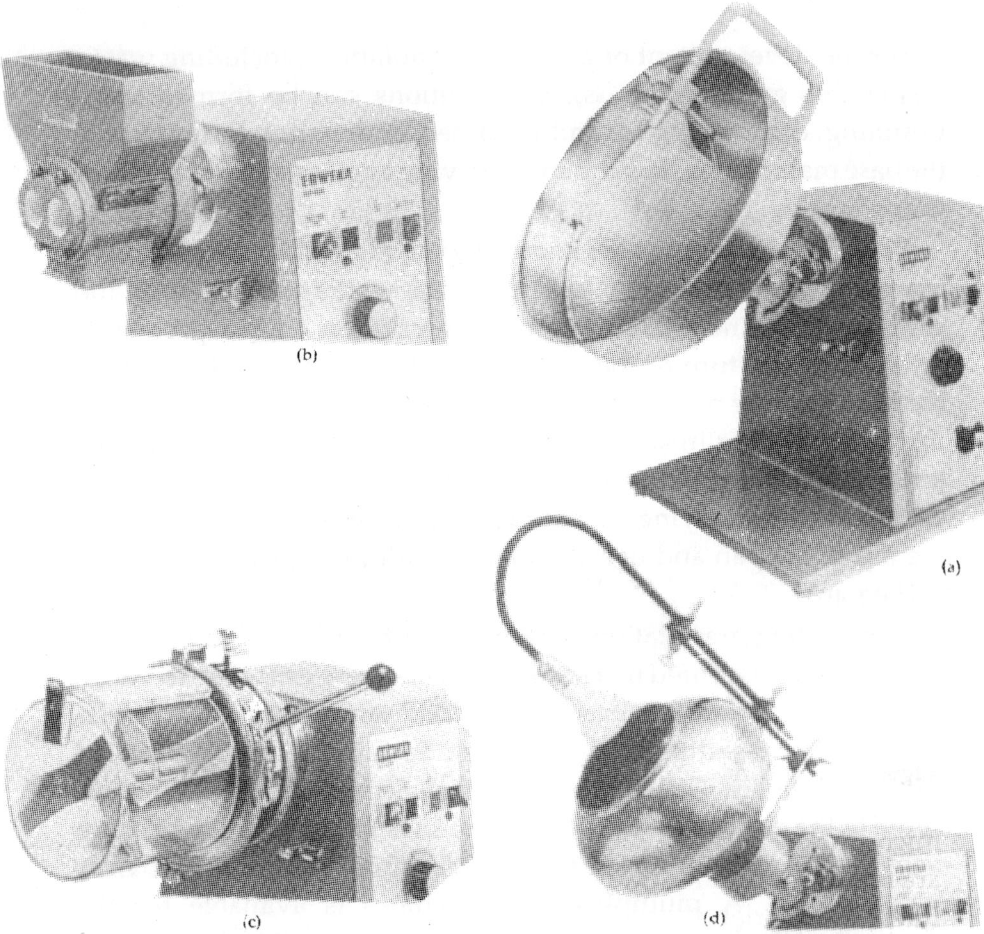

Figure 6.3 Erweka multi-purpose processor: (a) coating pan, (b) grinder (c) ribbon blender (d) tumbling and spray granulation (by courtesy of Erweka Apparatebau GmbH Heusenstamm, Germany)

formulation (Figure. 6.4). The equipment offers the advantage that, in addition to agglomeration and drying, operations involving impregnation, coating etc. can also be carried out in the equipment.

Similarly, spray drying is a well-known and well-tried technique for the preparation of agglomerated granules. It is used for a wide variety of food products such as soluble coffee and milk. The technique involves spraying a solution or a concentrated water suspension of the pesticide, with or without binder agent, into a conical cylinder in which hot air is circulated. Solid particles form from the spray droplets and agglomerate during the drying process, forming

Figure 6.4 Aeromatic fluid-bed granulator, laboratory unit STREA-1 (by courtesy of Aeromatic AG; Bubendorf, Switzerland)

microgranules. These particles collect in the bottom of the cone and are removed. A mini spray dryer for laboratory use is available from Buchi, Switzerland.

It is generally difficult to distinguish between fluid-bed and spray-dried granules, but they are clearly distinguishable from extrusion and pan granulation products. Other devices employed for granulation include the inclined rotating cylinder, used in making fertilizers, and Lödige mixers, which are used in food mixing.

At the developmental phase it is desirable to have a variety of laboratory equipment available to evaluate different techniques of granulation. However, selection of a process will depend upon the type of product required and the availability of the plant-size equipment for manufacture. The process equipment described for granulation is also useful for the development of slow-release and controlled-release formulations, which employ much the same techniques. A list of some of the more important laboratory process equipment for formulation development, along with the identity of their manufacturers/suppliers, is given in Appendix A.

3.2 Physical Testing Equipment

Formulation chemists have to employ a number of test methods for the evaluation of the physical characteristics of processed formulations. These methods can be classified into two categories. The first category constitutes methods used to correlate different parameters in order to improve certain properties of the compound that is to be formulated. The second category consists of methods used to test the products for their conformity to certain quality requirements and the selection of specification limit values (5).

Under the first category, a gamut of techniques is available for testing the wide range of physical characteristics that occur in formulation development. Examples range from light or electron microscopic methods through techniques for the measurement of rheological characteristics, interfacial tension, absorption isotherms etc. Equipment useful in these tests includes biological and stereo microscopes, supporting photographic facilities, a rotational viscometer and a tensiometer.

An extremely important property of most solid formulations and suspension concentrates is particle size. Critical sizes are in the range of 0.1 and 10 μm. The measurement of particle size generally requires sophisticated equipment, although some simple qualitative tests are available. The Hegman gauge is one such. However, more sophisticated means of measuring particle size, such as the Malvern, are desirable.

Although most of this laboratory equipment would be considered important for developmental purposes, procurement of the highest level of sophistication and sensitivity may be impeded by budgetary constraints. Unfortunately the more sophisticated and more sensitive equipment is usually more costly. In industrial research one must weigh the relationship between the data obtainable and the price of obtaining it. The formulation chemist will need to make a decision concerning the trade-off between the cost of the equipment and the quality of the data that will be obtained. In this connection, one can always concentrate on the most important properties of the formulation and first try to optimize them.

The second category consists of equipment that measures properties described in the specifications. The principle characteristics of the quality of a pesticide formulation are its biological activity, its tolerance by the crop and its ease of use. However as biological activities and crop tolerances are difficult to determine quantitatively in the laboratory, other parameters that can be measured readily by physical and chemical methods are specified in a specification. Often the method used has been tested in a collaborative study and has been widely approved. This includes equipment suitable for carrying out an assay of the active ingredient in a formulation and equipment able to assess both the chemical and physical stability of new formulations. Generally speaking, this involves, at its simplest, a UV spectrophotometer and chromatography equipment, i.e. a GC and/or HPLC. A tropical climate tester is desirable to study the stability of a formulation under conditions of accelerated storage and to check its compatibility with the packaging.

A suggested list of physical and analytical testing equipment, along with some of the identified sources, is given in Appendix B.

4 LABORATORY FOR BASIC FORMULATION RESEARCH

To set up a laboratory to carry out basic research on formulations and formulation adjuvants requires considerable expenditure of funds. In addition to the equipment cited earlier, it is essential to have available a very sophisticated means of measuring particle size with a Malvern c. a Coulter counter. The Malvern master sizer can measure pa. ticles in the range 0.1-600 μm; the Coulter counter covers the range 0.6-800 μm in diameter. One should note, however, that these two machines may produce different results if the particle shape differs substantially from the spherical.

Additional desirable equipment would include equipment to measure surface tension, zeta potential, absolute density etc.

Desirable mechanical equipment would include a spray granulator, a spray drier, a pan granulator, a colloid mill, a high-energy disperser and a drum blender. Such an array of equipment would give a good variety of approaches for the preparation of different formulations.

On the analytical side, equipment such as thermal analysers, including differential thermal analysis (DTA), differential scanning calorimetry (DSC) and thermal gravimetric analysis (TGA), and an atomic absorption spectrophotometer are particularly useful in evaluating the suitability of a mineral carrier as a pesticide diluent. These instruments yield information on the mineral characterization of clays, their crystallinity and the type of associated water (structural, bound or unbound), which may affect the stability of a formulated pesticide.

DTA is an excellent method for an accelerated evaluation of the compatibility of clays with an active ingredient. Here one compares the thermal curve when heating an active ingredient with the curve obtained when heating a mixture of the active ingredient with a potential carrier (6). Atomic absorption spectroscopy can generate information on heavy metal impurities present in the clay that may catalyse chemical decomposition of the active substance.

Equipment for GLC and HPLC, with standard detectors, is essential for a chemical assay of active ingredients. Some equipment is now available with an interface with a mass spectral detector (MSD). This detector offers an economical way to enable one to identify and analyse breakdown products of pesticides in the formulations and to analyse pesticide residues. However, equipment with an MSD is quite expensive and its acquisition has to be rationalized with the available budget.

Supercritical fluid chromatography (SFC), permits analysis of complex, high molecular weight or thermally labile mixtures that are difficult to separate by other techniques. It supplements the capabilities of gas and liquid chromatography, combining the solubility behaviour of liquids and the diffusion and viscosity properties of gases. The equipment can be coupled with an SFE, which rapidly extracts the components of interest from a complex mixture selectively and quantitatively for subsequent analysis. The technique is particularly useful for pesticide residue analysis.

Among the various spectrophotometric techniques, UV-visible and IR spectrophotometers are regularly employed for assaying pesticides in the formulations. A Fourier Transform Infra Red (FTIR) system provides a high resolution spectrum by means of a signal averaging of repeated scans at a desired signal-noise ratio. A photo diode-array UV-visible spectrophotometer offers the advantage of measuring full spectra (from 190 to 820 nm) very rapidly. Both devices can be interfaced with GC and HPLC to function as detectors.

Modern analytical equipment, by and large, is microprocessor-controlled, and often possesses its own dedicated computer. It is possible to network the dedicated computers of different pieces of equipment by ensuring their mutual compatibility at the time the equipment is ordered. This can facilitate the task of data storage and handling.

5 PILOT PLANT EQUIPMENT FOR THE PROCESSING OF FORMULATIONS

Ideally, pilot plant equipment for the processing of formulations will effectively simulate, on a larger scale, the processes observed in the laboratory. However, it does not always succeed in this. The situation becomes particularly complex when the equipment employed for laboratory-scale process development is not available in larger capacity models or when the larger equipment functions on a different principle. In such a situation, the laboratory process needs to be revalidated by changing process parameters such as temperature, time, feed rate, size, and number of recyclings.

Frequently the same pilot plant equipment will be used for processing different pesticides, such as herbicides, fungicides, and insecticides. Contamination problems may arise. Thus, versatile equipment should be selected which has easy clean-up and decontamination (7).

Many types of processing equipment, such as the size reduction mills, mixers and blenders and granulating systems described in Section 3.1, are available in larger models and are suitable for the pilot plant. There may be, however, scale-up problems. For example, in the size reduction process, the choice of the type of size reduction mill is dependent upon the physical properties of the feed material and the desired size distribution of the output. In some mills, such as a high-speed hammer mill or a pin mill, a temperature rise in the grinding chamber occurs due to heat generated during their operation. This temperature rise may cause the formation of sticky materials that will agglomerate or it may cause heat-sensitive materials to decompose. And of course, sticky agglomerated materials do not grind well. Some of these problems may be alleviated by processing the formulation in an air jet mill. On the other hand, a relatively efficient size reduction mill known as an air-swept ultra-fine vortex mill is widely used in Europe for the production of wettable powder formulations. The mill may provide a finely ground product (below 10 μm) at a much higher rate than a hammer mill and more economically than an air jet mill (7). The equipment is available in a pilot-plant size model from Ultrafine, France. However, sticky and heat-sensitive materials cannot be processed in it under normal conditions. For a wet grinding process, dynomills are available in several large-size models. Similarly, larger versions of Silverson mixers are also available. Both of these machines are commonly used in the pesticide industry for the processing of suspension concentrate formulations.

The availability of different types of size reduction mills in the pilot plant provides the flexibility to tailor operations with consideration of such factors as functionality, economy and safety.

For different granulation processes, a variety of equipment is available for pilot-plant scale operations, including sigma mixers, or kneaders, for dough making and screw extruders for making extrusion-type granules. Similarly, fluid-bed systems are offered by Aeromatic, Switzerland, and Alpine, Germany. The Aeromatic systems are available either as multi-processors or as a building-block system that can be assembled to suit the required use. Spray-drying equipment is available from manufacturers like Anhydro, Stork Friesland and Niro Atomizer. It comes in a wide variety of dryer shapes to suit different products and processing conditions. To improve the control of product dryness and raise the overall energy efficiency, it is possible to add a fluid-bed dryer to a basic spray dryer. Such spray dryer-fluid bed combinations are particularly suitable for heat-sensitive materials.

6 CONCLUSION

As rapid advances in technology and newer developments in highly sophisticated processes and analytical equipment have a symbiotic relationship, it is only to be expected that state-of-the-art equipment is soon superseded by even more versatile or sophisticated equipment. However, it is not necessary to have an extremely elaborate and fully equipped laboratory to do useful work on formulations and formulation development. As important to a good formulation laboratory as sophisticated equipment are innovative ideas drawn from day-to-day observations. Besides, many of the techniques used in pesticide formulations are drawn, either directly or indirectly, from food processing techniques. It is often possible to develop new formulations with very simple equipment. In this context, an appropriate approach for equipping a pesticide formulation laboratory might be the gradual addition of equipment as needed. Let the laboratory evolve in conformity with the exposure and training of the formulation chemists and in relation to the level of sophistication foreseen for various planned projects. In addition, the financial support available would have a significant influence on the development of the laboratory.

REFERENCES

(1) A.R. Woodford, "Setting up a formulation laboratory," Workshop in Pesticide Formulation Technology, New Delhi, Pesticide Develop. Programme in India, April 1989 (unpublished).

(2) D. Seaman, "Colloid and surface science and technology in the pesticide industry," *Chem. and Indust.*, 3 March 1979, p. 159.

(3) L.A Dombrowski, and M. Schieritz, "Dispersion and grinding of pesticides," *Advances in Pesticides Formulation Technology*, H.B. Scher, ed. (Washington, DC, American Chemical Society, 1984), p. 63.

(4) J.A. Polon, "Formulation of pesticidal dusts, wettable powders and granules," *Pesticide Formulations*, W. Van Valkenburg, ed. (New York, Marcel Dekker, 1973), p. 143.

(5) H.J. Niessen, "Physical test methods for formulation development, quality standards, pesticides science biotechnology," *Proceedings of the Sixth International Congress on Pesticide Chemistry*, R. Greenhalgh and T.R. Roberts, eds. (Oxford, Blackwell, 1986), p. 235.

(6) F.R. Rittig, G. Synnatschke and W. Guckel, "Differential thermal analysis as a screening technique for candidate carriers in pesticide formulations," *Proceedings of the Fourth International Conference on Thermal Analysis* 3,1974, p. 109.

(7) F. Fiedler, "Selected problems in the production of pulverised plant protectives," *Pesticide Formulation and Application in Developing Countries*, G. Fröhlich, W. Krammer and W. Rodewald, eds. (Berlin, Chemie-Export-Import, 3,1978), p. 126.

(8) C. Butcher, "Drying for quality: coffee with milk," *The Chemical Engineer*, 31 January, 1991, p. 17.

APPENDIX A

List of processing equipment for a pesticide formulation laboratory and some of the sources for their supply

	Equipment	*Source*
1.	Hammer mill (Condux type)	F.G. Bode & Co., GmbH Schlossmühlendamm 11 P.O. Box 900167 D-20457 Hamburg, Germany
2.	Laboratory grinding mill (Condux type)	
3.	Pin mill (Laboratory Universal Mill)	Alpine AG D-86199 Augsburg, Germany
4.	Air Jet mill (Micronizer)	Sturtevant International Sales Corp. 103 Clayton St. Dorchester Boston, MA 02122, U.S.A. Fluid Energy International Sales Corp. 153 Penn Avenue Hatfield, PA 19440, U.S.A.
5.	Air-swept micro vortex mill	Societe Ultrafine Usine de Villeneure surVerberie 60410 Verberie, France
6.	Agitated-type horizonal bead mill (Dynomill)	Willy A. Bachofen AG Machinenfabrik, 17 Utengasse Strasse CH-4005 Basel, Switzerland
7.	Agitated bead/sand mill (Eiger minimotor mill)	Eiger Engineering Ltd. 40 Hardwick Grange, Woolstone Warrington - Cheshire WA 1 4RF, England.
8.	High-sheer homogenizer (Silverson laboratory mixer)	Silverson Machines Ltd., Waterside, Chesham, Bucks HP5 1 PQ, England
9.	Multipurpose processor with (a) palletizer, (b) coating pan (c) extruder and (d) ribbon blender attachments	Erweka Apparatebau GmbH P.O. Box 1253 Ottostrasse 20-22 D-63150 Heusenstamm, Germany
10.	Fluid-bed granulator with Wurster accessory (Aeromatic Laboratory unit STREA-1)	Aeromatic AG Hauptstrasse 145 CH-4416 Bubendorf, Switzerland or Aeromatic Inc., Towaco, NJ 07082, U.S.A.
11.	Mini spray dryer	Buchi Laboratoriums Technik AG Flawil, Switzerland
12.	Ploughshare mixer (Lödige mixer)	Gebrüder Lödige Machinenbau GmbH Postfach 2050 Elsener Strasse 7-9 D-33102 Paderborn, Germany.

APPENDIX B

List of physical testing and analytical equipment and some of the sources for their supply

	Equipment	*Source*
1.	Sieving apparatus (Alpine Luftstrahlsieb)	Alpine AG, Postfach 629, D-86199 Augsburg, Germany
2.	Tropic Climate Tester (for accelerated storage studies)	Heraeus-Votsch GmbH. Beethovenstrasse 34 D-72334 Balingen - Frommern, Germany
3.	Biological and stereo microscopes	Olympus Optical Co. Ltd. San-Ei Building, 22-2 Nishi-Shinjuku 1-Chome Shinjuku-ku, Tokyo, Japan Nikon Corporation, Fuji Building, 2-3 Marunouchi 3-Chome, Chiyoda-ku, Tokyo 100, Japan
4.	Particle size analyser (Malvern master sizer to measure 0.1-600 μm particle sizes)	Malvern Instrument Ltd. Springlane South, Malvern, Worcestershire, WRI4 1 AQ, England
5.	Malvern Zeta Sizer IIc	
6.	Electronic particle counter (Coulter Counter TA-II)	Coulter Electronics Inc. Hialeah, FL, U.S.A.
7.	Brookfield LVT Viscometer	Brookfield Engineering Lab 240 Cushing Street Stoughton, MA 02072, U.S.A.
8.	Haake Rotovisko	MSE Scientific Instruments Crawley, Sussex, England.
9.	Deer rheometer	Integrated Petronic Systems London, England
10.	True denser	Sheshin Enterprises Co. Ltd. 5-27-7 Sadagaya Shivaya Ku, Tokyo, Japan
11.	Analytical balances	Sartorius Instruments 1430 Wankegan Road McGaw Park, IL 60085, U.S.A.
12.	Autotitrator/memotitrator	Mettler Instrumente AG CH-8606 Griefensee, Switzerland
13.	Gas-liquid chromatograph	Hewlett Packard Co. 3495 Deer Creek Road, Palo Alto, CA 94304, U.S.A. Perkin-Elmer Corporation 761 Main Ave. Norwalk, CT 06859-0012, U.S.A.

	Shimadzu Corporation 3, kanda-Nishikicho 1-Chome Chiyoda-ku, Tokyo 101, Japan.
14. High-pressure liquid chromatograph	Waters, Division of Millipore, Maple Street Milford, MA 01757, U.S.A.
15. Supercritical fluid chromatograph	Suprex Corporation SFC Research Centre 125 William Pitt Way Pittsburgh, PA 15238, U.S.A. Computer Chemical Systems Rt. 41 & Newark Road, P.O. Box 683, Avondale, PA 19311, U.S.A.
16. Infra-red spectrophotometer and Fourier Transform IR spectrophotometer	Perkin-Elmer Ltd., Norwalk, CT, U.S.A. Shimadzu Corporation, Tokyo, Japan Mattson Instruments Inc. 1001 Fourier Drive Madison, WI 53717, U.S.A.
17. Photodiode-array UV-visible spectrophotometer	Hewlett-Packard Co., Palo Alto, CA 94304, U.S.A. Beckman Instruments Inc. Alex Division, 23050 Camino Ramon, P.O. Box 5101 San Ramon, CA 94583, U.S.A.
18. Atomic absorption spectrophotometer	Perkin-Elmer Corporation, Norwalk, CT 06859, U.S.A. Pye Unicam Ltd., York Street, Cambridge, CB1 2 PX, England.
19. Thermal analyser	DuPont Company Instrument Systems Concord Plaza, Quillen Bldg., Wilmington, DE 19898, U.S.A Mettler Instrumente AG, Greifensee, Switzerland.

CHAPTER

7

Economics and Technologies of Controlled-Release Formulations of Pesticides

Baruch S. Shasha

Plant Polymer Research
Northern Regional Research Center
*United States Department of Agriculture**
Agricultural Research Service
1815 N. University Street
Peoria, IL 61604, United States

1 INTRODUCTION

The control of pests is crucial since the rapid growth of the world population makes the need for more efficient agricultural production paramount. The world population is soaring at a faster rate than at any time in history: 0.8 billion in 1760, 2.5 billion in 1950, and 5.25 billion today. According to a United Nations projection it will reach over 8 billion by 2025 (1). This rapid growth comes at a time when inefficient pest control techniques are causing significant losses in crop yields costing billions of dollars

Three factors determine the rate at which a pesticide is applied (2):

(a) The minimum amount of pesticide to control the pest;
(b) The maximum amount of pesticide the crop can tolerate;
(c) The length of time the pesticide is desired to remain at an effective concentration.

*The mention of firm names or trade products does not imply that they are endorsed or recommended by the Department of Agriculture over other firms or similar products not mentioned.

A conventional wettable powder, emulsifiable concentrate or soluble formulation is fully active when applied under adequate environmental conditions but can quickly lose activity due to leaching, volatility, adsorption to clay and organic matter and chemical and microbial degradation. To compensate for these losses and to obtain effective control, a conventional formulation must be applied at a rate much higher than that needed to control the pest. This excess not only increases the cost of pesticide application but also poses potential hazards to the environment and non-target organisms.

An effective approach to solving these problems associated with a current formulation is through controlled-release technologies. Such technologies have been pioneered by the drug industry over the past 30 years. The purpose of controlled release was to maintain an effective level in the body without undesirable side effects caused by high dosage. By applying the same principles to pesticides, formulators were able to make products that were capable of maintaining an effective level in the soil or on foliage and at the same time to reduce the rates of application and minimize the levels of pesticide in the environment. These technologies are the subject of intensive research, and a complete review of the literature cannot be given here. However, many excellent publications are available (2-7).

Some examples of slow-release formulations in agriculture are as follows:

(a) Microcapsules of the insecticides methyl parathion and diazinon enclosed in a nylon-like material;
(b) Encapsulated algicide copper sulphate to be used in water;
(c) Encapsulated endothal in an inorganic matrix;
(d) Encapsulated insecticide carbaryl in a matrix of polymerized beta-pinene;
(e) Encapsulated rodenticide warfarin;
(f) Encapsulated herbicide alachlor.

2 ECONOMIC FUTURE

Due to the high cost of research and more stringent regulatory requirements, few new pesticides are entering the market and many old ones that were once considered safe cannot survive the re-registration criteria of environmental protection agencies. Therefore, the economic future of the pesticide manufacturer will be largely dependent upon new formulations. It is obvious that people of all nations have become less tolerant to hazards, especially environmental ones. The general perceptions are that pesticides are overused and that the cost/benefit ratios in many cases are not favourable. The people of the United States, for example, use over 1.1 billion pounds of pesticides yearly (over 4 pounds of toxic material per capita) and still the loss due to pests is estimated to be 30%. Thus, the economic threshold should always become a prime consideration in choosing the type of formulation needed. This threshold is defined as the minimum pest level at which pest control measures should be instituted to prevent pest populations from reaching a damage level greater than the control costs (8).

3 SUCCESSFUL PESTICIDES/SUCCESSFUL FORMULATIONS

In searching through the patent literature one can find hundreds of chemicals and groups of chemicals that are claimed to be effective pesticides. Unfortunately, very few are brought to market profitably. Common denominators among the most successful are the following: the products are efficacious, easily and economically produced and have found their own niches in the marketplace. An example of such a product is the herbicide 2,4-D, which has remained popular in combating broadleaf weeds for over four decades.

Controlled release technology offers the opportunity to enhance efficacy by protecting the pesticide against most losses. The object is to deliver a bioactive substance to a target site at a constant effective rate over a specified period of time. The concentration must be high enough to control pests without harming the environment. Such formulations should be stable and compatible with granular and spray application technologies. They should provide prolonged concentrates for continuous control without damaging the crop. The formulation should be effective only during the growing season, with no carryover to the next season. The economics of preparing the formulation should include both low cost encapsulating material, preferably biodegradable, and simple manufacturing technology. The formulation should have the potential to create new markets for existing pesticides, and improve overall efficacy while increasing the safety of handling and decreasing environmental pollution. Also these formulations should be suited for modern integrated pest management systems, which can reconcile the conflicting needs of controlling pests and preserving the quality of the environment. The formulation should also have the potential to overcome the environmental instability of pesticide chemicals such as insect growth regulators and pheromones by incorporating them into a matrix for a slow release to prolong biological activity. Other chemicals such as aquatic herbicides and molluscicides cannot be used on a practical scale unless incorporated into a controlled release formulation.

4 CRITERIA FOR THE SELECTION OF CONTROLLED RELEASE FORMULATIONS

Due to the added costs of controlled release versus conventional formulations, the following criteria should be considered (9):

(a) The physical characteristics of the technical product, such as volatility and water solubility;

(b) The chemical characteristics of the technical product such as corrosiveness or oxidative instabilities;

(c) Instability due to photodegradation, hydrolysis, breakdown by microorganisms or sensitivity to drastic pH changes;

(d) Unacceptably high dermal or oral toxicity;

(e) Physical or chemical incompatibility of the pesticide in conventional form with other products;

(f) The pesticide in conventional form is available too fast, resulting in toxicity to crops, followed by non-performance, and/or is a hazard to non-target organisms and to the surrounding environment;

(g) The pesticide in conventional form is too expensive because it needs repeated application.

5 THE ENVIRONMENTAL IMPACT OF CONTROLLED RELEASE FORMULATIONS

The environmental impact of controlled release formulations has been studied and reviewed (5). It was concluded that inert organic polymers making up the wall material cannot have any significant deleterious effect on non-targets in the environment. That is especially so since the polymeric material must degrade before it can have any impact. However, certain wall materials such as chlorinated natural rubber, which is used in antifouling coatings, may be toxic to some microorganisms. The polyamides that are used in the microencapsulation of methyl parathion (Penncap-M) degrade very slowly under ambient conditions, which may extend the life of active ingredients and in turn cause residues in the food crops. Another concern is that certain formulations might act on non-target organisms in an unexpected manner. For example, the environmental hazards of encapsulated methyl parathion to honeybees was first reported in 1976. The encapsulated product, a nylon-type wall material, is 30-50 μm in diameter. Because this size is in the range of certain pollen grains, bees could not distinguish between them. Furthermore, it was found that the capsules easily adhered to bee hairs; the real hazard appeared to be the lasting effect of the pesticide liberated from the microcapsules. When bees were confined to colonies that had been killed 19 months earlier, presumably from methyl parathion, the bees were poisoned. A precaution to prevent unnecessary bee kills is mentioned on the label of the pesticide container: "Do not apply to weeds in bloom on which an economically significant number of bees are actively foraging". When Penncap-M won registration in the United States in 1984, it was the first major controlled-release pesticide approved for agricultural use. The manufacturer conducted environmental and toxicological studies on Penncap-M and the polymer that made up the formulation. The results compared very favorably with the EC formulation. Only after extensive field studies was the encapsulated product found to be hazardous to pollinating insects such as honeybees.

6 MECHANISMS

Controlled-release mechanisms can be categorized on the basis of the release-rate-controlling step into four general types (10):

(a) Diffusion through a nonporous polymer membrane, where the migration of a molecule from a reservoir through an enveloping homogeneous polymer membrane normally occurs by an activated diffusion process. The permeant first dissolves in the surface layers, migrates through the bulk material under a concentration gradient and then evaporates or desorbs from the low concentration surface;

(b) Diffusion through a nonporous, solid polymer matrix with two variations of the diffusional process in which a permeant migrates through an enveloping, homogenous matrix. In the first system the permeant is completely dissolved in the surrounding polymer. In the second the permeant is dispersed in the matrix at a concentration greatly exceeding its solubility in the matrix;

(c) Diffusion through a porous, solid polymer matrix where the active agent is leached out from the matrix by the bathing fluid, which is able to enter the matrix through pores. The active agent is presumed to dissolve slowly into the permeating fluid phase and to diffuse from the system along the cracks and capillary channels filled with extracting solvent;

(d) Surface hydrolysis or dissolution of a polymer matrix where the rate of release depends simply on the surface area exposed. A controlled-release formulation with a constant output rate, such as water evaporation from a cake tin in which the rate of evaporation remains constant, is a zero-order rate. A first-order rate occurs when the rate declines with time and is proportional to the amount of active agent still in the reservoir. A second order occurs when the output rate is proportional to the square of the amount of agent remaining in the reservoir.

If the total amount of active agent is M_∞ and the amount that will be released in unit of time is M_t, then the release rate dM_t/dt is

Zero order: $$\frac{dM_t}{dt} = \text{constant}$$

First order: $$\frac{dM_t}{dt} = \text{constant} \times (M_\infty - M_t)$$

Second order: $$\frac{dM_t}{dt} = \text{constant} \times (M_\infty - M_t)^2$$

A controlled release formulation with a zero-order release is an ideal formulation, providing the release duration that has no deleterious effect on non-target biota. Such a release is illustrated in Figure 7.1, where curve (a) represents standard release and (b) represents controlled release.

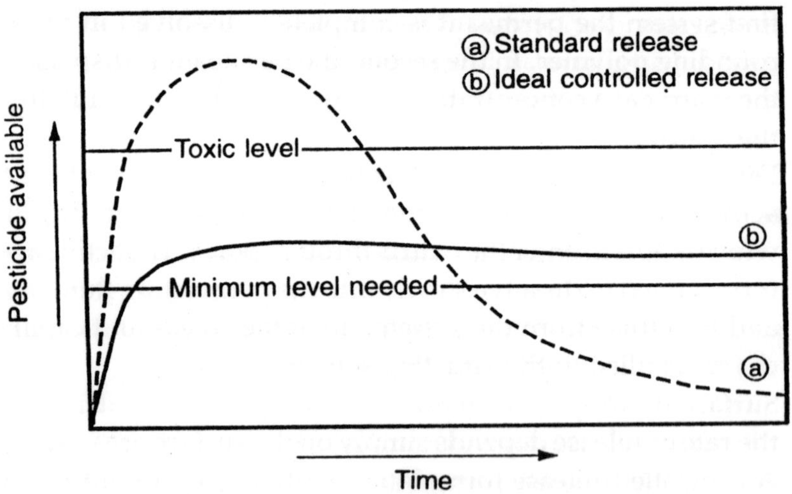

Figure 7.1 Theoretical curves comparing pesticide concentrations available from (a) standard spray application and (b) ideal controlled release devices

7 TYPES OF CONTROLLED RELEASE SYSTEMS

A pesticide can be either chemically or physically attached to a polymer. The resulting combination will contain the pesticide in a manner capable of being released upon hydration, hydrolysis, erosion, biodegradation, diffusion, osmosis, mechanical rupture or other suitable means.

7.1 Chemically bound

Chemically bound pesticides, mainly herbicides, are of two types:

(a) Those which are prepared by attaching a polymerizable site to the active ingredient, followed by polymerization of the new derivative:

$$\begin{array}{c} \text{Herbicide} - \text{COOH} \\ + \\ \text{CH}_2 = \text{CHOCOCH}_3 \end{array} \xrightarrow[\text{H}_2\text{SO}_4]{\text{HgSO}_4} \begin{array}{l} \text{CH} = \text{CH}_2 \\ | \\ \text{OCO} - \text{herbicide} \end{array} \longrightarrow \begin{array}{l} (-\text{CH} - \text{CH})_n \\ | \\ \text{OCO} - \text{herbicide} \end{array}$$

(b) Those which are prepared by chemically binding derivatives of active ingredients to a suitable polymer. For example, herbicides containing carboxylic functionalities have been reacted to form acid chlorides, which in turn were attached through hydroxyl groups of natural polymers. Herbicides containing primary amino functionality were reacted with phosgene to form isocyanates, which in turn were attached through the hydroxyl group of natural polymers:

$$\begin{array}{c} \text{Herbicide} - \text{COOH} \\ + \\ \text{SOCl}_2 \end{array} \longrightarrow \text{Herbicide} - \text{COCl} \xrightarrow{\text{Cellulose}} \text{Cellulose} - \text{OCO-herbicide}$$

$$\begin{array}{c} \text{Herbicide-NH}_2 \\ + \\ \text{COCl}_2 \end{array} \longrightarrow \text{Herbicide-NCO} \xrightarrow{\text{Cellulose}} \text{Cellulose} - \text{OCONH-herbicide}$$

Such chemically bound combinations have found application in forestry and agronomic crops. The rate of release can be increased by lowering the molecular weight or increasing the hydrophilicity of the polymer carrier. The rate of release also depends upon the degree of substitution of the herbicide moiety within the polymer, the pH of the hydrolysis medium and the size of the particles.

7.2 Microencapsulation

Microencapsulations are processes in which tiny particles or droplets are surrounded by a coating to give small capsules. The material inside the capsule is referred to as the core, internal phase or fill, whereas the coating is known as the wall, shell or membrane. Microcapsule diameters vary between a few micrometers up to a few millimeters. Advantages of microencapsulated products include ease of handling, reduced volatility and toxicity, stability, sprayability and controlled release of active ingredients. There are a half dozen prominent methods for their preparation.

7.2.1 *Coacervation*

This process disperses the pesticide in a suitable vehicle containing a coating material that will deposit around the pesticide. The coating-coacervate is then hardened by chemical cross-linking and solvent evaporation to form discrete microcapsules. Himel and Cardarelli (11) developed a sprayable system based upon coacervation that encapsulates herbicides "in flight". That is, the encapsulation process takes place as the dispersion of herbicide with vehicle and coating material is sprayed and dried.

In coacervation techniques the wettability of the core material with the coating is crucial. Properly wetted solid particles are easier to coat than liquid cores. If a liquid core material is highly insoluble in the coacervate-forming solution, proper wetting may be difficult. It is also critical that the core not be soluble in the polymer solvent and that the polymer not partition into a liquid core.

7.2.2 *Interfacial polymerization*

This method (12) necessitates the use of at least a two-phase system: (a) an organic phase containing the pesticide along with a reactive monomer and (b) an aqueous phase emulsifier along with another reactive monomer. The reactive monomers undergo a polycondensation reaction at the interface of the organic and aqueous phases to form a polymer that coats the pesticide. Microcapsules formed with this process contain as much as 90% pesticide, which may be used in sprayable formulations in conjunction with thickening or suspending agents. In interfacial polymerization, the solubilities of the reactants in the phases permits a choice of solvents and polymers.

7.2.3 *Air-suspension coating*

This method (12) involves coating particles as solutions or melting them to give better control and flexibility. They are coated while suspended in an upward-moving air stream. They are supported by a perforated plate having different patterns of holes inside and outside a cylindrical insert. To fluidize the settling particles, air is permitted to rise through the outer annular space. Most of the rising air (usually heated) flows inside the cylinder, causing the particles to rise rapidly. At the top, as the air stream diverges and slows, they settle back onto the outer bed and move downward to repeat the cycle. As the particles start flying, they encounter a fine mist of the coating solution. Only a small amount of solution is applied in each pass. Many different coating materials have been used with this process, including compounds and water-soluble polymers.

7.2.4 *Molecular complex formation*

By virtue of their spatial-geometrical structures, certain chemicals are capable of entrapping pesticides at the molecular level (2). One of these chemicals is the Schardinger dextrins, which originate mostly from starch. These dextrins are six-, seven- or eight-member rings built of D-glucose units with a hydrophobic cavity in which the pesticide can be bound. Another class that can entrap chemicals is the liposomes, which are cell-like structures derived from phosphatidyl cholines and phosphatidyl ethanolamines. They are triglycerides which can form a series of closed concentric membranes that entrap other substances. Triazine herbicides have been encapsulated within such structures.

7.2.5 *Centrifugal extrusion*

Liquids can be encapsulated using a rotating extrusion head containing concentric nozzles (12). In this process, a jet of core liquid is surrounded by a sheet of wall solution or melt. As the jet moves through the air it breaks into droplets, each being coated with the wall solution. While the droplets are in flight, the molten wall material may be hardened or a solvent may be evaporated from the wall solution. This process is excellent for forming particles 400-2,000 μm in diameter.

7.2.6 *Spray drying*

The main advantage of using this technique is the ability to handle labile materials (12). The active ingredient is dissolved or suspended in a melt or polymer solution and becomes entrapped in the dried particle. In modern dryers the viscosity of the solutions to be sprayed can be as high as 300 cps. In a modified technique the core material is incorporated into a meltable fat or wax and the emulsion or suspension has to be chilled only below its melting point to form particles.

7.3 Matrix Encapsulation

Products obtained here lack a distinctive wall surrounding each individual particle of active ingredient (2). The pesticide is dispersed within a polymer and becomes entrapped within many small cells of a continuous matrix. The active ingredient may be dissolved or suspended in various polymers to yield ribbons, sheets or granules. Often, an excipient or porosogen is added to such formulations. Such excipients may be inorganic filler particles or water-soluble polymers that provide points at which the surrounding medium can penetrate the product in order to regulate the rate of release of active ingredients:

(a) *Encapsulation within an alginate.* Alginic acid is a linear polysaccharide composed of polymannuronic and polyglucuronic acid units and is isolated from brown seaweeds. Its sodium salt is water-soluble and forms a tough gel when treated with calcium or barium cations. Alginate beads (2) are made by dispersing the pesticide into a solution of sodium alginate and adding the dispersion to an aqueous gellant solution of calcium or barium chloride. Beads prepared this way are outwardly similar to many prepared by microencapsulation but differ in that they have a matrix structure rather than a central core of active ingredient;

(b) *Encapsulation within a lignin matrix.* The process involves emulsifying the active ingredient with an alkali lignin, followed by acidification to precipitate the composites, which are isolated and dried. Upon immersion in water the composites develop cracks and fissures. The release of active ingredient is mainly by diffusion;

(c) *Encapsulation within inorganic matrices.* Herbicides have been entrapped in plaster of paris matrices for use with container-grown ornamentals (2). Most are formulated as tablets of various sizes and shapes. Herbicides have also been entrapped within a silicate matrix using sodium silicate and calcium chloride in a manner similar to the alginate method.

7.3.1 Encapsulation within a starch matrix

Starch is one of the most abundant biopolymers. It is found mainly in the plant kingdom, where it occurs as the principal food reserve. Currently, the United States produces over 200 billion lb a year from the corn crop alone. It is also a very inexpensive polymer. It consists solely of glucose units and it can be easily fractionated to yield amylose, a straight-chain polymer, and amylopectin, a branched-chain polymer. While amylose is capable of forming tough, flexible films, amylopectin does not possess this property. Starch reacts with an iodine solution to give a dark-blue color. Because of the abundance of free hydroxyls (three hydroxyls for each glucose unit) starch is easily derivatized. The degree of derivatization or substitution (D.S.) is defined as the number of substituents per glucose unit. Thus, theoretically, a D.S. of 3 is possible.

Simple and economical procedures are described in the literature to entrap water-insoluble pesticides within starch. The main methods (13-17) are the xanthate, the calcium and the borate procedures, which all involve chemical modification of starch. Another method using extrusion procedure without chemically modifying the starch was described recently (18). The pH of particles formed by the different procedures varies. Particles made by the xanthate process have a pH of about 4. With the calcium, a pH of about 11 is obtained; with the borate, a pH of about 9.5. With the extrusion process a pH of about 6 is obtained. In both the calcium and borate procedures, modifications can be made to yield final products with pHs close to neutral.

7.3.1(a) *Xanthate procedure*: The xanthate procedure is based on the reaction of starch with carbon disulphide and sodium hydroxide to yield starch xanthate, which is thoroughly mixed with the pesticide. The relative amount of carbon disulphide and sodium hydroxide to starch can be varied to different degrees of substitution. The xanthate mixture is neutralized and cross-linked to yield xanthide or treated with a salt such as ferric chloride to yield a metal xanthate. The pesticide particles are dispersed and entrapped within the starch xanthate or xanthide matrix:

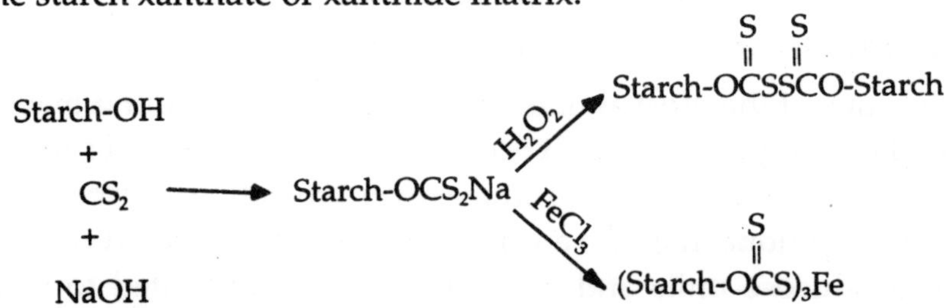

Unmodified corn starch (162 g, dry basis) is slurried in water (11) and mixed with carbon disulphide (40 ml) followed by the addition of sodium hydroxide solution (40 g, in 350 ml water). Gelation occurs immediately. The mixing is continued until a homogeneous mixture is obtained. The xanthate thus obtained is kept at ambient temperature for 1 hour to yield a product with a D.S. of 0.3. Although the product can be used for up to 30 days when kept at 5°C, it is advisable to prepare a fresh batch as needed to minimize the formation of by-products. To encapsulate butylate (s-ethyl diisobutylthiocarbamate) using hydrogen peroxide as a cross-linking reagent, starch xanthate (200 g) is cooled to 5°C and mixed thoroughly with butylate (6.7 E, 10 g), followed by the addition of glacial acetic acid (8 ml) and hydrogen peroxide, 30% (5.5 ml). The mixture is allowed to stand for 15 min, filtered and pressed under a rubber dam to remove most of the water. The product is ground with a blender and dried under a fume hood. Substituting the peroxide with sodium nitrite yields a product with a faster rate of release of active ingredient. Microscopic examination of granular particles containing butylate shows that products obtained with the nitrite are porous and have cracks while those with peroxide have smooth surfaces (19) (Figure. 7.2).

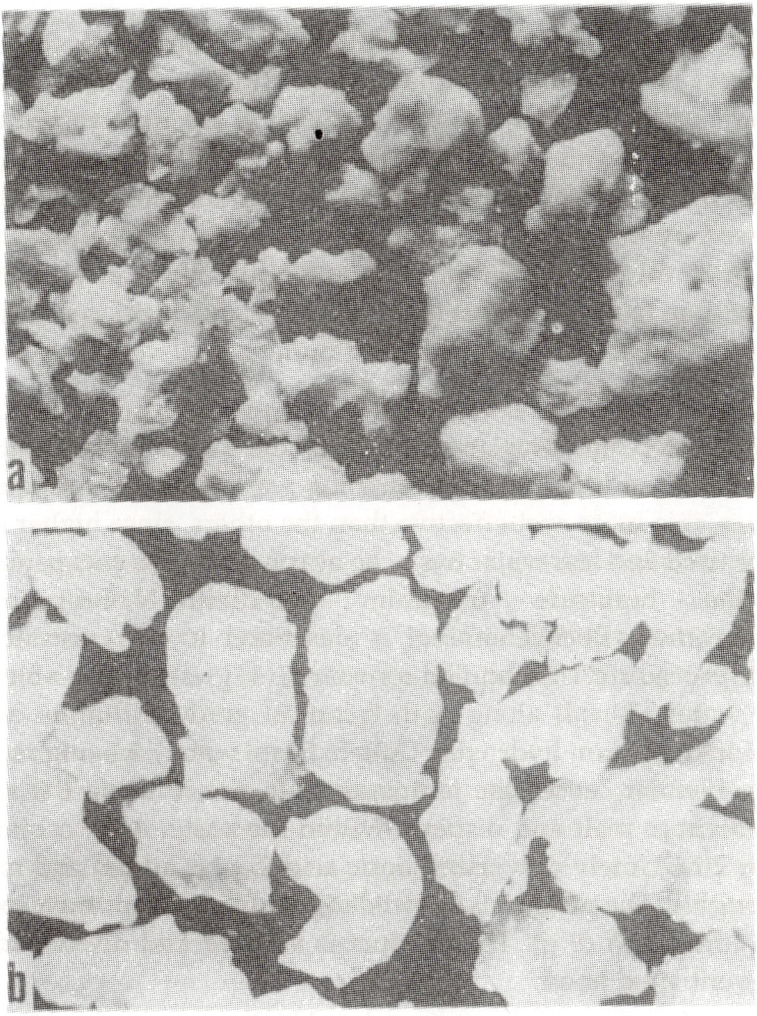

Figure 7.2 Microscopic pictures of granular particles containing butylate: (a) particles prepared with nitrite, (b) particles prepared with peroxide

These pores and cracks are probably due to nitrous acid gas produced during the cross-linking. Pilot plant processes have been developed (14) and cost estimates have been made (15).

7.3.1(b) *Calcium procedure:* The calcium procedure (16) and the borate procedure that follows avoid the use of toxic, flammable carbon disulphide. The encapsulation of the herbicide vernolate (S-propyl dipropylthiocarbamate) is described to demonstrate the calcium procedure. Unmodified cornstarch (45 g) is mixed with water (250 ml) and technical vernolate (20 g). The starch is gelatinized by the addition of sodium hydroxide solution (10%, 50 ml), followed by the addition of calcium chloride solution (dihydrate, 10 g in 25 ml water). The mixture is filtered and pressed using a rubber dam to remove most of the water. The product is sheared in a blender to pass 10 mesh and dried under a well-ventilated hood.

7.3.1(c) *Borate procedure:* The borate procedure (17) avoids the filtration step that was required with the xanthate and calcium procedures, so it is useful in retaining water-soluble compounds. Besides, it is more cost-effective than the others, since higher solid can be used and less water has to be evaporated. The encapsulation of the herbicide trifluralin [2,6-dinitro-N,N-dipropyl-4-(trifluoromethyl)benzenamine] is described to demonstrate the borate procedure. Unmodified cornstarch (45 g) is mixed in a blender with water (70 ml) along with technical grade trifluralin and a solution of sodium hydroxide (3.3g in 50 ml water). Mixing rapidly in the blender causes the temperature to rise, causing the active ingredient to melt and disperse within the paste. After cooling to below 20°C, finely pulverized boric acid (5 g) is added and mixed thoroughly. The product is allowed to stand for 15 min and is coated with cornstarch (8 g). The product is sheared and dried under a well-ventilated hood.

Examples of pesticides encapsulated by the borate procedure are listed in Table 7.1.

Table 7.1 Encapsulation of pesticides by the starch-borate procedure

Commercial name	*Chemical name* [a]	*Formulation*	*Proportion recovered (%)*
Chlorthaldimethyl	Dimethyl-2, 3, 5, 6-tetra-chloro-1, 4- benzenedi-carboxylate	75% WP	99

(Contd...)

Table 7.1 *(Contd...)*

Commercial name	*Chemical name* [a]	*Formulation*	*Proportion recovered (%)*
Chlorothalonil	2, 4,5,6-Tetrachloro-1, 3-benzenedicarbonitrile	75% WP	100
Terbutryne	N-(1, 1-dimethylethyl)-N'-ethyl-6-(methylthio)-1, 3, 5-triazine-2, 4-diamine	80% WP	98
Fonofos	O-ethyl-S-phenyl ethyl-phosphonodithioate	Liquid, 95-99% tech.	81
Diazinon	O, O-diethyl-O-6-methyl-2-(1-methylethy1) -4-pyrimidinyl phosphorothioate	Liquid, 8 EC	96
Butylate	S-ethyl bis(2-methylpropyl)-carbamothioate	Liquid, 98% tech.	91
Butylate+		Liquid, 6, 7 EC	100
EPTC	S-ethyl dipropylcarbamo-thioate	Liquid, 98% tech.	86
EPTC+		Liquid, 6, 7 EC	97
Vernolate	S-propyl dipropyl-carbamothioate	Liquid, 7 EC	94
Pebulate	S-propyl butylethyl-carbamothioate	Liquid, 6 EC	96
Cycloate	S-ethyl cyclohexyl-ethyl carbamothioate	Liquid, 6 EC	95
Molinate	S-ethyl hexahydro-l-H azepine-l-carbothioate	Liquid, 98% tech.	84
Triallate	S-(2, 3, 3-trichloro-2-propenyl)-bis-isopropyl thiocarbamate	Solid, 92% tech	95
2, 4-D	2, 4-Dichlorophenoxy-acetic acid, propylene glycol butyl ethyl ester	Liquid, 73% EC	94
Trifluralin	2, 6-Dinitro-N, N-dipropyl-4-(trifluoromethyl) benzenamine	Solid, 95% tech	91
Benfluralin	N-butyl-N-ethyl-2, 6-dinitro-4- (trifluoromethyl) benzenamine	Solid, 95% tech	100

(Contd...)

Table 7.1 *(Contd...)*

Commercial name	*Chemical name* [a]	*Formulation*	*Proportion recovered (%)*
Ethalfluralin	N-ethyl-N-2 (2-methyl-2-propenyl)-2, 6-dinitro-4- (trifluoromethyl) benzenamine	Solid, 97% tech	100
Metain-Sodium (Vapam)	Sodium methylcar-bamodithioate	Liquid, 32.7% aq	100

(a) Chemical abstracts nomenclature.

7.3.1(d) *Extrusion procedure:* Unlike the xanthate, calcium and borate procedures, which require the use of batch processes to encapsulate pesticides, the extrusion procedure makes the use of a continuous process possible. Unmodified cornstarch is gelatinized in the extruder in the presence of a water-herbicide mixture at starch concentrations ranging from 20 to 65%. A co-rotating, intermeshing, twin-screw extruder was used to encapsulate liquid or solid herbicides to yield products with encapsulation efficiencies exceeding 90%.

7.3.1(e) *Encapsulated pesticides within starch:* Finely powdered solid or liquids with high boiling points can be encapsulated by any of the above procedures to yield particles 400-1,600 μm in diameter. Particles in this range are suitable for use with conventional farm equipment. Although a smaller particle size would allow better uniformity of coverage (Table 7.2), the encapsulation efficiency is reduced.

Regardless of the procedure used to encapsulate, shelf-life is good and there is no appreciable loss on storage for at least one year. Furthermore, volatile encapsulated products do not escape upon placement in an open container for several weeks. However, when they are wetted or immersed in water, the active agent diffuses from the matrix. A screening test was developed for judging the release properties of thiocarbamate-containing products to assist in

Table 7.2 Relationship of capsule diameter and number of capsules per gram

Capsule diameter (μm)	*Approx. no. of capsules per gram (assume density of 1)*
5	15,279,000,000
25	122,230,000
100	1,909,800
200	238,730
400	29,841
500	15,277
600	8,842
700	5,568
800	3,730
900	2,620
1000	1,910

the selection of formulations for subsequent bioassay. Among the important factors that affect the rate of release are the physical characteristics of the entrapped chemical. In an aqueous medium the higher the solubility in water of the pesticide, the faster it will diffuse out of the starch matrix. Thus, the herbicide EPTC (S-ethyl dipropylthiocarbamate) will be released faster than the herbicide butylate (S-ethyl diisobutylthiocarbamate) since the water solubility of EPTC at 20°C is 370 ppm and butylate's is 45 ppm. Scanning electron micrographs that depict the surface and internal structure of several starch-encapsulated granular formulations of EPTC and butylate reveal a different structure for fast-and slow-release formulations. The fast-release formulation appears to have a porous surface with little distinctive internal structure. Diffusion would be rapid as few barriers are present. The slow-release formulation has a smooth, hard surface with a honeycomb internal structure (Figure 7.3). The scanning electron microscope would be an excellent quality control technique in the preparation of starch-encapsulated herbicides.

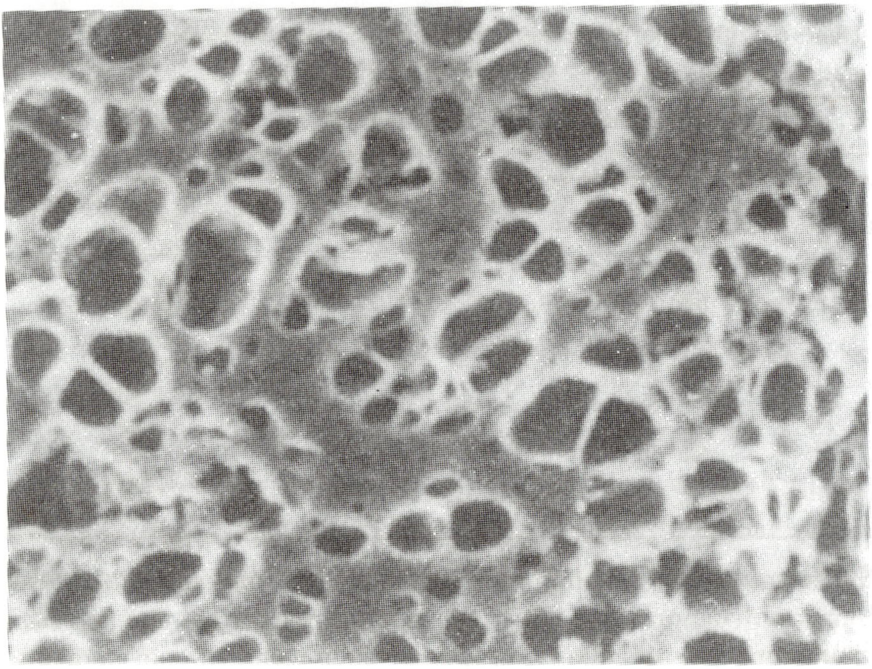

Figure 7.3 Scanning electron micrograph of fractured granule of encapsulated butylate

8 QUALITY TESTING

While bioassay tests are the preferred tests for encapsulated products, they are time-consuming and unavailable in most laboratories. Thus, quick and indicative tests were developed to screen large numbers of samples for their comparative rate of release. The tests developed fall mainly into two broad categories (20): (a) those that examine the rate of breakdown of starch by enzymatic means and hence the release of an active ingredient and (b) those that examine the rate of diffusion of an active ingredient from the starch matrix.

The rate of breakdown is related to the degree of cross-linking of starch. It was determined by measuring the swelling of the encapsulated product in water and by measuring the amount of free glucose released under standard conditions upon treatment of the encapsulated product with diazyme (an amylase-type enzyme that hydrolyzes starch to free glucose). A good correlation was found between swelling and diazyme tests, and they are complementary to each other.

If the encapsulated agent is a solid with low solubility in common solvents, then the particle is examined through a magnifying lens with 30-50 times magnification to observe whether loose particles of the agent are present unencapsulated. A second test, which is more meaningful and easier to carry out, involves immersion of the product in a dilute aqueous iodine solution for 1-2 minutes. Since only the starch part will stain with iodine, the observation of the immersed sample through the lens will reveal the entrapment quality.

If the encapsulated agent is a solid or liquid soluble in common organic solvents, then, besides the iodine test, a portion of the product (about 10 g containing 15% encapsulating agent) is suspended in the appropriate organic solvent (50 ml) for 1 hr. The portion of the agent extracted reflects the part that is not encapsulated or that is close to the surface of the encapsulated granule.

If the encapsulated agent is a volatile liquid, a portion of the product is exposed to an air current (a well-ventilated fume hood will suffice) and the remaining product is analyzed periodically. Another simple test, the "wet test", consists of placing several 1g portions of the product in 2 ml of water. The water is allowed to evaporate, and after 4 hr another 2 ml of water is added and again allowed to evaporate. Addition of water and evaporation is repeated a total of four times, and the dried product is analyzed after each wetting-drying cycle for loss of the encapsulated agent.

9 DISCUSSION

The next decade will bring an increased focus on environmental issues, and controlled release technology for agrochemicals can play an important role. The technology has many distinctive advantages and has grown steadily in sophistication. Among the advantages are the reduction in the amount of active ingredient needed for fewer applications while at the same time reducing volatility and toxicity.

Most pesticides are toxic to people and wildlife. However, their rates of release can be controlled upon entrapping the toxic material, which decreases their toxicity. Current problems are in two areas:

(a) *Economic.* Traditionally, agrochemicals are sold in bulk quantities and farmers are extremely sensitive to any increase in cost, especially when many countries have huge surpluses of agricultural commodities;

(b) *Education.* The consumer needs to be educated not only about the advantages of these formulations but also the change in performance of active ingredients. The agrochemical industry needs to be responsive about the protection of clean air and water not only by developing safer pesticides but also by improving the safety of existing ones. Some manufacturers are reluctant to invest in products that will have a higher production cost and that the farmer will use less of.

To develop and market any new technology successfully, many issues have to be assessed and resolved, such as customer needs, product performance, competing products and consumer and environmental safety.

For research to be effective one must adopt "strategic defence viewpoints", which consist of identifying flaws as early as possible in the development stage and trying to modify the product accordingly or moving on to something else with better potential. Some of the factors that are especially important are the following:

(a) *Customer need.* If a product does not satisfy a real need, the probability of failure is very high. It has been said (21) that "effective marking strategy depends upon good information about customer needs and wants and the extent to which they are being served by current suppliers — the most common error is the producers overestimate the uniqueness of their product and its ability to satisfy customer needs." Assessing the real needs can be especially hard for technical people since they are too often enamored of the benefits of a new technology without addressing the real concerns of the customer;

(b) *Performance properties.* If the product does not perform well, no one will buy it. In the field of controlled-release formulations, one must be careful how performance is assessed. Often the first assessment is a simple plug-in for an existing product. This is usually doomed to failure since most formulations are highly optimized. As was mentioned earlier, controlled-release products behave differently from conventional formulations since the availability of active ingredient at any given time is different.

(c) *Fit to current operations.* Most firms do best in their own field, so the probability of success goes up if the product fits a company's current operations. It is important to remember that people resist change in their normal practices;

(d) *Cost.* The cost, of course, has to be competitive but initially it is high per unit volume because most manufacturing is affected by economies of scale: as volume goes up, capital and operating costs go down;

(e) *Relative advantage.* The product has to be perceived as superior to existing products or practices, preferably creating its own niche in the market place.

Since understanding customer-consumer needs is so crucial, one of the best ways to do this is to form a partnership with the manufacturer as early as possible in the development stage. This allows the research to be tailored to the eventual needs.

REFERENCES

(1) M. Macioti, "How to save the planet," *Times Higher Education Supplement*, 15 February, 1990, *ISI Press Digest* 17, p. 19.

(2) M. M. Schreiber, B. S. Shasha, D. Trimnell and M. D. White., "Methods of applying herbicides", *Controlled Release Herbicides*, WSSA Monograph 4, Weed Society of America, 1987, p. 177.

(3) S. D. Bruck, *Controlled Drug Delivery* (CRC Press, 1983).

(4) A. F. Nydoniens, *Controlled Release Technologies: Methods, Theory and Applications* (CRC Press, 1980).

(5) N. G. Das, *Controlled Release Technology Bioengineering Aspects* (Wiley, 1982).

(6) B. Cross and H. B. Scher, *Pesticide Formulations*, ACS Symposium Series 371, 1987.

(7) M. H. Gutcho, *Microcapsules and Other Capsules* (Noyes Data Corporation, 1979).

(8) H. H. Shorey, *Handbook of Pest Management in Agriculture,* Vol. 2, 1981, p. 307.

(9) D. J. Lawatsch, *Proceedings of the Controlled Release Symposium,* 1977.

(10) F. W. Harris, *Proceedings of the Controlled Release Symposium,* 1974.

(11) C. M. Himel and N. F. Cardarelli "Flight encapsulation of particles," U.S. Pat. 4,286,020, 1981. CIP, U.S. Pat 4,353,962, 1982.

(12) R. E. Sparks, *Microencapsulation* (East Brunswick, New Jersey, Center for Professional Advancement, 1980).

(13) B. S. Shasha, W. M. Doane and C. R. Russell, "Starch encapsulated pesticides for slow release," *J. Polymer Science,* 14,1976. p. 417.

(14) E. I. Stout, B. S. Shasha and W. M. Doane, "Pilot plant process for starch xanthide encapsulated pesticides," *J. Applied Polymer Science,* 24, 1979, p. 153.

(15) G. E. Hamerstrand, "Starch encapsulated pesticides: a preliminary cost estimate," *Eighth International Symposium on Controlled Release of Bioactive Materials,* 1981.

(16) B. S. Shasha, D.Trimnell and F. H. Otey, *Encapsulation of Pesticides in a Starch-Calcium Adduct,* 19, 1981, p. 1891.

(17) D. Trimnell, B. S. Shasha, R. E. Wing and F. H. Otey, "Pesticide encapsulation using a starch-borate complex as wall material," *J. Applied Polymer Science,* 27,1982, p. 3919.

(18) R. E. Wing, Private communication.

(19) M. M. Schreiber and M. D. White, "Granule structure and rate of release with starch-encapsulated thiocarbamates," *J. Weed Science,* 28,1980, p. 685.

(20) B. S. Shasha, "Methods for testing and factors affecting rate of release of encapsulated products prepared via starch xanthate," *Fifth International Symposium on Controlled Release of Bioactive Materials,* 1978.

(21) R. G. Schafermeyer, "Developing new products more effectively," Conference on Commercializing Industrial Uses for Agricultural Commodities, 1990.

CHAPTER

8

Suspension Concentrates

Th. F. Tadros

Zeneca Agrochemicals
Jealott's Hill Research Station
Bracknell, Berkshire, RG 12 6EY, United Kingdom

1 INTRODUCTION

The formulation of agrochemicals as dispersions of solids in liquids (to be referred to as suspension concentrates) has attracted considerable attention in recent years. Such formulations offer a number of advantages for the user, being easy to measure, not dusty and readily dispersible in water. Moreover, being aqueous-based they produce less hazard to the operator in handling and they have no problems of flashpoint. They also cause less damage (phytotoxicity) to the plant. These aqueous suspension concentrates (SCs) are also suitable for optimization of biological efficacy. Firstly, one may control the particle size by controlling the milling conditions and proper choice of the dispersing agent. Secondly, it is possible to incorporate high concentrations of surfactant in the formulation. This is sometimes essential to enhance wetting, spreading and penetration. Stickers may also be added to enhance adhesion and in some cases to provide slow release. In recent years, there has been considerable research into the factors that govern the physical stability of suspension concentrates (1,2). The theories of colloid stability could be applied to predict the stability of these systems. In addition, analysis of the problem of sedimentation at a fundamental level has been undertaken (1,2). Since the density of the particles is larger than that of the medium (water), SCs tend to separate as result of sedimentation. The sedimented particles tend to form a compact layer, sometimes referred to as clay, which is very difficult to redisperse. It is, therefore, essential to reduce sedimentation and prevent the formation of clays. In this chapter, attempts will be made to address the above-mentioned phenomena at a fundamental level. The chapter will start with a section on the preparation of SCs and the role of the dispersing agent. This is followed by a section on the control of the physical stability of SCs. The various theories of colloid stability will be briefly summarized. The next section will deal with the problem of sedimentation and prevention of claying. The various methods that may be applied to reduce sedimentation and prevent the formation of hard clays will be summarized. The last section of this chapter will deal with the methods that may be applied for the assessment of the physical stability of SCs. For the assessment of flocculation and crystal growth, particle-size analysis techniques are commonly applied. The bulk properties of the suspension, such as sedimentation and dispersion on dilution, may be assessed using rheological techniques. The latter will be summarized with particular emphasis on their application in prediction of the long-term physical stability of suspension concentrates.

2 PREPARATION OF SUSPENSION CONCENTRATES AND THE ROLE OF THE DISPERSING AGENT

Suspension concentrates are usually formulated using a wet milling process, which requires the addition of a dispersing agent. The latter should satisfy the following criteria:

(a) Be a good wetting agent for the agrochemical powder (both external and internal surfaces of the powder aggregates or agglomerates must be spontaneously wetted);
(b) Be a good dispersing agent to break such aggregates or agglomerates into smaller units and subsequently help in the milling process (one usually aims at a dispersion with a volume mean diameter of 1-2 μm);
(c) Provide good stability in the colloid sense. This is essential for maintaining the particles as individual units once formed. Powerful dispersing agents are particularly important for the

preparation of highly concentrated suspensions. Any flocculation will cause a rapid increase in the viscosity of the suspension, and this makes wet milling a difficult job.

In this section the wetting of agrochemical dispersions and their subsequent dispersion and milling will be discussed. The subject of colloid stability will be dealt with in the next section.

2.1 Wetting of Agrochemical Powder

Dry powders of organic compounds usually consist of particles of various degrees of complexity, depending on isolation stages and the final drying process. Generally, the particles in a dry powder form either aggregates, in which the particles are joined together at their crystal faces, or agglomerates, in which the particles touch only at edges or corners, thereby forming a looser, more open structure. It is essential in the dispersion process to wet the external as well as the internal surfaces and displace the air between the particles. This is usually achieved by the use of surface-active agents or macromolecules of the ionic or non-ionic type. Wetting of a solid is usually described in terms of the equilibrium contact angle, θ, and the appropriate interfacial tensions, using the classical Young's equation,

$$\gamma_{sv} - \gamma_{sl} = \gamma_{lv} \cos\theta \qquad (1)$$

or,

$$\cos\theta = (\gamma_{sv} - \gamma_{sl})/\gamma_{lv} \qquad (2)$$

where γ represents the interfacial tension and the symbols s, l and v refer to the solid, liquid and vapour respectively. It is clear from equation (1) that if $\theta < 90°$, a reduction in γ_{lv} improves wetting. Hence, the use of surface-active agents or macromolecules that reduce both γ_{lv} and γ_{sl} to aid wetting is clear. However, the process of wetting of particulate solids is more complex and involves at lease three distinct types of wetting (3,4): adhesional wetting, spreading wetting and immersional wetting. This is illustrated in Figure 8.1, which describes schematically the wetting of a cube of solid by a liquid. Assuming the surface area of the cube to be unity, then the work of dispersion (W_d) is given by the following expression:

$$W_d = W_a + W_i + W_s = -6\,\gamma_{sv} - 6\,\gamma_{sl} = -6\,\gamma_{lv}\cos\theta \quad (3)$$

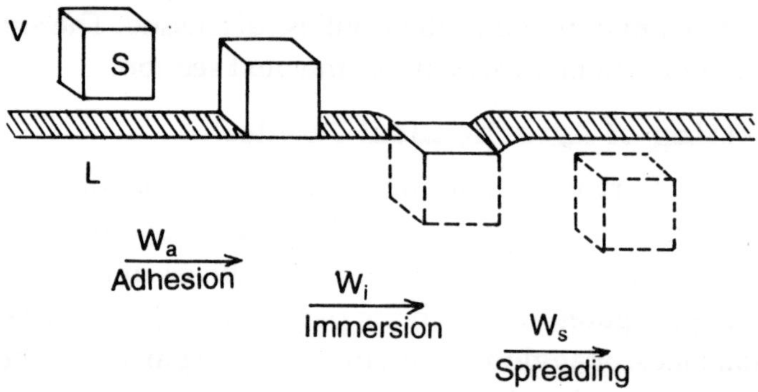

Figure 8.1 The three stages involved in the complete wetting of a solid cube by a liquid

As the wetting of a solid by a liquid depends on two measurable quantities, γ_{lv} and θ, the above equation may be used to predict whether the process is spontaneous, as it is when W_d is negative.

The next stage to consider is the wetting of the internal surface, which implies the penetration of the liquid into channels between and inside the agglomerates. This process is governed by the capillary pressure, P, which is given by the following equation,

$$P = -2\gamma_{lv}\cos\theta/\gamma = -2\;(\gamma_{sv} - \gamma_{sl})/r \quad (4)$$

Equation (4) shows that to increase penetration, θ and γ_{sl} have to be made as small as possible, e.g. by sufficient adsorption of surface-active agent. However, when $\theta = 0$, P is proportional to γ_{lv}, i.e. a large surface tension is required. These two opposing requirements show that the proper choice of a dispersing agent is not simple. In most cases a compromise has to be made between minimizing θ and not having too small a surface tension to aid penetration.

2.2 Dispersion and Milling

For the dispersion of aggregates and agglomerates into smaller units, one requires high-speed mixing, e.g. a Silverson mixer. In some cases the dispersion process is easy and the capillary pressure may be sufficient to break up the aggregates and agglomerates into primary units. This process is aided by the dispersing agent, which becomes adsorbed on the particle surfaces. However, for reduction of particle size it is necessary to apply a milling process (sometimes referred to as comminution). The most common procedure for such a process is bead milling, whereby glass or ceramic beads are mixed with the suspension. The suspension is then rotated at a high speed using various types of mills, e.g. a Dynomill. During this rotation process, impaction between the beads and the particles takes place, fracturing the particles. This process is aided by the dispersing agent, which becomes adsorbed at the solid/liquid interface. Rehbinder and co-workers (5,6) suggested that adsorption of the surfactant at a newly formed surface during a fracture process facilitates crack propagation and thereby the formation of smaller particles. This effect is sometimes referred to as the Rehbinder effect.

3 CONTROL OF THE COLLOID STABILITY OF SUSPENSION CONCENTRATES

The dispersing agent used for the preparation of a suspension concentrate should provide repulsion between the particles that is sufficient to overcome the everlasting van der Waals attraction. In other words, it is essential to maintain the particles that are formed in a milling process as individual units. Aggregation and flocculation should be prevented over a long period of time (2-3 years) at various temperatures (from, say, -10° to +50°C). This requires the use of powerful dispersing agents that provide an energy barrier against aggregation. Generally speaking, two types of stabilization mechanisms may be applied. The first type depends on the creation of an electrical double layer at the solid/solution interface. This is achieved by adsorption of an ionic surfactant or polyelectrolyte. Charge separation occurs as a result of dissociation of these molecules. The second type of stabilization depends on the adsorption of a non-ionic surfactant or macromolecule. The adsorbed layers prevent the approach of the particles to a distance of separation where van der Waals forces of attraction may be significant. This effect is usually referred to as steric stabilization. Following is a discussion of the two types of stabilization mechanisms, including a summary of the factors that control the stability of the suspension in the colloid sense.

3.1 Electrostatic Stabilization

Surfactants of the anionic or cationic type may be used for the stabilization of suspension concentrates. The most commonly used materials are the alkyl and alkyl benzene sulphonates and the cationic alkyl trimethyl ammonium surfactants. When these surfactants are added to the continuous phase, adsorption of the surfactant ion on the particle surface occurs, and the counter-ions produce an extended layer in bulk solution. Part of the counter-ions may approach closely to the particle surface, whereas the rest produce a diffuse layer that extends to a significant distance from the particle surface. This arrangement of surfactant ions and counter-ions is referred to as an electrical double layer. The counter-ions that are present in the first layer near the surfactant ions are referred to as the Stern plane, whereas the remaining diffuse layer is referred to as the Gouy diffuse layer. This picture of the double layer has been developed by Gouy-Chapman (7), Stern (8) and Grahame (9). The potential at the surface resulting from the adsorbed surfactant ions is referred to as the surface potential, ψ, whereas the potential at the locus of the first layer of counter-ions is referred to as the Stern potential. The latter is usually taken to be equal to the measured zeta potential, ξ.

The double layer extension is determined by the electrolyte concentration C. This extension is given by the reciprocal of the Debye-Huckel parameter $(1/\kappa)$, which for a 1:1 electrolyte is given by the following expression:

$$(1/\kappa) = \left(\frac{\varepsilon \varepsilon_o kT}{4\pi e^2 C Z^2} \right)^{1/2} \qquad (5)$$

where ε is the permittivity of the medium, ε_o is the permittivity of free space, e is the electronic charge and Z is the valency of the ion.

For 1:1 electrolyte, $(1/\kappa)$ is equal to 100 nm, 10 nm and 1 nm in 10^{-5}, 10^{-3} and 10^{-1} mol dm^{-3}, respectively. In other words, the double layer extension increases with a decrease in electrolyte concentration. This is reflected in the repulsive energy, which increases with a decrease in electrolyte concentration.

When two particles with their extended double layers approach to a separation distance h that is smaller than twice the double layer thickness, overlap of the double layers occurs, resulting in a repulsive force V_R that is given by the following expression,

$$V_R = 2\,\pi\, a\, \varepsilon_r\, \varepsilon_o\, \psi_o^2 \ln\,[1 + \exp-(\kappa h)] \tag{6}$$

where a is the particle radius, ψ_o is the surface potential, which may be replaced by the ξ potential. It is clear from equation (6) that V_R decreases exponentially with distance of separation h between the particles. The rate of decrease of V_R with h depends on the electrolyte concentration as determined by the value of κ. At any given h, the value of V_R increases with decrease of electrolyte concentration.

For a description of the stability of a suspension, it is essential to consider the balance between repulsion and attraction at various distances of separation. The attractive force between particles was considered by Hamaker (10), who arrived at the following expression for the attractive force V_A:

$$V_A = -\,\mathrm{Aa} / 12h \tag{7}$$

where A is the net Hamaker constant that is given by the following expression:

$$A = \left(A_{11}^{\;1/2} - A_{22}^{\;1/2}\right)^2 \tag{8}$$

where A_{11} is the Hamaker constant of the particles and A_{22} that of the medium. The Hamaker constant of any material, A_{ii}, is given by the following expression:

$$A_{ii} = \pi\; q^2 \beta_{ii} \tag{9}$$

where q is the number of atoms or molecules per unit volume and

β_{ii} is the London dispersion constant.

It can be seen from equation (7) that V_A is inversely proportional to the distance of separation between the particles. Thus, V_A increases rapidly with decrease of separation distance h and it reaches very high values when h is very small.

The net interaction between two particles is obtained by summing V_R and V_A at various h values, *i. e.*,

$$V_T = V_R + V_A \quad (10)$$

This forms the basis of the theory of colloid stability introduced by Deryaguin and Landau (11) and Verwey and Overbeek (12) (the DLVO theory). A schematic representation of the variation of V_T with h is shown in Fig. 8.2 which shows two extreme cases of high and low values of V_R (curves A and B respectively).

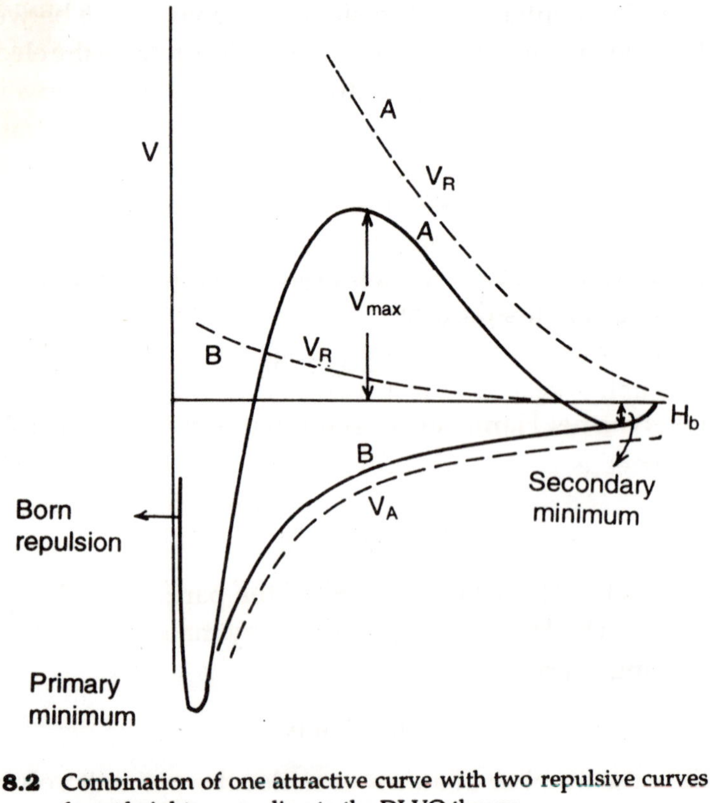

Figure 8.2 Combination of one attractive curve with two repulsive curves of different heights, according to the DLVO theory

When the repulsive force is small, as in curve *B*, the net force is attractive and the suspension undergoes rapid aggregation. On the other hand, when the repulsive force is appreciable, as in curve *A*, one obtains the net force-distance curve represented by *A* in Figure 8.2. This curve shows two minima and one maximum. At large distances of separation, the attractive force is larger than the repulsive force. This is due to the fact that the repulsive force decays exponentially with distance of separation and at large distance approaches a very small value. On the other hand, the attractive force decays as a power law with h, and at large distances, it will have a value that is larger than the repulsive force. This results in a residual attraction at large distances resulting in a secondary minimum, whose depth depends on the particle size, Hamaker constant and electrolyte concentration. At very short distances of separation, V_A is much larger than V_R, resulting in a deep primary minimum. The rapid increase in V_T at extremely short distance is due to the Born repulsion. At intermediate distances of separation, V_R is now larger than V_A, resulting in an energy maximum whose height depends on electrolyte concentration, particle size and zeta potential. V_{max} increases with decrease in electrolyte concentration and increase in the zeta potential.

The stability of a suspension that is stabilized by an ionic surfactant can be explained on the basis of the DLVO theory summarized above. Adsorption of the surfactant ions on the particle surface usually results in a reasonably high zeta potential (50-100 mV). If the electrolyte concentration is kept low, a high energy barrier is produced, preventing any particle aggregation. Such a barrier is usually greater than 25 kT (where kT is the thermal energy of the particles), so the suspension remains stable for a long time. Thus, ionic surfactants can produce a highly deflocculated system provided the electrolyte concentration is kept to a minimum. However, in many industrial preparations, the water used may contain a significant concentration of an electrolyte, in particular Ca^{+2} and Mg^{+2}, which results in a compression of the double layer, i.e. a significant reduction of V_R. Under these conditions, the energy maximum may be significantly reduced or eliminated altogether, resulting in particle aggregation. This shows the limitation of using ionic surfactants for formulating suspension concentrates.

An alternative and more efficient dispersing agent should provide stability under the practical conditions encountered, e.g. in the presence of electrolyte. Polyelectrolytes provide such alternatives, of which the naphthalene formaldehyde sulphonated condensates and the lignosulphonates are the most widely used in the formulation of suspension concentrates. The polyelectrolyte nature of these materials make them less susceptible to moderate concentrations of electrolyte. These dispersing agents operate by a combination of electrostatic and steric stabilization mechanisms. Any compression of the double layer by the addition of electrolyte is compensated by steric stabilization since these molecules produce relatively thick adsorbed layers.

3.2 Steric Stabilization

Nonionic surfactants and macromolecules are commonly used for formulation of suspension concentrates. The most commonly used non-ionic surfactants are those of the ethoxylate type, of which alkyl and alkyl phenyl ethoxylates are the most familiar. These surfactants consist of a hydrophobic part, namely the alkyl or alkyl phenyl, and a hydrophilic part, which is the polyethylene oxide chain. The molecule adsorbs with the hydrophobic part on the particle surface, leaving the polyethylene chain dangling in solution. It is necessary to have strong adsorption of the hydrophobic portion and usually the alkyl chain length is 12 or more. This may provide sufficient anchoring to the particle surface. In some cases the adsorption is enhanced by the introduction of a phenyl group and/or polypropylene oxide chain. The polyethylene oxide chain, which provides the steric barrier, is chosen with sufficient length to provide sufficient repulsion. In most cases, a chain with more than 10 ethylene oxide units is chosen. However, one should be careful in the choice of the optimum chain length, since a very long ethylene oxide chain, although providing a better steric barrier, may cause reduced adsorption. A useful empirical index for the choice of the non-ionic surfactant is the hydrophilic-lipophilic balance (HLB), which is simply the proportion of hydrophilic components to lipophilic groups in the surfactant molecule. In general, an HLB number between 8 and 18 is chosen, depending on the nature of the agrochemical solid. Once the optimum HLB number surfactant is chosen, it is useful to use the one with the highest molecular weight possible. One should be extremely careful in applying such an empirical approach, since the surfactant chosen may not satisfy all the criteria required for effective steric stabilization. These criteria will be discussed below after consideration of the theories of steric stabilization.

With non-ionic polymers, the situation is more complicated. These molecules consist of a large number of segments and hence adopt different configurations, depending on the structure of the molecule. The simplest configuration to consider is that of a homopolymer, which adsorbs with a few segments in direct contact with the particle surface (referred to as trains), leaving the rest of the chains dangling in solution as loops and tails (13). However, as we will see later, homopolymers are not the most effective in steric stabilization and more complex structures of block and graft copolymers are normally used. The most commonly used block copolymers are those based on blocks of ethylene oxide and propylene oxide. Chains consisting of a central propylene oxide with two ethylene oxide chains on either side are used. These polymers (referred to as A-B-A block copolymers) are available commercially under the trade name Pluronic by BASF and Synperonic PE by ICI. They cover a wide range of compositions where the chain length of the propylene oxide and ethylene oxide chains is varied. More recently, ICI has prepared a highly effective dispersing agent of the graft type, which it sells under the trade name Hypermers. An example of these agents is a graft consisting of a backbone of polymethyl methacrylate with several polyethylene oxide side chains (Hypermer CG-6 ex ICI Surfactants). Recent investigations (14) have demonstrated the strong adsorption of the graft copolymer on pesticide particles (this was confirmed by the lack of desorption of the chain once adsorbed). The multiplicity of the polyethylene side chains provides an effective steric barrier.

The origin of the steric stabilization produced by the above-mentioned non-ionic surfactant, block and graft copolymers can be understood from a consideration of the interaction forces that exist when two particles containing adsorbed surfactant or polymers approach each other in a suspension. Two main repulsive energies may be distinguished on such approach (15). When the particles, with an adsorbed layer of thickness δ, approach to a distance that is smaller than 2δ, interpenetration and/or compression of these chains occur. Both effects increase the osmotic pressure in the overlap region, since these chains (mainly polyethylene oxide) are in good solvent conditions (the chains are very soluble in water). As a result of the increase in osmotic pressure, solvent molecules diffuse to the overlap region, thus separating the particles. This repulsive effect is referred to as the mixing free energy term, G_M. It implies that mixing of the highly solvated chains of polyethylene oxide is highly unfavourable. For two spheres with radius a, G_M is given by the following expression:

$$\frac{G_M}{kT} = \frac{4\pi\phi_2^2}{3V_1}\left(\frac{1}{2}-\chi\right)\left(\delta-\frac{h}{2}\right)^2\left(3a+2\delta+\frac{h}{2}\right) \qquad (11)$$

where ϕ_2 is the volume fraction of polymer in the adsorbed polymer layer and χ is the chain-solvent interaction parameter (usually referred to as the Flory-Huggins parameter). It can be seen from equation (11) that if χ is less than 1/2, G_M is positive and the mixing interaction is repulsive. This is the case when the stabilizing chain is in good solvent conditions. However, when χ is greater than 1/2, G_M is negative and the mixing interaction is negative. This will, in most cases, lead to catastrophic flocculation of the suspension, a process that is referred to as incipient flocculation. There is one point at which χ is equal to 1/2, which is referred to as the theta point, and it marks the flocculation point. It is clear that one should keep the suspension well away from the theta point by maintaining the chains in good solvent conditions. Most dispersants based on polyethylene oxide are stable under normal conditions, e. g. within a reasonable temperature range and at moderate electrolyte concentrations. In cases where a high electrolyte concentration has to be used, flocculation may occur, particularly at high temperatures. Under these conditions the medium may no longer be a good solvent for the chains. It is, therefore, essential to ensure that the conditions to which the suspension is subjected do not bring the chains to worse than theta conditions. One way of checking this is to measure the cloud point of the surfactant or polymer in the medium of the suspension. In most cases, there is a relationship between the cloud point and the theta point.

The second type of interaction that occurs with particles containing adsorbed surfactants or polymers results from the entropy loss on significant overlap of the chains. Under these conditions, the chains lose configurational entropy, which amounts to an increase in the free energy. This effect is referred to as the entropic or elastic effect and is given by the following expression:

$$G_{e_l} = 2 \ \nu \ kT \ \ln \ (\Omega/\Omega_\infty) \tag{12}$$

where ν is the number of chains per unit area, Ω is the number of configurations on overlap, whereas Ω_∞ is the number of configurations at infinite separation distance. It is clear from equation (12) that G_{el} is always positive and it increases very rapidly with a decrease in h, particularly when $h < \delta$. Figure 8.3 shows schematically the variation of G_M, G_{el} and G_T with h. G_T is given by the sum of the steric G_s and van der Waals G_A energies:

$$G_T \quad = G_s + G_A \ = \ G_M + G_{el} + G_A \tag{13}$$

It can be seen from Figure 8.3 that G_M increases rapidly with a decrease of h as soon as $h < 2\delta$. On the other hand, G_{el} increases rapidly with a decrease in h when the latter is of the order of δ. The total energy of interaction shows only one minimum, G_{min}, whose depth depends on the adsorbed layer thickness, particle size and Hamaker constant. At $h < 2\delta$, G_T increases rapidly with a decrease in h. Unlike the DLVO theory, the energy distance curve does not show any maximum, and G_T goes almost to infinity at separation distances of δ. This implies that sterically stabilized suspensions may approach thermodynamic stability. This clearly shows the advantage of using

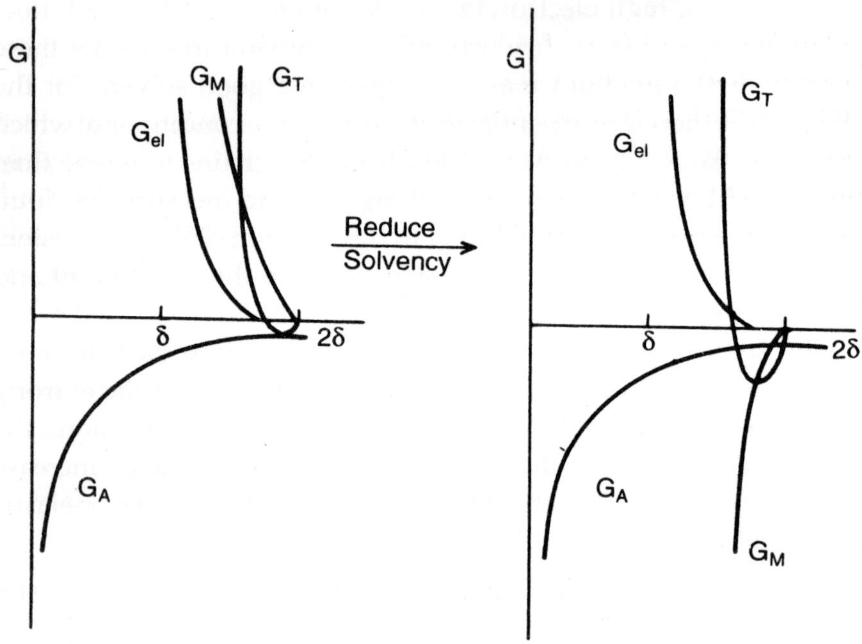

Figure 8.3 Total interaction distance curve for sterically stablized suspensions

non-ionic block and graft copolymers for the stabilization of suspension concentrates. The criteria for effective steric stabilization are summarized below:

(a) The dispersing agent should be strongly adsorbed to the particle surface. For this purpose, the anchoring groups (the B portion in an A-B-A block) is made to have a high affinity to the particle surface and a low affinity to the medium;

(b) The dispersant should fully cover the particle surface. Any bare patches may lead to flocculation, either through van der Waals attraction or bridging;

(c) The stabilizing chain should be of sufficient length to provide an effective barrier. The adsorbed layer thickness should be greater than 5 nm in order to screen the van der Waals attraction;

(d) The stabilizing chain should remain in good solvent conditions at all the storage temperatures.

A special case of steric stabilization is produced with polyelectrolytes such as lignosulphonates. In this case both double layer and steric repulsion operate. Any reduction in electrostatic repulsion, say as a result of the presence of electrolyte, will be compensated by the steric repulsive force. Unfortunately, most commercially available polyelectrolytes do not contain sufficient anchor groups and hence the adsorption of these materials is usually not very strong. With time, some desorption may take place, leading to flocculation. In addition the polyelectrolytes may be incompatible with other adjuvants or with other pesticide formulations which may be mixed with the SC on application.

4 CONTROL OF SEDIMENTATION AND PREVENTION OF CLAYING

As mentioned in the introduction, the particle size of SCs is not sufficiently small (average 1-2 μm) for Brownian motion to overcome gravitational sedimentation. Under these conditions, the particles tend to settle to the bottom of the container, forming a hard sediment (referred to as clay). The particles in a sediment tend to rotate and approach each other to small distances of separation. This results in a dilatant sediment (shear thickening), which is very difficult to redisperse. It is, therefore, essential to reduce sedimentation of suspension concentrates and prevent the formation of dilatant clays. Before discussing the methods that may be applied for reduction of sedimentation it is essential to analyze the sedimentation of suspensions and how it is affected by the concentration (volume fraction ϕ) of the system (16).

4.1 Sedimentation of Suspensions

For a very dilute suspension with rigid, non-interacting particles, the sedimentation velocity, v_o, can be described by Stokes' law, whereby the hydrodynamic force is equated with the gravitational force:

$$6\pi\eta R V_o = (4/3)\pi R^3 \Delta\rho g \tag{14}$$

$$V_o = \frac{2R^2 \Delta\rho \ g}{9 \ \eta} \tag{15}$$

where $\Delta\rho$ is the density difference between the particles and the medium ($\rho_2 - \rho_1$) and η is the viscosity of the medium. The Stokes' velocity and the time needed for complete sedimentation of a dilute suspension in a 0.1 m container is given in Table 8.1 for three different particle sizes, with a density difference of 0.2.

Table 8.1 Particle size and rate of settling

Particle radius (μm)	$\Delta\rho$	*Stokes' velocity (ms^{-1})*	*Time needed for complete separation in 0.1 m container*
0.1	0.2	4.4×10^{-9}	250 days
1.0	0.2	4.4×10^{-7}	60 hours
10.0	0.2	4.4×10^{-5}	40 minutes

It can be seen from the above table that the sedimentation of a very dilute suspension is very rapid when the particle size exceeds 0.1 μm. However, for more concentrated suspensions the sedimentation rate is considerably reduced as a result of interparticle interaction. For moderately concentrated suspensions ($\phi < 0.2$), hydrodynamic interaction reduces the sedimentation velocity. In this case, the sedimentation velocity, v, may be related to the Stokes velocity, v_o, by the following equation,

$$v = v_o (1 - k\phi) \tag{16}$$

where k is an empirical constant that is equal to 5 - 6.

Bachelor (17) developed a theory for the sedimentation of suspensions with $\phi < 0.2$ that took into account hydrodynamic interaction between the particles:

$$v = v_o (1 - 6.55\phi)$$

For a more concentrated suspension ($\phi > 0.2$), sedimentation is a complex function of the volume fraction. As ϕ increases, v

decreases, and when ϕ exceeds 0.4, one enters the hindered settling regime, whereby v becomes independent of the particle size. Usually v decreases exponentially with ϕ in the same manner as the viscosity of the suspension increases with ϕ. This is illustrated in Figure 8.4, which shows the variation of v and $\eta_r = \eta/\eta_o$, where η is the viscosity of the suspension and η_o that of the medium.

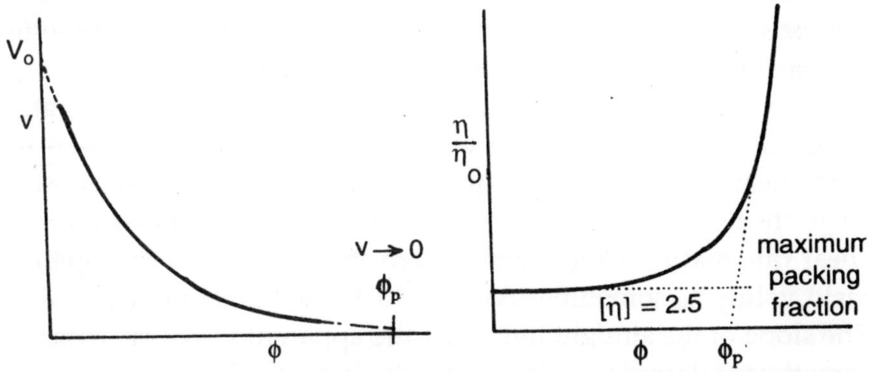

Figure 8.4 Variation of v and η/η_o with the volume fraction of the suspension

It can be seen from Figure 8.4 that v approaches zero and η_r approaches infinity when ϕ approaches ϕ_p, the so-called maximum packing fraction. The latter is equal to 0.64 for random packing and 0.74 for hexagonal packing when the particles are monodisperse. Dougherty and Krieger (18) derived an equation relating η_r to ϕ:

$$\eta_r = \eta/\eta_o = (1 - \phi/\phi_p)^{-(\eta)\phi_p} \qquad (17)$$

Since $v/v_o = \eta_o/\eta$ or $v/v_o = \alpha\,(\eta_o/\eta)$, where α is a constant,

$$v/v_o = (1 - \phi/\phi_o)^{\alpha(\eta)\phi_p} = (1 - \phi/\phi_o)^{k\phi_p} \qquad (18)$$

4.2 Reduction of Sedimentation and Prevention of Formation of Dilatant Sediments

Several methods may be applied to reduce the sedimentation of SCs, most of which are based on the addition of an inert material in the continuous phase. Such materials should cause a large increase in viscosity at low shear rates. However, the viscosity at moderate and high shear rates should be sufficiently low for ease of dispensing and dispersion of the SC on dilution. In some cases the SC may be applied without dilution, e.g. in seed dressing. Again, the viscosity of the suspension should be high at low shear rates to reduce sedimentation and prevent claying but sufficiently low for application in the seed dressing operation, for easy flow and adequate dressing of the seeds. Therefore, the materials that are added to reduce sedimentation should give the required flow characteristics (rheology). The most common flow behaviour required is that of shear thinning suspensions, whereby the viscosity is very high at low shear rates but decreases rapidly with increase in shear rate. This is illustrated in Figure 8.5, which shows the variation of shear stress and viscosity with shear rate. Such behaviour is sometimes referred to as pseudoplastic flow. The flow curve shows a linear region above a critical shear rate that if extrapolated to zero shear rate gives an apparent yield value, τ_B, sometimes referred to as the Bingham yield value. The slope of the straight line gives the apparent viscosity, which is sometimes referred to as the plastic viscosity, η_{pl}. The viscosity-shear rate curve shows two Newtonian regions at low and high shear rate ranges, with a shear thinning region at intermediate shear rates.

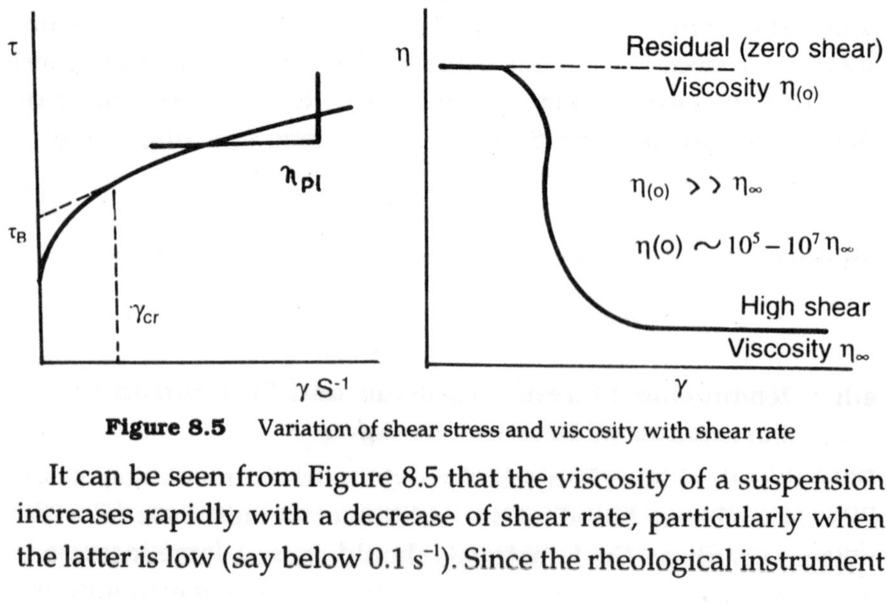

Figure 8.5 Variation of shear stress and viscosity with shear rate

It can be seen from Figure 8.5 that the viscosity of a suspension increases rapidly with a decrease of shear rate, particularly when the latter is low (say below 0.1 s^{-1}). Since the rheological instrument can carry out measurements at values below 0.1 s^{-1}, one usually obtains a limiting viscosity when the shear rate approaches zero. This limiting viscosity is referred to as the residual (or zero) shear viscosity, $\eta_{(o)}$. As schematically shown in Figure 8.5, $\eta_{(o)} >> \eta_{(\infty)}$. Usually $\eta_{(o)}$ is ~ $10^5 - 10^7$ greater than $\eta_{(\infty)}$. Such low shear values can be measured using constant stress measurements, as will be shown in the next section. To reduce sedimentation one requires such high residual viscosity. As we will also see in the next section, it is necessary to measure the elastic modulus of the suspension, which gives a measure of the overall separation of the suspension. By following Figure 8.5, one can find a summary of the methods that can be applied to reduce sedimentation and prevent the formation of dilatant clays.

4.3 Balance of the Density of the Disperse Phase and Medium

It can be seen from equation (15) that if $\rho_1 = \rho_2$, i.e. $\Delta\rho = 0$, then $v = 0$. Thus, if one can increase the density of the medium to a value that is equal to that of the particles, it is possible, in principle, to eliminate sedimentation. This approach is, however, of limited application. Firstly, the density of the medium can only be increased by a small amount (perhaps by dissolving an inert substance such as sugar in water). Thus, only in cases where the density of the particles is only slightly greater than that of water is this possible. Secondly, matching density is only possible at one temperature since the variation of density with temperature is different for solids and liquids.

4.4 Use of Water-Soluble Polymers

High molecular weight water-soluble materials, such as hydroxyethyl cellulose and polysaccharides (sometimes referred to as thickeners), are commonly used in agrochemical suspensions to reduce sedimentation and prevent claying. These polymers produce non-Newtonian solutions above a certain concentration, C^*, which depends on the molecular weight of the polymer. This concentration, C^*, sometimes referred to as the semi-dilute value, marks the onset of polymer coil or polymer rod overlap. Above C*, the viscosity of the solution increases more rapidly with a further increase in its concentration. The polymer solution shows viscous and elastic behaviour (usually referred to as viscoelasticity). The particles that are suspended in such viscoelastic solutions show a significant reduction in sedimentation as a result of the very high residual viscosity and the elastic nature of the polymer solution, which behaves like a gel. The most commonly used polymers are those sold under the trade name Kelzan or Keltrol (Kelco Co., United States) and Rhodopol (Rhone-Poulenc, France). These are polysaccharides produced by polymerization of waste sugars using micro-organisms. These polymers show significant non-New- tonian behaviour at fairly low concentrations (>0.05%). This results from their high molecular weights ($>10^6$), which causes polymer coil or rod overlap to occur at low concentrations. In most cases, a concentration of 0.1 - 0.2% is sufficient to prevent claying of suspension concentrates. In some cases, higher concentrations may be required to prevent separation of the suspension. However an increased viscosity at high shear may prevent spontaneous dispersion of the suspension on dilution, thereby requiring considerable agitation in the spray tank.

4.5 Use of Inert Fine Particles

When suspended in aqueous media, clays such as sodium montmorillonite (bentonite) produce viscoelastic gels as a result of interaction between the clay platelets. The latter consist of negatively charged faces at all pH values. The negative charge on the clay faces results from isomorphic replacement of ions of higher valency such as Si^{+4}, by ions of lower valency such as Al^{+3}. This substitution in the crystal lattice of the clay results in a deficiency of positive charges, which equates to the acquisition of negative charges on the faces of the clay particles. Thus, an electrical double layer is produced around the clay particles (the counter-ions being Na^+ or other monovalent ions). At low electrolyte concentrations, the double layers are expanded and hence strong repulsion between the clay particles occurs. This results in a viscoelastic system (gel). However, van Olphen (19) suggested a different mechanism for gel formation for sodium montmorillonite. This is based on the interaction between the double layers of the faces and edges. The latter may produce a double layer of opposite sign to that on the face as a result of the following mechanism: the bonds of the oxide layer at the edges may become broken and hence the edges behave as an oxide, i.e. they become covered with OH groups. In most cases the edges are populated with Al-OH groups that may attract a H^+ or OH^- ion, depending on the pH of the solution. At pH values below the iso-electric point of alumina (<7-9), the hydroxyl groups attract a proton and thereby form a positive surface charge. This produces a double layer with a sign opposite to that at the face (which always has a negative surface charge). Face-to-edge attraction results, producing a three-dimensional structure (sometimes referred to as a house-of-cards structure). This structure or gel is viscoelastic and hence it can be used to reduce sedimentation in SCs. For the gel to be produced it is essential to have a low electrolyte concentration. This ensures adequate swelling of the clay and expansion of the double layers to maximize interparticle interaction.

Finely divided oxides such as silica can also produce a gel when adequately dispersed in an aqueous solution. The most commonly used oxide is fumed silica (trade name Aerosil, ex Degussa, Germany) which can produce a gel at 1-2 w/v% in aqueous solution at pH values above the isoelectric point of silica (pH > 3). Thus, by dispersing silica in an aqueous SC, a gel may be produced in the continuous medium that prevents sedimentation and claying of the SC.

4.6 Use of Mixtures of Polymers and Finely Divided Solids

As mentioned above, polymers may increase the high shear viscosity of the SC to an unacceptable level that makes it difficult to dilute the concentrate without considerable agitation. A way to avoid this high shear viscosity is to use a mixture of a polymer such as Kelzan and sodium montmorillonite. By adding 1-2% sodium montmorillonite, the amount of polymer needed to produce an acceptable gel is considerably reduced. Usually a polymer concentration not exceeding 0.1% is sufficient. The gel produced will have a sufficiently high residual (zero shear) viscosity and a low high-shear viscosity, which makes it easy to disperse the SC in water with minimum agitation. Similar gels may be produced by using a mixture of silica and Kelzan. Such anti-settling systems are possibly more robust than those produced by the individual components. The interaction between the particles and the polymer chains produce such an optimum structure (possibly as result of the synergetic effect of bridging or depletion flocculation; see below).

4.7 Controlled Flocculation

As shown in Figure 8.1, the energy distance curve for an electrostatically stabilized suspension shows weak attraction at large distances of separation (secondary minimum). If this attraction is made sufficiently strong by making the secondary minimum sufficiently deep (1-5 kT units), it is possible to produce a weak gel as a result of interparticle interaction. Such a gel may be sufficient to prevent sedimentation and claying of the suspension. The depth of the secondary minimum is controlled by particle size and electrolyte concentration. This has been recently demonstrated (20) by the addition of electrolytes (NaCl, $CaCl_2$ or $AlCl_3$) to a suspension concentrate stabilized by an anionic polyelectrolyte. Above a certain electrolyte concentration (which was lower the higher the valency of the electrolyte), weak flocculation of the suspension occurred and a weak gel was produced. This resulted in a relatively high sediment volume and it was easy to redisperse the suspension by gentle agitation.

Sterically stabilized suspensions can also be flocculated by controlling the depth of the minimum in the energy-distance curve (Figure 8.3). This may be achieved by controlling the adsorbed layer thickness. If the latter is made sufficiently small, the energy minimum becomes sufficiently deep for a weak attraction to occur. The gel produced may be sufficiently strong to prevent claying and the SC can be redispersed by shaking the container.

4.8 Depletion Flocculation

Addition of a free (non-adsorbing) polymer to sterically stabilized suspension concentrates can, under ceratin conditions, produce weak flocculation. The mechanism of this flocculation may be understood by considering of the interaction between the particles when the free polymer does not adsorb on the particle surface. Under these conditions, a polymer-free zone around the particle surface is produced since any reduction in entropy, as a result of the approach of the polymer chain to the surface, is not compensated by an adsorption energy. The thickness of the polymer-free zone (depletion zone), Δ, is of the order of the radius of gyration of the free polymer, R_g. When two particles with two depletion zones approach to a separation distance h that is smaller than 2Δ, the free polymer chain becomes "squeezed out" from in between the particles. As a result, the osmotic pressure in bulk solution becomes higher than that in the overlapping depletion zone, which consists of free solvent. This results in a weak attraction between the particles, usually referred to as depletion flocculation. One may define a free polymer volume fraction, ϕ_p^+, at the onset of depletion flocculation. ϕ_p^+ is of the same order of magnitude ϕ_p^*, the semidilute volume fraction at which polymer coils begin to overlap. ϕ_p^+ decreases with an increase in polymer molecular weight. The magnitude of the attraction free energy due to depletion, G_{dep}, is proportional to the osmotic pressure of the free polymer solution. Therefore, at any given ϕ_p, G_{dep} increases with an increase in the molecular weight of the free polymer. This phenomenon of depletion flocculation has been recently investigated using an agrochemical suspension concentrate (21). The suspension was sterically stabilized using a graft copolymer containing polyethylene oxide (PEO) side chains. The free polymer used was either PEO or hydroxyethyl cellulose (HEC) . In both cases, flocculation occurred at a polymer concentration that decreased with an increase in the molecular weight of the free polymer. It was also shown that depletion flocculation reduced sedimentation and prevented the formation of dilatant clays.

5 ASSESSMENT OF THE LONG-TERM PHYSICAL STABILITY OF SUSPENSION CONCENTRATES

Several methods have to be applied for the assessment of the long-term physical stability of suspension concentrates and, in some cases, to predict their shelf-life. These methods may be classified into two categories: (a) those based on dilution of the suspension and measurement of its particle size distribution, flocculation and crystal growth and (b) those based on measurement of the bulk properties of the suspension, such as its sedimentation, claying and rheology. The various methods are summarized below:

5.1 Particle-Size Distribution Measurements

The most common methods for particle-size analysis include microscopy, the Coulter counter and light diffraction techniques. Both optical and electron microscopy may be applied for particle-size distribution measurements, although these are by far the most tedious. Optical micrographs of the particles in a diluted suspension may be used to obtain the particle-size distribution. This allows one to obtain a size analysis for particles that are greater than 0.6 μm (this is perhaps the lower limit for resolution by optical microscopy). The size distribution may be obtained manually by determining 500-1,000 particles in photomicrographs taken at random in the field of view of the optical microscope. Alternatively, the particle-size distribution can be obtained automatically using image analysis. Electron microscopy allows one to measure most of the size distribution since there is practically no lower limit for size determination.

The Coulter counter is perhaps the most rapid technique for determination of the particle-size distribution of suspensions. The basis of the Coulter counter is fairly simple. A dilute suspension is allowed to flow through an orifice in to an electrolyte solution (1 molar NaCl). Two electrodes are placed at each side of the orifice and the conductivity of the electrolyte is monitored. As each particle passes through the orifice, it displaces an equivalent volume of electrolyte. By monitoring the change in conductivity as the particles pass through the orifice, one can obtain an equivalent volume distribution. A multichannel analyzer is used to monitor several thousand particles passing through the orifice. The method is fairly accurate, but it has a lower limit for size measurement, usually 0.6 μm.

More recently, diffraction methods have been applied for particle-size distribution measurements, but these methods have a larger lower limit, usually of the order of 2 μm. By combining light diffraction with forward scattering, one may be able to extend size measurement to lower values. Some manufacturers claim that a size analysis as low as 0.1 μm may be possible.

Another technique for particle-size analysis is the photosedimentation method, which can be applied to obtain a size distribution in the range 0.1-5 μm within a reasonable measurement time. Basically the particle-size distribution is followed at a fixed interval of time using a disc centrifuge fitted with an optical device for monitoring turbidity during spinning.

5.2 Measurement of Flocculation

The simplest method for measuring the flocculation of suspensions is by direct microscopic counting of the number of particles at various intervals of time. This method is obviously tedious and can be applied only for a dilute suspension. For dispersions stabilized by non-ionic surfactants and non-ionic polymers containing hydrophilic groups of the PEO or polyvinyl alcohol type, flocculation of the suspension usually occurs above a critical temperature that is referred to as the critical flocculation temperature (c.f.t.). The latter may be determined for a dilute suspension using a simple turbidimetric technique. A sample of the suspension is placed in a spectrophotometer containing a device for heating the sample cell at a defined rate. By monitoring the turbidity as a function of temperature, one can locate the c.f.t., since this will be the temperature at which there is a rapid increase in turbidity. The c.f.t. may also be determined using rheological measurements without having to dilute the suspension. In this case, a rheological parameter such as the yield value or modulus is followed as a function of temperature. Again the c.f.t. is the point at which the rheological parameter shows a rapid increase.

5.3 Measurement of Crystal Growth

For measuring the rate of crystal growth, the particle-size distribution is measured as a function of time. This can be done using the Coulter counter, which is able to measure particles above 0.6 μm. A more sensitive method is to use the optical disc centrifuge which allows one to obtain the particle-size distribution down to 0.1 μm. Plots of the average particle size as a function of time allow one to obtain the rate of crystal growth, which in many cases follows first-order kinetics. Any change in crystal habit should be followed by optical microscopy, which should be carried out at intervals during storage. Temperature cycling could also be applied as an accelerated test for studying crystal growth. The rate usually increases with temperature changes, particularly when temperature cycling is carried out over wide intervals. When an SC undergoes crystal growth on storage, methods should be designed to reduce such a process. The most common procedure is to use crystal growth inhibitors, which are normally found by a trial and error procedure. In some cases, crystal growth inhibition may be achieved by proper choice of the dispersing agent. Polymers such as poly(vinyl alcohol) (22) and some graft copolymers may be effective in reducing crystal growth, since they strongly adsorb on the particle surface, thus blocking some of the active sites on the surface of the crystals. Molecules that have the same crystal structure as the agrochemical may also reduce crystal growth by adsorbing on active sites and crystal dislocations. In some cases, dyes may also inhibit crystal growth, since these molecules are strongly adsorbed on the crystal faces.

5.4 Rheological Measurements

For the quantitative assessment of the long-term physical stability of suspension concentrates, e.g. its sedimentation, claying and dispersion on dilution, various rheological measurements should be carried out simultaneously. Two main types of measurement are essential: (a) steady-state shear stress-shear rate measurements and (b) measurement under small deformation (transient and dynamic). Below the two types of measurements are described briefly:

5.4.1 Steady state measurements

As mentioned before, most "structured" SCs (i.e. containing an antisettling system) SCs give a pseudoplastic flow, as illustrated in Figure 8.5. The results of the shear stress τ - shear rate γ curves can be fitted to the Bingham (23) or the Casson (24) models. In the Bingham model,

$$\tau = \tau_\beta + \eta_{pl}\gamma \tag{19}$$

In the Casson model,

$$\tau^{1/2} = \tau_c^{1/2} + \eta_c^{1/2}\gamma^{1/2} \tag{20}$$

One should follow the variation of τ_β or τ_c, and η_{pl} or η_c, with time. These measurements give useful information for good application of the SC-type of formulations. For example, for adequate flow in pipes (as in a seed dressing type of application) or for spontaneity of dispersion on dilution, τ_β or τ_c should not exceed an optimum value. Moreover, any increase of τ_β or τ_c on storage usually indicates some progressive flocculation or gelling. Thus, steady-state measurements can provide some guide-lines for optimizing the formulation and ensuring its shelf-life. However, for prediction of the long term physical stability of suspension concentrates, these measurements must be combined with low deformation measurements, which are discussed below.

5.4.2 Low deformation measurements

There are three types of measurements that give quantitative information on long-term physical stability. Such low deformation measurements give valuable information on the structure of the system on standing.

5.4.2 (a) *Transient measurements:* In these measurements a constant stress τ is applied on the system and the strain or deformation is measured as a function of time. One usually follows the creep compliance, J (m^2/N), with time. As an illustration, Figure 8.6 shows typical creep curves for a pesticidal SC (containing 250 g/dm^3) structured with bentonite clay (30 and 45 g/l) (25). The creep curves are characterized by three main regions: (a) directly after application of stress one observes a rapid elastic deformation resulting in an elastic compliance J_o, (instantaneous shear modulus, $G_o = \tau/\gamma = 1/J_o$); (b) a slow elastic deformation, i.e. a mixed

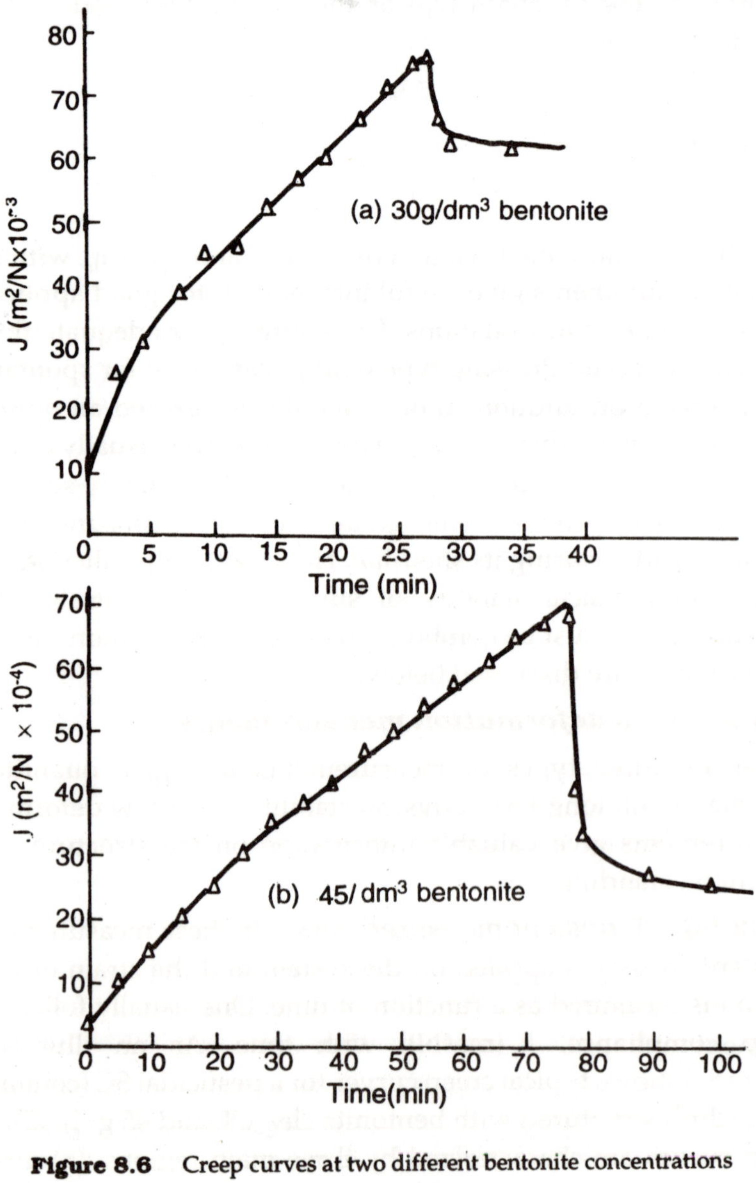

Figure 8.6 Creep curves at two different bentonite concentrations

viscoelastic region (in this region bonds are broken and reformed, but not at the same rate, thus giving a spectrum *i* of retarded compliances); (c) a region of viscous deformation, i.e. a region where the strain changes linearly with time. In the latter region individual aggregates flow past each other, since the time required to restore broken bonds is longer than the test period. The reciprocal of the slope of the linear portion of the creep curve gives the viscosity η_τ at the applied stress.

Thus, the above creep curves allow one to obtain two parameters, G_0 and η_τ. In the above example, a value of G_0 can only be obtained with some accuracy when J_0. is measurable. This is the case with the 30 g/dm^3 bentonite suspension, which gives a J_0 of 1.1×10^{-2} m^2/N or a G_0 of 93 N/m^2. With the 45 g/dm^3 bentonite suspension, J_0 is smaller and hence G_0 is much larger, as expected. The η_τ calculated from the slope of the linear portions of the curve amounts to 3.3×10^2 and 9.0×10^3 Pa for the formulations containing 30 and 45 g/dm^3 bentonite, respectively. This clearly illustrates the very high viscosities obtained at low shear stresses. It also shows that by increasing the bentonite concentration from 30 to 45 g/dm^3, η_τ increases by more than one order of magnitude, whereas the high shear viscosity (at 600/s) only increases by a factor of 2.

Clearly, η_τ is not the residual viscosity $\eta_{(0)}$, which is the limiting value as $\tau \rightarrow 0$. To obtain $\eta_{(0)}$, one should measure creep curves at successively smaller stress values and then plot η_τ versus τ, extrapolating to $\tau = 0$. This is illustrated in Figure 8.7. This is usually carried out automatically using a constant stress instrument, e.g. Bohlin CS (Bohlin Reologie, Sweden), allowing one to obtain the residual viscosity. The latter is the viscosity at zero shear rate and it is the parameter that determines sedimentation in the container.

5.4.2 (b) *Dynamic (oscillatory) measurements:* In these measurements one applies a sinusoidal strain with amplitude γ_o (or stress

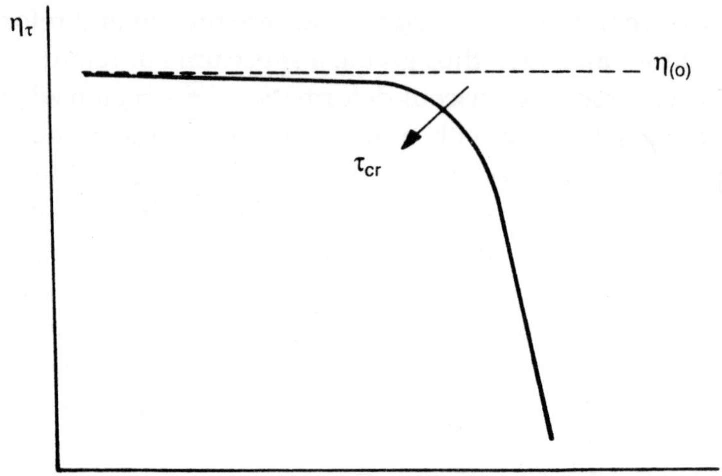

Figure 8.7 Variation of viscosity with shear stress

with amplitude τ_o) at a frequency ν in Hz (or ω in radians, where $\omega = 2\pi\nu$). This can be carried out using a Bohlin VOR (Bohlin Reologie, Sweden) or a fluid rheometer (Rheometrics, United States). For a viscoelastic system such as that of an SC, the stress (or strain) oscillates with the same frequency but out of phase with the strain (or stress). This is illustrated in Figure 8.8, which shows the variation of strain and stress with time. It can be seen that the stress is shifted by a period Δt from the strain. The phase angle shift δ is equal to $\Delta t\,\omega$. For a viscoelastic system, $0 < \delta < 90°$.

The ratio of the stress amplitude to the strain amplitude gives the complex modulus G^*:

$$G^* = \tau_o/\gamma_o \tag{21}$$

The complex modulus is formed from a storage component, G' (the real part of G^*), and a loss component, G''(the imaginary part of G''). G' is a measure of the energy stored in a cycle and hence it gives the elastic part of the complex modulus. On the other hand G'' gives a measure of the energy dissipated as heat in a cycle and hence it gives a measure of the viscous part of G^*.

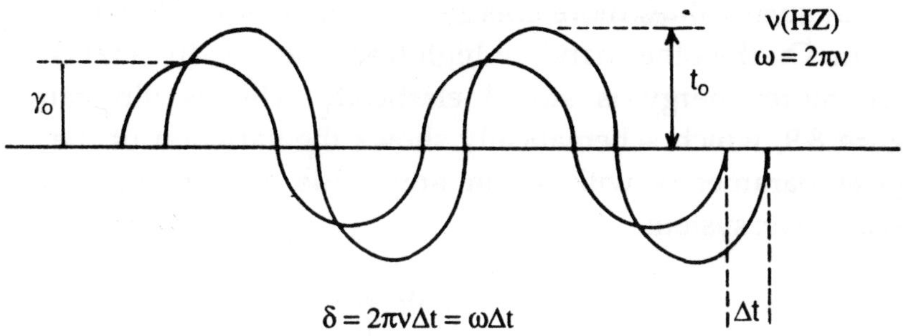

Figure 8.8 Corresponding stress and strain wave forms for a viscoelastic system

G' and G'' are related to G^* by the following equations,

$$G' = G^* \cos\delta \tag{22}$$

$$G'' = G^* \sin\delta \tag{23}$$

$$G^* = G' + i\,G'' \tag{24}$$

The dynamic viscosity η' is obtained from G'' by the following equation,

$$\eta' = G'/\omega \tag{25}$$

To carry out dynamic measurements, the suspension is placed in a concentric cylinder platen and a sinusoidal strain is applied to the cup, while the corresponding stress is measured on the bob. The latter is connected to torsion bars of different strength and the angular displacement is measured using a transducer. Initially the frequency is kept constant and the strain amplitude is gradually increased (strain sweep measurements). The rheological parameters G^*, G', G'', δ and η' are measured as a function of applied strain in order to establish the linear viscoelastic region where all these parameters are independent of the applied strain amplitude.

Once the linear region is established, measurement of the rheological parameters is made as a function of the frequency in the linear region. The trend obtained, which depends on the system, can be used to obtain information on the "structure" of the suspension. For example, at low frequencies ($<10^{-2}$ Hz) one usually finds $G'' > G'$, since at such values more energy is dissipated than stored by the system. On the other hand, at high frequencies ($>10^{-1}$ Hz) $G' > G''$, since more energy is stored elastically. This is illustrated in Figure 8.9, which schematically shows the variation of the rheological parameters with strain amplitude and frequency for a visco-elastic system.

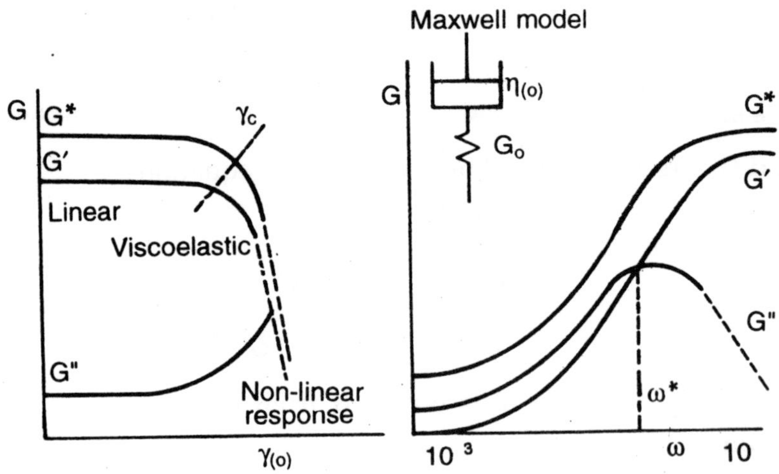

Figure 8.9 Schematic representation of the variation of rheological parameters with strain amplitude and frequency for a viscoelastic suspension concentrate

The magnitude of the various rheological parameters and their variation with strain amplitude and frequency may be used for optimization of the SC. Firstly, one should aim at a structure that produces a linear viscoelastic region up to strain values of few per cent. If the rheological parameters decrease at low strain values, this is usually associated with gross flocculation. Secondly, the suspension should be considerably more elastic than viscous at moderate and high frequencies. In this case also, the storage modulus shows little dependence on the applied frequency. However, the optimum values of G^* and G' depend on the system and the application procedure. Values in the region 10-30 Pa at a frequency of 1 Hz for G' are commonly quoted for adequate storage stability.

It should be mentioned, however, that for prediction of the

long-term physical stability of suspension concentrates, one has to combine the various rheological parameters. For example, sedimentation may be predicted from the residual (zero shear) viscosity, which has to be high enough (>10^5 Pa). On the other hand, separation and subsequent consolidation may be predicted from a knowledge of G' and its dependence on frequency. For prediction of ease of flow and dispersion on dilution, the yield value and plastic viscosity offer useful parameters. The best procedure is to measure the above parameters as a function of time during the early stages of the formulation and then establish the optimum parameters needed for adequate control of stability and shelf-life. This is by no means an easy task and it requires experience and good knowledge of the basic principles of rheology.

REFERENCES

(1) Th. F. Tadros, *Adv. Colloid Interface Sci.*, 12, 1980, p. 141.

(2) Th. F. Tadros, *Adv. Colloid Interface Sci.*, 32, 1990, p. 205.

(3) T. C. Patton, *Paint Flow and Pigment Dispersion* (New York, Interscience, 1964).

(4) G. D. Parfitt, *J. Oil Col. Chem. Assoc.*, 50, 1967, p. 822.

(5) P. A. Rehbinder, *Colloid J. USSR*, 20, 1958, p. 493.

(6) E. D. Shchukin and P.A. Rehbinder, *Colloid J. USSR*, 20, 1958, p. 601.

(7) G. Gouy, *J. Physique* (4), 9, 1910, p. 457; D. L. Chapman, *Phil. Mag.*, (6), 25, 1913, p. 475.

(8) O. Stern, *Z. Elektrochem.*, 30, 1924, p. 508.

(9) D. C. Grahame, *Chem. Rev.*, 41, 1947, p. 441.

(10) H. C. Hamaker, *Physica* (Utrecht), 4, 1937, p. 1058.

(11) B. V. Deryaguin and L.D. Landau, *Acta Physicochim. USSR*, 14, 1941, p. 633.

(12) E. J. W. Verwey and J. Th. G. Overbeek, *Theory of Stability of Lyophobic Colloids* (Amsterdam, Elsevier, 1948).

(13) Th. F. Tadros, *The Effect of Polymers on Dispersion Properties* (London, New York, Academic Press, 1982), p. 246.

(14) D. Heath, R. D. Knott, D. A. Knowles and Th. F. Tadros, *ACS Symposium Series, 246*, 1984, p. 254.

(15) D. H. Napper, *Polymeric Stabilisation of Colloidal Dispersions* (London, New York, Academic Press, 1983).

(16) Th. F. Tadros, *Solid/Liquid Dispersions* (London, New York, Academic Press, 1987).

(17) G. K. Bachelor, *J. Fluid Mech.*, 52, 1972, p. 245.

(18) I. Krieger and M. Dougherty, *Trans Soc. Rheol.*, 3, 1959, p. 137 I. Krieger, *Adv. Colloid Interface Sci.*, 3, 1972, p. 111.

(19) H. Van Olphen, *An Introduction to Clay Colloid Chemistry* New York, Interscience Publishers, 1963.

(20) Th. F. Tadros, *Colloids and Surfaces*, 18, 1986, p. 427.

(21) Th. F. Tadros and A. Zsednai, *Colloids and Surfaces*, 43, 1990, p. 105.
(22) Th. F. Tadros, "Particle growth in suspensions", *SCI Monograph No 28*, A. L. Smith, ed. (London, Academic Press, 1973).
(23) R. W. Whorlow, *Rheological Techniques* (Chichester, Ellis Horwood, 1980).
(24) N. Casson, *Rheology of Disperse Systems*, C.C. Hill, ed. (Oxford, Pergamon, 1959), pp. 84-104.
(25) D. Heath, Th. F. Tadros, P. K. Thomas and R. P. Warrington, *ACS Symposium Series* 254, 1984.

CHAPTER

9

Water–Dispersible Granules

A. R. Woodford

Rhone-Poulenc Agriculturé
Fyfield Road Ongar
Essex CM5 OHW, United Kingdom

1 INTRODUCTION

The use of chemical products to help to control the pests that cause loss of yield in arable crops is of comparatively recent origin. In fact, although some chemicals were used before the Second World War, their widespread use only started after the War. However, with the advance of our scientific knowledge of plant biology and chemical processes, the compounds used for the control of pests have become more and more sophisticated as well as more and more effective.

Following on this increase in the use of chemicals on the land, more attention was paid to the question of their safety and effect on the environment. At first this concern was for the safety of the public who consumed the foodstuff that was being treated with these chemicals, and great emphasis was placed on residues in crops. More recently this concern has been directed to residues in soils and in run-off water, which could find its way into our drinking supply. Only limited attention was devoted to other environmental effects.

The effect of the spray on the operator, has in recent years received much more attention and concern is now being directed to the form in which the product is presented to the spray operator and to safety when handling formulations in field situations.

Early formulations were mainly wettable powders or aqueous solutions, followed later on by emulsifiable concentrates and then suspension concentrates. Testing these formulations during use showed that the emulsifiable concentrates clearly presented a hazard due to the risk of contact with the solvent-based formulations, while wettable powders presented a hazard because of their dusty nature.

The first solid formulations were developed particularly for those active ingredients that were water-soluble. Variations of this technology were used to present non-water-soluble products in a similar form. In recent years a wide variety of different techniques have been used to produce these as granules. Some of these techniques are described in the rest of this chapter as well as some of the techniques used to evaluate the products and to test their performance characteristics.

1.1 General Description

The GIFAP coding (1) for these products is WG, which is the shorthand form for water-dispersible granules or a concentrated granules for dispersion in water. As this name implies the formulations that come under this heading have two main characteristics: they are water-dispersible and they are granules. It is important to distinguish these granules from the more common granular presentations of pesticides, which are for direct application.

In order to be quite clear about the terminology that is used, the following definitions are used:

(a) *Granules.* For the purposes of this chapter, the term granule will be used for any particle greater than 150 μm and not greater than about 10 mm, although they are generally between 250 μm and 2 mm. The reason for making this clarification is that several of the products on the market are relatively small granules and could be almost called powders and it is some times difficult to distinguish between granules and powders. However the essential difference is that granules are dust-free.

(b) *Water-dispersible granules.* These products disperse quickly and completely when added to water and do not need any special dispersion techniques. The dispersion rate expected is about the same as that of a wettable powder but not as rapid as that of a suspension or emulsifiable concentrate.

1.2 Advantages

The main advantage of these granules is that they avoid the use of solvents and thus reduce the risk to operatives during manufacture and to farm workers during spraying operations. Compared to wettable powders, they are not dusty, they are easily wetted and because they can be made to be free-flowing powders, they can be measured by volume. On farms this is a considerable, advantage over powders, which almost always have to be measured by weight. A further advantage is that cheaper and more easily disposable packaging can be used. Fibreboard or paper are quite suitable for these granules so that expensive and difficult-to dispose-of packaging, such as plastic containers and tin plate drums, are not necessary.

1.3 Disadvantages

Because of the way these granules have to be made, it is almost inevitable that in the primary processing they will be more expensive than wettable powders and flowables. The same first-stage operations, such as mixing and milling, whether wet or dry, have to be carried out for all these types of formulation. However, in addition the water added to the powder pre-mix or used in the slurry pre-mix has to be removed to dry the granules. There is, therefore, always this added cost in making granules. An additional problem is that the bulk density of these granules is usually lower than that of either the wettable powder or the equivalent flowable. Thus the packaging may be more expensive and offset some of the gains.

2 PROCESSING

The manufacture of water dispersible granules is more specialized than the manufacture of most other formulation types. Emulsifiable concentrates and aqueous solutions are produced by simple mixing and wettable powders require only powder milling. Flowables are more difficult to produce due to the wet milling stage. However, for flowables, a wide range of equipment can be used without making major alterations to the process. This is not the case with water dispersible granules. The granulating plant is highly specific and it is rarely possible to transfer the technology from one type of granulation plant to another.

2.1 Advantages

The main advantage of these granulation processes lies in the wet-route processes, where the dust hazards associated with the handling of powders, both during and after milling, are very significantly reduced.

2.2 Disadvantages

The main disadvantages of these processes is their lack of versatility. In broad terms, a particular granulating plant produces one type of granule and requires the formulation to have certain characteristics. Thus to have a range of formulation types it is necessary to have more than one granulator.

A second disadvantage is that in most of the processes, water has to be added and subsequently removed. As a result there will, always be the add-on-cost of the removal of water. The more water used, the greater is this additional cost.

3 PROCESSES

A multiplicity of processes can be used to produce water-dispersible granules of insoluble active ingredients (2). Despite their great diversity, they can be reduced to two basic routes: (a) those in which the starting materials are essentially dry and are subsequently granulated by making them wet and then redrying and (b) those in which the starting materials are wet and are granulated and dried.

The compositions of the formulations used for granulation are almost independent of the particular route chosen for the manufacture. Following is a typical composition for such products:

(a) *Active ingredient.* This can be a mixture of more than one active;
(b) *Dispersing agent.* Most commonly a surfactant;
(c) *Suspending agent.* Can be either a surfactant or a mineral or organic swelling agent;

(d) *Wetting agent*. Most commonly a surfactant;
(e) *Binder*. Usually a lignosulphonate but can be a gum or similar product;
(f) *Filler*. Most often a mineral filler is used, but inert soluble salts have been tried.

From this list it can be seen that the formulation composition is very similar to that of a wettable powder. The function of each of these components is self-explanatory and, apart from the binder, very similar to their function in a normal wettable powder. This is because the product is used in the same way as the normal wettable powder and must show at least as good or better properties of suspensibility and wettability.

The function of the binder is to assist in the aggregation of the powder to yield the required granule size. A binder is not always necessary, and simple wetting with water followed by drying is sufficient to granulate some products. However, it is well known that just adding water and drying can lead to a type of hydration bonding that is usually very firm, and the particles will not subsequently redisperse.

Thus, if a binder is necessary, it must be carefully chosen as it is important that any binding agent used can overcome this effect of water. It must aid the binding together of the primary particles but must not cement them together to such an extent that they do not readily redisperse once they are dropped into water.

Compounds used for this purpose are often polymeric. Various lignosulphonates have given very good results. Sugars and molasses have also been used as have water-soluble cellulosics and other gums, but they are less commonly used than the lignosulphonates. In common with all the other formulation components, binders must not cause any chemical degradation of the pesticide and must be non-phytotoxic.

The remaining components of the formulation require no special mention, except to say that in procedures that start by blending the dry ingredients together, it is better to use all powdered solids than to have to incorporate small quantities of liquids uniformly into the powder. Although mixing takes place in all the granulation processes, it is usually not sufficient to ensure a homogeneous blend.

3.1 Dry Route Processes

As indicated above, all these processes involve mixing dry ingredients, adding water or a similar granulating aid in the processing plant, granulating the mixture and redrying. In all the examples there is no further size reduction of the active ingredients during the granulation stage. The basic plant diagram for most of the dry- route processes is the same (Figure. 9.1).

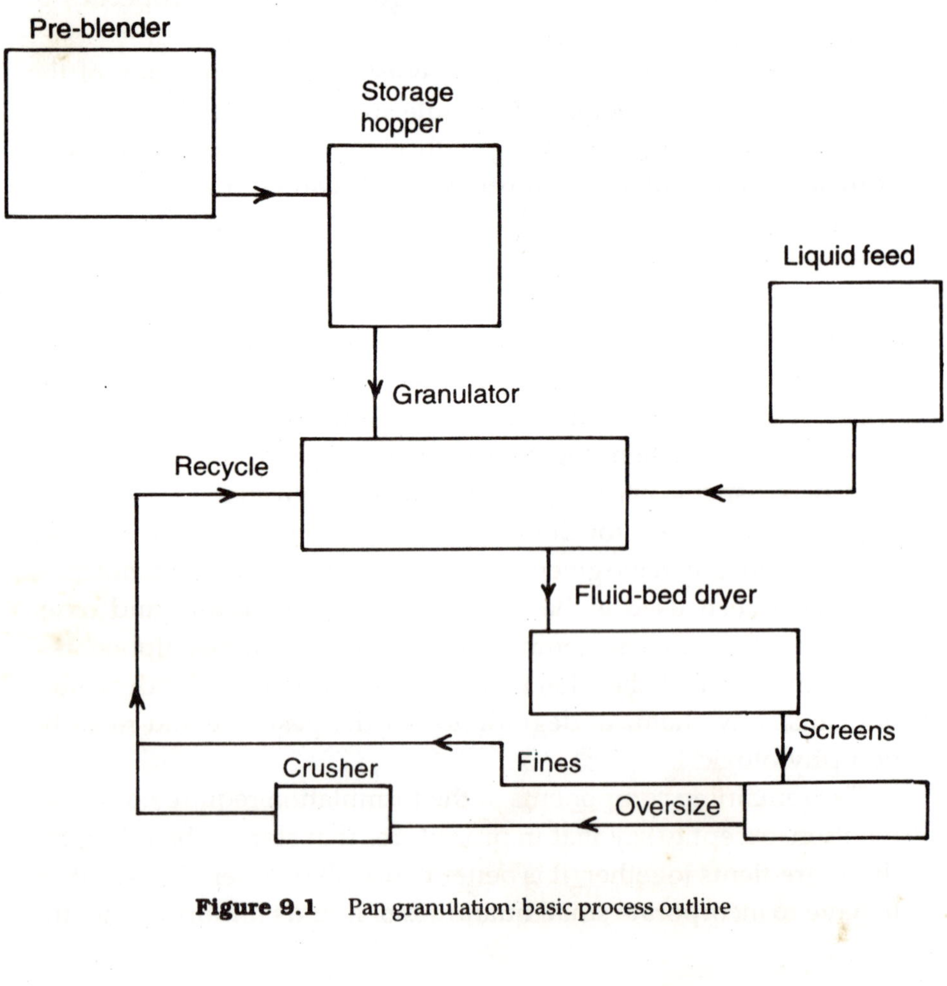

Figure 9.1 Pan granulation: basic process outline

The dry ingredients are mixed together in a ribbon or similar blender and passed to a feed hopper. This feed hopper discharges to the granulator, where the liquid is sprayed on to the powder. After granulation, the product is passed to a fluid bed or other drying unit and the dried granules are screened. The screens are chosen to split the product into three size ranges: the undersize (or fines), the oversize and the required product. The fines are usually fed straight back into the granulator as are the oversize after they have been ground to a suitable size.

Although the precise pieces of equipment used in an ancillary plant vary, the basic principles in the process are the same. The most important variable lies in the granulator itself. A number of typical granulators are described in the following paragraphs.

3.1.1 Pan granulators

3.1.1(a) *Inclined disc or pan agglomerator:* In these granulators (Figure 9.2), the dry powder is continuously fed onto the pan and at

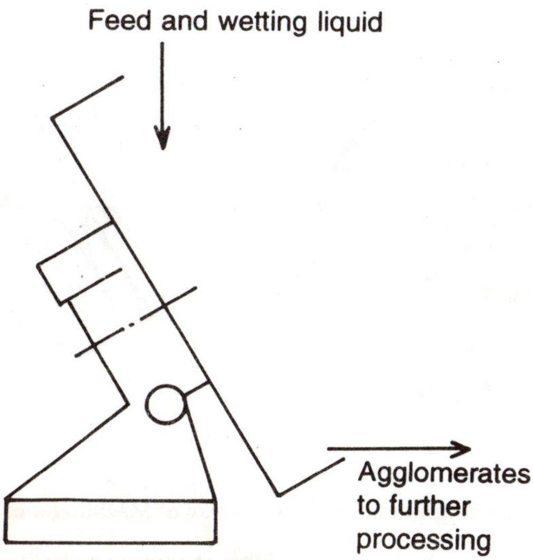

Figure 9.2 Disc granulation

the same time water is sprayed onto the powder(3). The rotation of the pan causes the powder to aggregate and to accumulate in the lower part of the pan.

As in all agitated beds of this type, the larger particles accumulate on the surface of the powder in this region and as more powder is added, the larger granules flow over the edge of the pan and are directed to the dryer. The speed of rotation and the angle of tilt of the pan can be varied to modify the size and yield of granules.

To improve the ease of operation and yield of these granulators, a number of modifications have been made to the pans. These include modifying the pan contours to give different depths within the pan and adding scraper blades to help to keep the powder surface within the pan level.

3.1.1(b) *Eirich granulator:* This type of pan granulator (Figure 9.3) has a number of important differences from the normal pan granulators(4). The main difference is that it is now a batch process and therefore is rather more labour-intensive. On the

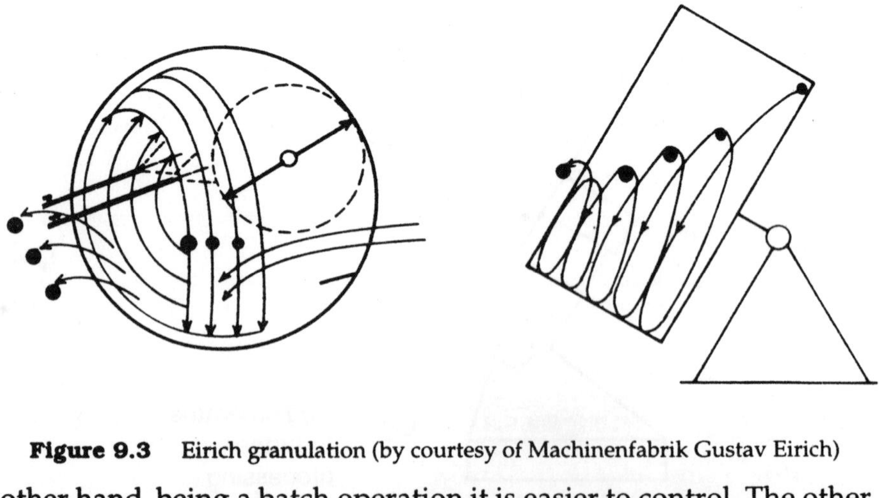

Figure 9.3 Eirich granulation (by courtesy of Machinenfabrik Gustav Eirich)

other hand, being a batch operation it is easier to control. The other difference is that within the pan there is a small rotating blade that acts to break up any large aggregates and thus, in principle, gives a higher yield of useful granules.

The mechanical parameters that can be varied are the speed of rotation of the pan and the speed of rotation of the lump breaker. However, as it is possible to vary not only the composition but also the amount of water and the granulation time, there is a considerable amount of flexibility in the process.

3.1.2 *Fluid-bed granulator*

The basic principle in these granulators (Figure 9.4) is that the powder is fluidized and water sprayed on in such a way that the solid aggregates(5).

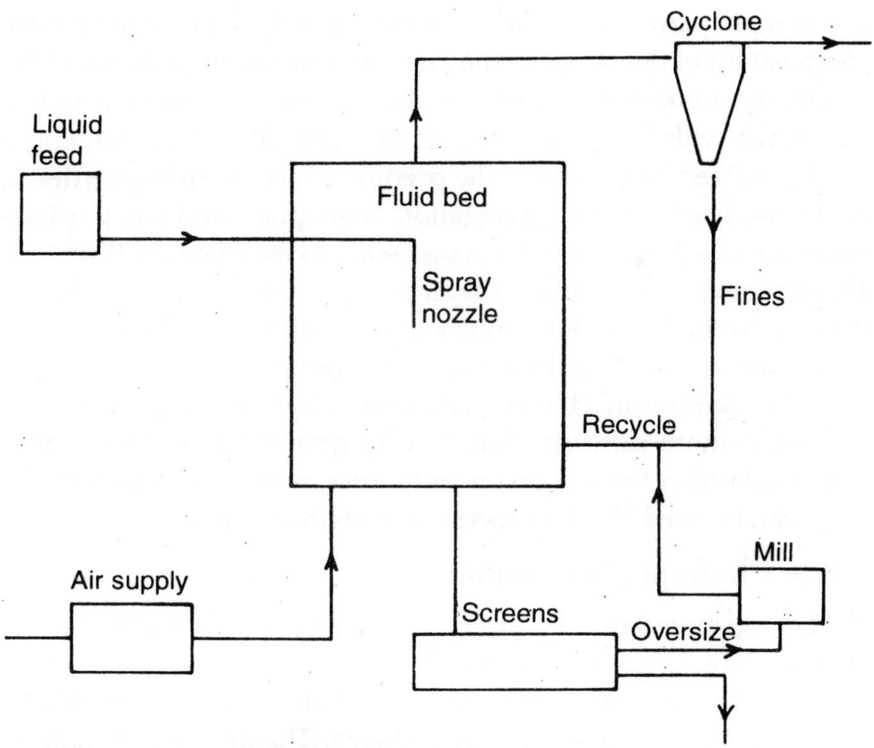

Figure 9.4 Fluid-bed spray dryer

These granules are then dried in the same unit. In this type of process the two stages, granulation and drying, referred to in the description of pan granulators are combined into one.

In its simplest form this is a batch process, but there are more sophisticated plants that can operate in a continuous manner. A number of commercial units are available which, as far as the granulator is concerned, are self-contained. The Glatt and the Aeromatic are such units.

In the continuous units (Figure 9.5) the fluid bed is divided into sections and the powder passes along the granulator as it increases in size until it is finally discharged. The main advantage of this type of process is that once equilibrium conditions have been established a high yield of good product can be obtained.

The main parameters that can be varied in these processes are the composition of the powder, the amount and rate of addition of the water, the temperature and the dwell time. The main problem associated with batch processes is the difficulty of maintaining a good fluidized bed. Initially the premix, which is a fine powder, is being fluidized. But as aggregation takes place and the particles become more dense, more air is necessary to maintain the fluid bed. This means careful control of the air pressure is necessary at all times. It has to be adjusted continuously as the nature of the bed changes. This becomes less of a problem in the continuous processes once the bed is at equilibrium. However, the initial start-up of the continuous process is more difficult. This type of granulator will be further discussed under the section on wet routes, where the same process plant can be used but the process starts with a slurry.

3.1.3 Schugi granulator

The Schugi granulator (Figure 9.6) offers a unique procedure for the continuous granulation of powders. The device has been developed into a self-contained unit including all the ancillary operations necessary to start from basic ingredients. The dry raw ingredient starting materials are metered into a mixing unit . From here, the mixed powders are fed to the Schugi agglomerator, where the liquid, usually water, is sprayed in. The granules are then fed directly to the fluid-bed dryer and screened. The fines and oversize are treated as in all the other processes.

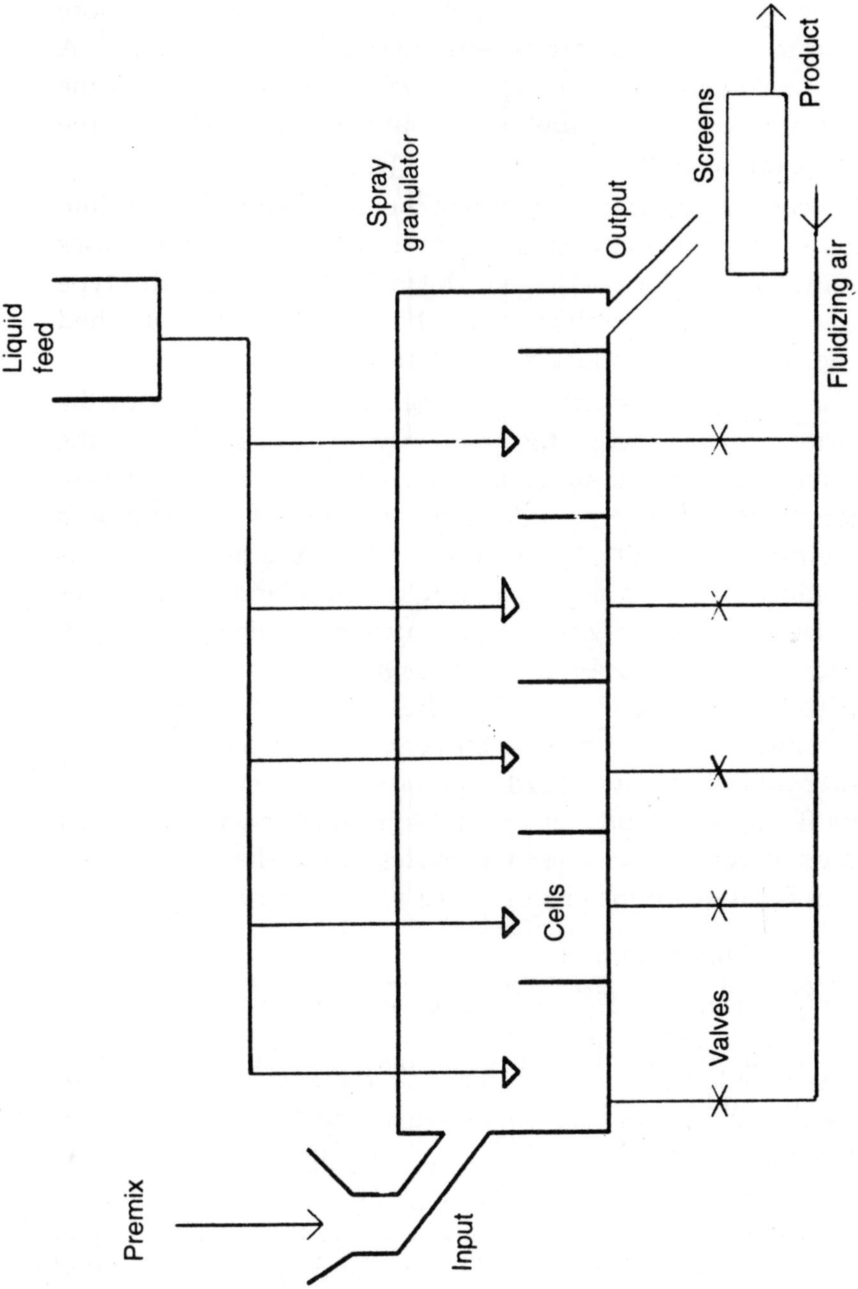

Figure 9.5 Multiple-cell spray granulator

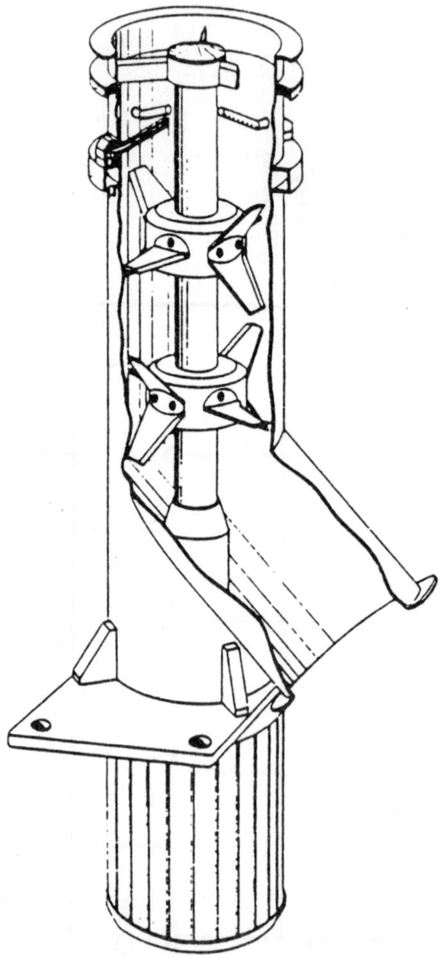

Figure 9.6 Schugi granulation (by Courtesy of Ritez Division of Bepex Corporation) (2)

The most important feature of this procedure is the Schugi granulator and the fluid bed. The powder agglomeration is a two-stage process. The first stage occurs in the Schugi, where the main agglomeration takes place, and is completed in the fluid bed. As the dwell time in the Schugi is only a matter of a few seconds, it is important that there is no delay before the granules pass to the fluid bed since little or no drying takes place inside the Schugi. The main parameters in this procedure are the formulation composition, the concentration of powder in the Schugi and the amount of liquid added. It is also possible, if necessary, to vary the rotor speed. With these variables it is possible to achieve a good range of granules.

This procedure has many advantages. Pre-blending is continuous and thus there is no need to store the blended pre-mix. Equilibrium conditions are quickly established and the particle size distribution remains fairly constant.

Because of the very short dwell time in the Schugi itself, the control of both the solids feed and the water flow is very important. The higher the ratio of solids to water, the more important is this control. This means that the mechanism for feeding these components into the Schugi must be efficient and easily controlled. One additional feature of this procedure is that all the solid raw materials need to be free-flowing powders if the feed mechanisms are to work efficiently and yield a uniform blend. If possible, it is preferable not to have any important biologically active ingredients in the water phase because of the problems this may bring in the reworking of the out-of-specification granules.

3.1.4 Miscellaneous Dry-Route Processes

There are a variety of other dry-route processes that are sufficiently different from the preceding ones to merit separate mention.

3.1.4(a) *Extrusion:* There are several different forms of extruders but they all use the same basic principle. The mixed powdered ingredients are wetted with water to produce a very thick slurry. This slurry is then forced under pressure through a perforated plate. After passing through the plate the granules either break off of their own accord or are cut off with some form of knife. Usually these granules are then dried as in the previous examples. Two basic types of extruders are most commonly used: (a) screw extruders (Figure 9.7), in which the mixed paste or powder, before being forced out of a perforated plate, is compressed in the steadily reducing spaces between the flights of a rotating screw and (b) a direct roller compression extrusion (Figure 9.8) (2).

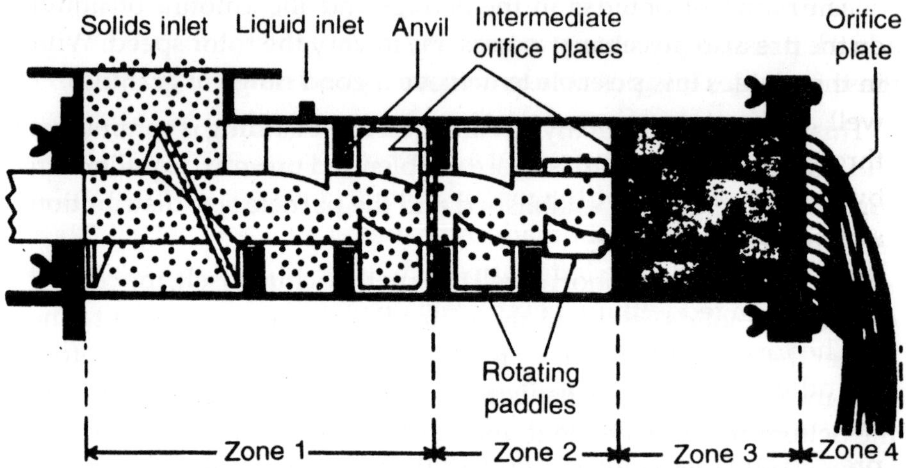

Figure 9.7 Ritez extructor (by Courtesy of Ritez Division of Bepex Corporation

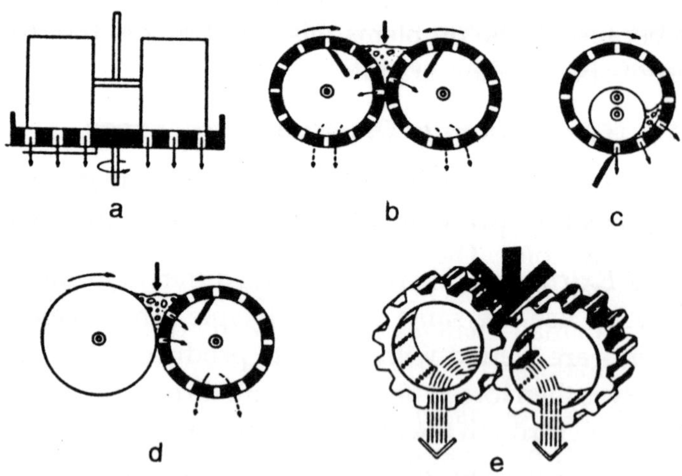

Figure 9.8 Roller-type extrusion process

In both these processes the granule diameter is decided by the diameter of the holes in the perforated plate. Their length is determined by the speed at which they are cut off. Granules can be produced with a wide variety of diameters and lengths. But in practice, it is usual to keep the granules small so that they disperse quickly when used by the pesticide applicator.

The hardness of the granules is largely, if not entirely, dependent on the pressure necessary to force the formulation through the holes in the plate. This is dependent on the hole diameter and length as well as the state of compaction of the formulation as it is forced through the hole. In the case of a pellet mill this is mainly controlled by the composition of the formulation. In the case of the extruder, some changes can also be made by changing the screw profile and/or the screw speed. The latter approach is the more useful since this can be changed without completely dismantling the extruder.

The formulation composition is largely the same as that used in the other procedures but more care is necessary in the selection of the surfactants. In an extrusion process, the formulation is compressed and finally forced through the holes in the plate. These holes can be as small as 1 mm, so the pressure necessary to force the formulation through these holes can be very high. This means that the granules become very highly compacted and often the surface becomes hardened and glazed. Both phenomena lead to difficulty in dispersing the granules and often to poor dispersibility.

For the above reasons, the additives in these formulations must be carefully chosen. It is important to avoid any tendency for the damp paste to be dilatant as this leads to an increase in viscosity under pressure and greater force is required to push the paste through the holes. The composition of the powdered formulation must be such that it is not highly sensitive to the water content. In such cases small changes in water content significantly alter the viscosity, making it nearly impossible to get consistent granules.

3.1.4(b) *Peg or pin granulator:* This granulator consists of a rotating shaft carrying a large number of pegs arranged spirally around the shaft. The speed of rotation and the intensity of the mixing action lead to rapid agglomeration. By feeding at one end and discharging at the other, the process can be made continuous.

The pre-mix is continuously fed in at one end of the granulator and water added in the first part of the unit. As more pre-mix is added, the contents of the mixer are driven down towards the other end, where the granules are discharged. This type of granulator is used more commonly in the ceramics industry but it has been used for direct granulation of herbicides. The procedure is also used to produce granules that can then be fed to another type of granulator such as a roller granulator or an extruder. The advantage of this type of plant is that it can handle very fine powders, which, due to their lightness and dustiness, would present serious handling problems in the other types of granulators.

In all these dry-route processes one of the main problems is associated with the water addition. Where plain water alone can be used there is no problem. However, where there is some additive in the water problems arise. When material is being reworked, the reworked material will receive a second application of water containing the additive when it is regranulated. In such an instance, the content of this additive in the granules will slowly increase.

A further problem with these processes is that the yield of the required granules is often quite low and in some instances can be as low as 30%. This means that some of the material may be treated through the granulator many times. Obviously, the more efficient the granulator the less rework is necessary. One of the main pressures on manufacturers is to design granulators that have greatly improved yields.

3.2 Wet-Route Processes

In all of these processes, the feed stock to the granulator is wet, either as a slurry or as a paste. Whatever pre-processing is necessary can be either wet or dry, but at some stage the formulation is converted to a slurry before being fed to the granulator.

3.2.1 *Spray drying*

The basic principles of spray drying (Figure 9.9) are as follows (7) :

(a) The liquid feed is broken up into droplets in some form of spray device at the top of a tall tower;

(b) These droplets are mixed with hot air in the column;

(c) As the droplets pass down the tower, the water evaporates and the droplets dry;

(d) The dried droplets are finally separated from the air.

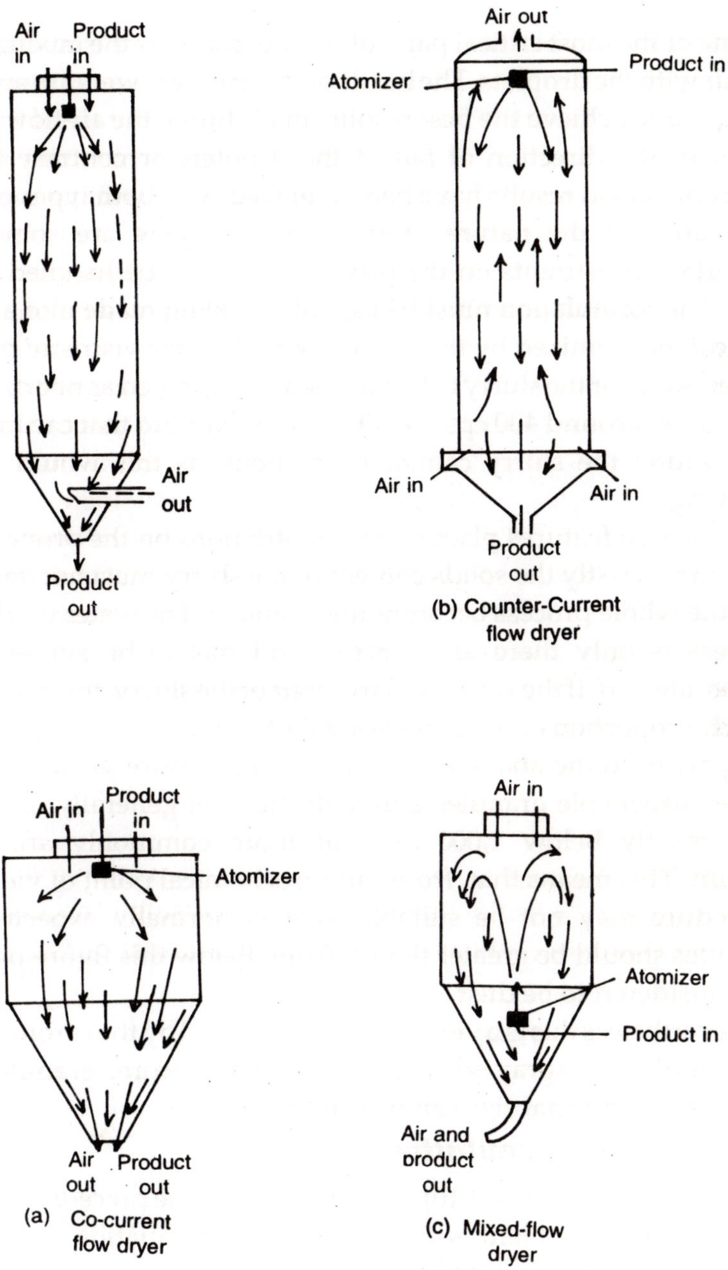

Figure 9.9 Spray dryers (7)

One of the most critical parts of this operation is the mixing of the hot air with the droplets. The heads of the drying towers are specially designed to achieve the best results. In addition, the air flow can be either in the direction of fall of the droplets or contrary to that direction. Good results have been achieved with both types of flow.

Because of the nature of this process, there are some very important constraints on the products that can be handled in this way. The formulation must be capable of being made into a slurry that can be atomized by the spray head. Thus the viscosity must be low enough for the slurry to be pumped and sprayed as fine droplets. Viscosities around 400 cps are often the maximum that can be used. In addition the slurry cannot be dilatent, as this would hinder spraying.

These two features place serious restrictions on the properties of the slurry. Firstly the solids content of the slurry must not be so low that the whole process becomes uneconomic. The water used in the process is only there as a carrier and has to be subsequently evaporated off. If the water is a large part of the slurry, this represents a high proportion of the processing cost.

Apart from the above problems, this procedure yields excellent water-dispersible granules although they are generally of a small size, mostly below 1,000 μm but more commonly are below 100 μm. This means that, from the agrochemical point of view, this procedure may not be suitable as it is normally expected that granules should be greater than 150 μm. Below this figure particles are considered to be dust.

Nevertheless this is a very useful process and by the careful choice of formulation, spray generator and temperature, granules of a suitable commercial size can be produced.

3.2.2 Spray granulation

This procedure (Figure 9.10) is an extension of the preceding process (8). The main problem with direct spray granulation is that it is difficult, if not impossible, to get large granules. As a result, equipment designers have successfully included a granulation stage within the spray-drying process. The inclusion of this granulation

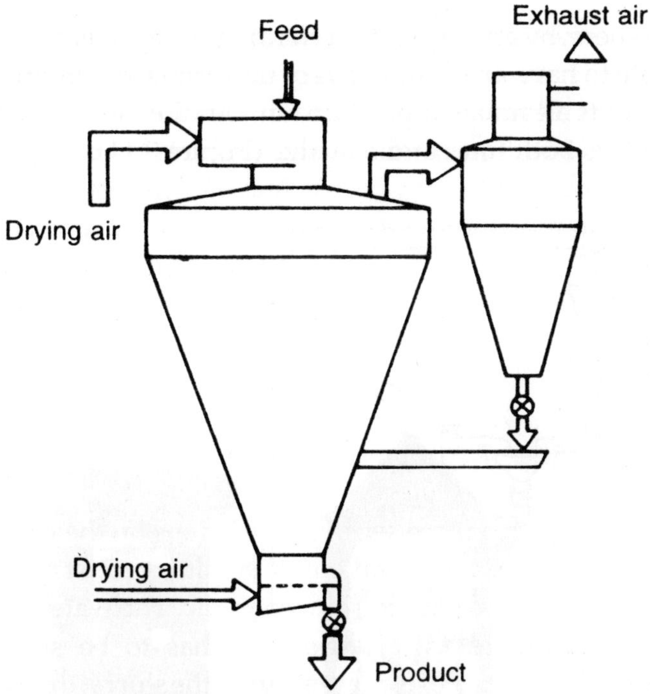

Figure 9.10 Niro-type spray granulator (schematic)

stage enables granules of a larger average size to be obtained and to reduce, if not eliminate, the risk of there being particles that could be considered as dust.

The main difference between this process and conventional spray-drying is that in the spray granulators the granules are retained for a short time in the tower before being discharged. Thus the freshly formed droplets are encouraged to agglomerate at this stage in the lower part of the tower. The degree of agglomeration that takes place depends on the time the granules remain in the tower and their stage of dryness when they reach the bottom of the unit.

Several types of these granulators are made where the main variations in the equipment are in the way the slurry is sprayed in and the way the air is introduced. In some of these granulators the slurry is sprayed in from the bottom of the tower. This is analogous to the procedure used in normal spray-dryers to help reduce the height of the towers without reducing the flight path of the spray. In the granulating version of these machines, one model has been designed where, by adjusting the feed rate and the retention volume, it is possible to have an output of very uniform-sized granules. This machine yields all finished product without the need for screening and eliminates both fines and oversize (Figure 9.11).

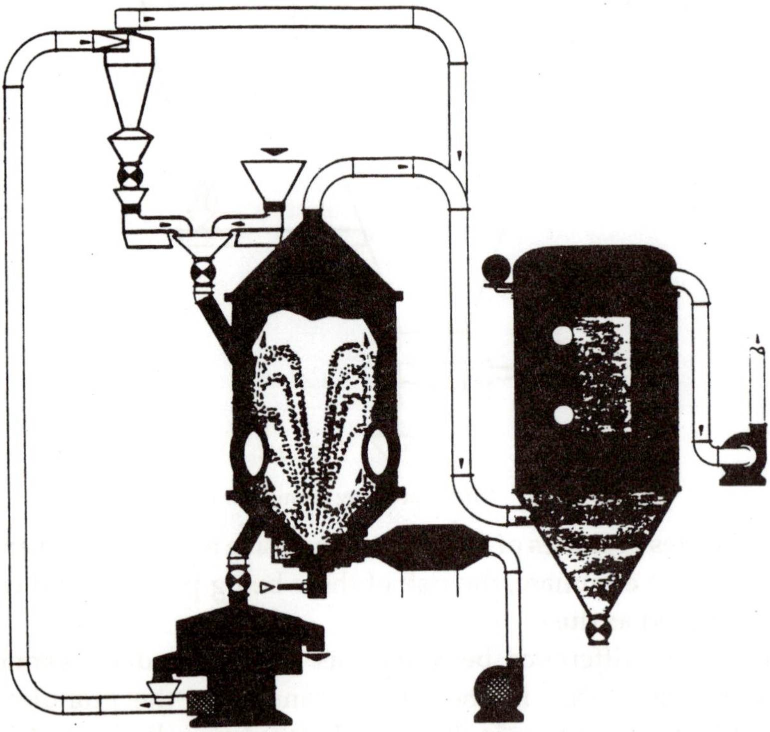

Figure 9.11 Alpine spray granulator (by Courtesy of Hosokawa Micron Europe B.V.)

In all these spray granulation processes, the composition of the formulation is of the utmost importance as the ultimate performance of the manufacturing plant is very dependent upon the ease with which the slurry is sprayed.

The process is not suitable for all active ingredients. It is essential that the active can be made into a sprayable slurry or a solution. Thus, low melting point actives cannot, in general, be used. Because of the need to evaporate off the water, high concentration slurries with a low viscosity are an advantage. Thus, solids that are difficult to wet and that yield very thick slurries are not easy to use.

In the case of solutions a problem also arises involving the temperature effect of the solubility of the actives. If the active ingredient solubility increases faster than the rate of evaporation of the water, the spray droplet remains tacky for long enough to stick to the sides of the granulator as the droplets fall down the tower. However with these small exceptions, if an active can be made into a slurry, then it can be granulated by this method.

3.2.3 Miscellaneous methods

There are very few other true wet-route granulating procedures that are being used for agrochemicals. There are other procedures that use a solution as starting material, but they are generally only used for water-soluble actives and so far do not seem to have been used with slurries. However, it may be possible to adapt them to handle such materials.

These procedures include drum-drying, freeze-drying and other procedures for evaporating water as well as prilling methods, which use melts as a starting material. Some considerable adaptation is necessary to use these processes with non-soluble actives.

4 TESTING

In the development of water-dispersible granules, it is important to have test methods yield meaningful information at all stages of development (9). The purpose of these tests varies depending upon the particular stage of development. Thus, tests used in the development stage have to be more searching than those used for quality control. The former have to distinguish between formulations to meet the end-use, whereas the latter only have to ensure that the final product has been manufactured to the required standard.

It is for this reason that the tests described below have been divided into three sections. There is, of course, no clear dividing line between the sections and very often the same tests are used at all three stages.

4.1 Tests Used in Formulation Development

The choice of tests in this area must be based on two main principles:

(a) The end-use to which the product will be put.

(b) The main parameters of the manufacturing process.

Some typical tests are described next.

4.1.1 Granule dispersibility

This parameter is most easily measured by pouring a known weight of granules into a cylinder and measuring the number of inversions necessary to give complete dispersion. The following is a typical procedure: About 5 g of granules are poured into a 250 ml measuring cylinder containing 250 ml CIPAC standard water. The cylinder is then inverted end-on-end once a second. After five inversions, the contents are examined for undispersed granules. If any persist, inversion is continued, examining at intervals, until all the solid is dispersed. The number of inversions is then noted. For normal use all the granules should be dispersed after 10 inversions.

4.1.2 Friability

Granule friability is a most important parameter since the more friable the granule the more dust is generated during the packaging, handling and transport of the granules.

There are two aspects to friability: the effect of pressure on the granules, which can cause crushing, and shaking or vibration, which can cause inter-granule attrition. The following simple tests are useful in assessing these properties:

4.1.2 (a) *Crushing* **:** The principle of this test (Figure 9.12) is that a measured quantity of granules is placed in a cylinder and a weight placed on them to simulate the load they would experience if they were on a pallet.

The friability in this test is assessed by measuring the amount of dust (i.e. particles passing 150 μm) at the start of the test and again after a given period of time. This test can be made more severe by subjecting the cylinder to vibration during the test.

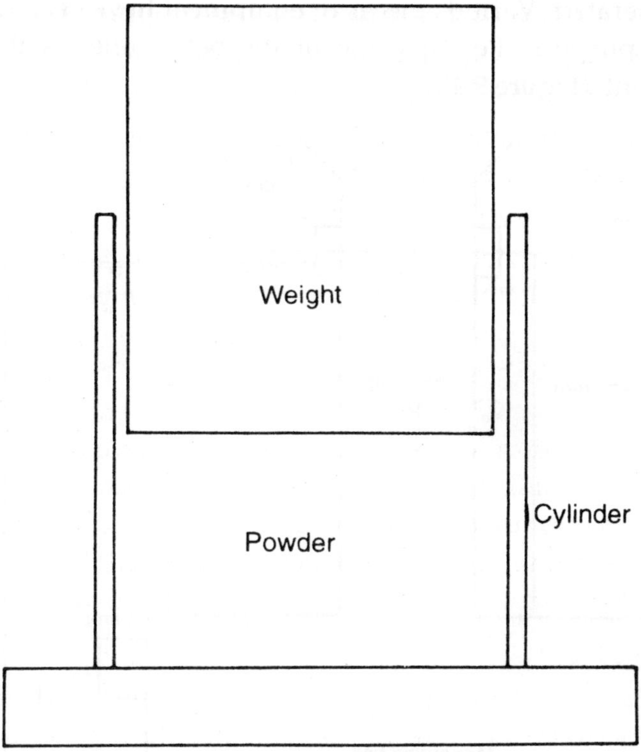

Figure 9.12 Static compression test

4.1.2 (b) *Attrition:* A simple form of attrition test that gives a good indication of the friability of granules is to place a weighed quantity of granules in a closed jar and rotate end-over-end for a given length of time. As in the previous case, the amount of dust generated during the test is a measure of the friability of the granules.

Neither of these tests is an absolute measure of friability. However, they do give the formulator a very good indication of their suitability and an ability to determine whether any changes need to be made in the additives.

4.1.3 Dustiness

This is perhaps the most difficult test to select because of the difficulty of obtaining consistent results. Most of the tests used are based on the principal of dropping, under controlled conditions, a quantity of granules onto a surface and then measuring, by some means, the dust generated. Various pieces of equipment have been developed for this purpose. Perhaps one of the better ones is the Casella instrument. (Figure 9.13).

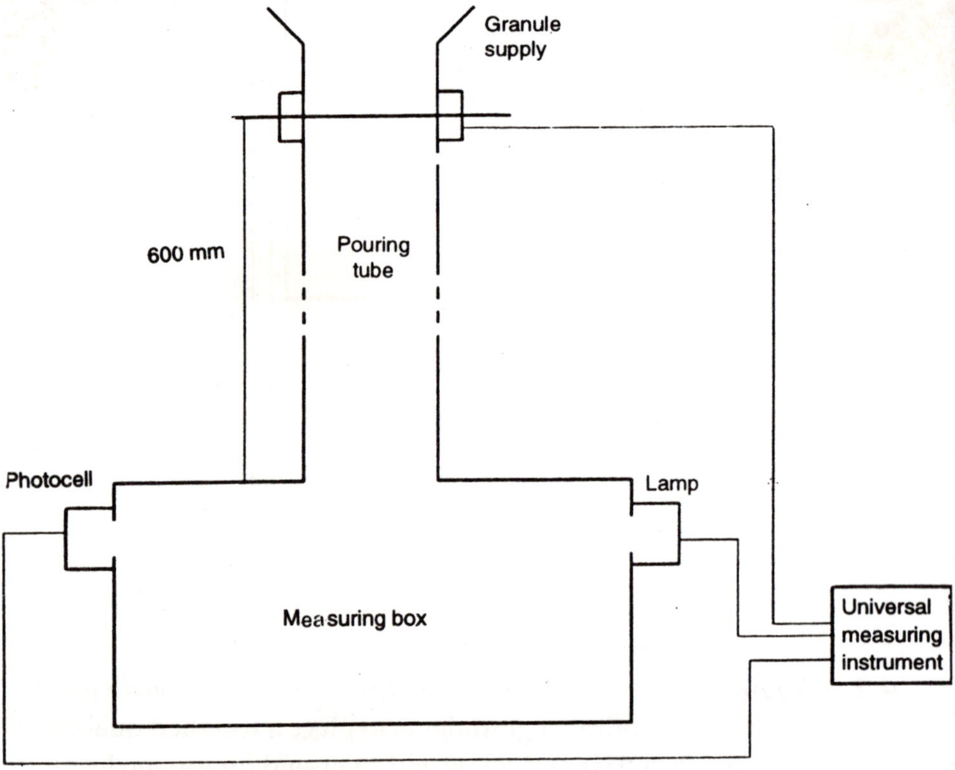

Figure 9.13 Casella dust measurement

This instrument consists of a measuring box with a removable case fitting inside the box. On top of the box is a supply pipe and funnel down which the sample is dropped. At the side of the box and at right angles to it is a photoelectric measuring device, which measures the light passing across the box both before and after the granules have been dropped into the box. The advantage of this equipment is that measurements can be taken immediately and at various time intervals so that the nature of any dustiness can be assessed without disturbing the atmosphere in the box.

In a second piece of equipment (Figure 9.14) the granules are dropped down a 3 litre cylinder that has a small side-arm about 1 cm from the base. To this arm is attached a means of drawing air from the cylinder through a millipore filter on which any dust is collected.

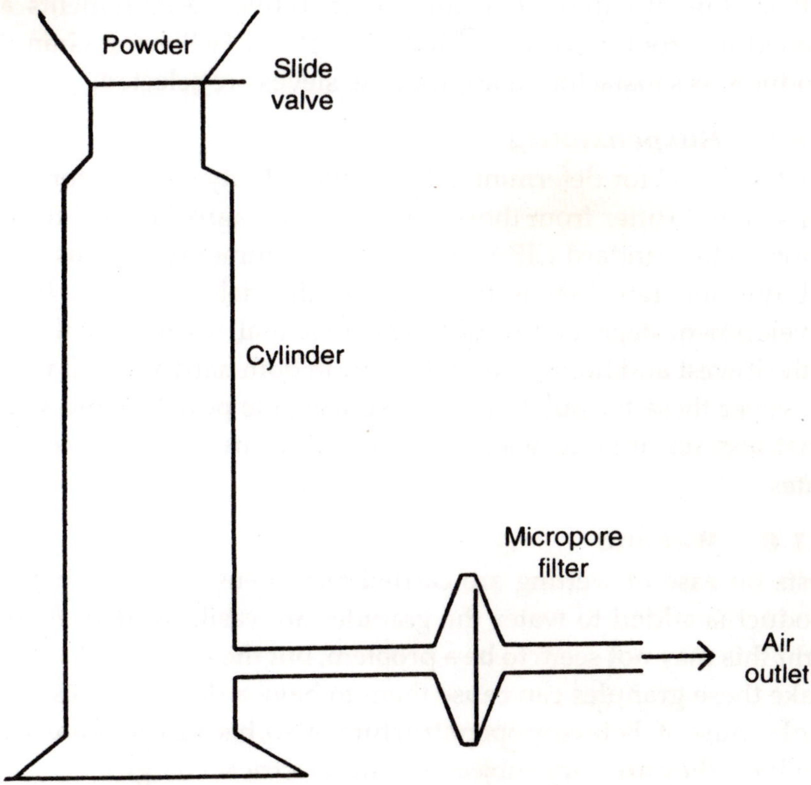

Figure 9.14 Cylinder method for dust measurement

Although this is a very simple test procedure it does suffer from the fact that the air in the cylinder is disturbed when the sample is taken and it is difficult to get reproducible results unless the extraction air is very well controlled.

In both cases the results are expressed as a percentage of dust in the sample. This is by direct measurement in the case of the cylinder test, but the Casella has to be calibrated.

4.1.4 *Granule size distribution*

The size distribution of granules is important from two aspects. Firstly the product must be consistent from batch to batch as it comes out of the production process and secondly its appearance must meet marketing needs.

It is thus apparent that only comparative measurements are needed, and for this purpose CIPAC MT 59.2, Dry Sieving-Granular Products, is satisfactory if appropriate sieves are selected.

4.1.5 *Suspensiblity*

The tests used for determining the suspensibility of these formulations do not differ from those used for other agrochemical formulations. The standard CIPAC test is most commonly used using an MT dilution rate that is the same as the field use rate. In the development stage it is useful, but not essential, to carry out this test at the lowest and highest use rates and in both hard and soft water. However these formulations are like wettable powders and somewhat less sensitive to water hardness than are emulsion concentrates.

4.1.6 *Wetting*

Tests on ease of wetting are carried out to ensure that when the product is added to water the granules are easily wetted. At first sight this may not seem to be a problem, but the techniques used to make these granules can cause them to have a density of less than one because of their very open structure. Also, because they are dried products they are very subject to static electrical charges.

No special tests are necessary and the standard CIPAC MT 53.3, Wetting of Dispersible Powders, is a perfectly satisfactory test.

4.1.7 *Ultimate particle-size distribution*

The ultimate particle size must be measured on the product when diluted and as used in practice, as this is the form in which the pesticide will be applied. There is a wide range of methods for determining particle-size distribution, and the choice should be based on the level of sophistication needed. The methods include sedimentation under normal gravity (as in the case of CIPAC MT 42 and the EEL sedimentometer) and various types of centrifugal sedimentometers that yield data on sedimentation under simulated increased gravity. Another instrument for measurement of particle sizes is the Coulter counter, which determines the change in electrical properties as a suspension passes through an orifice. Yet another type is based on light scattering through the suspension, and the most sophisticated examples (such as the Malvern Nanosizer and the Coulter Autosizer) are based on laser scattering.

4.2 In-Process Testing

The level of in-process testing can only be assessed against the background of the particular manufacturing plant to be used. The two most important characteristics of granules are the particle size distribution and the moisture content. In some processes, particularly those that are completely self-contained, like the Schugi, it is also useful to determine the assay when mixtures of actives are used.

4.2.1 Moisture content

This can be determined by measuring the weight of a sample before and after drying in an oven. (Some typical conditions are heating for 24 hr at 105°C.) A more modern method is the infra-red balance, in which an infra-red lamp provides the heat source and the apparatus includes a balance so that the weight loss can be measured at any time interval without disturbing the sample.

4.3 Product Testing for Quality Control

The routine quality control testing of these granules follows the normal principles used for any other formulation. Tests must include the following: active ingredient content, impurities, water content, bulk density, wetting and suspensibility.

Other tests can be included. For example, CIPAC is developing a model specification that can form a good starting point for the development of a general specification.

The tests used are not different from those used above. They include water content (CIPAC MT 30.1/2), bulk density (CIPAC MT 58.4), wettability (CIPAC MT 53.3 Modified) and suspensibility (CIPAC MT 15.1 Modified).

Active ingredient tests are specific for each active and no general methods are available that can be used other than those which appear in the FAO specifications for a particular ingredient. In all these tests the normal FAO specification limits are applied.

5 CONCLUDING COMMENTS

As shown in the preceding paragraphs, the wide range of granulating equipment and the types of granules obtained indicates the general versatility of water-dispersible granules. They have introduced a new dimension into the area of agricultural pesticide formulations and at the same time have improved the ease of handling and safety of this type of formulation. With the ever-increasing need to reduce the impact of pesticides on the environment, these granules offer several advantages. They are dust-free, they permit a wide range of packaging types to be used and they enable containers to be used that can be easily destroyed, such as cartons. Further, by packaging them in water-soluble packaging they can be converted into a closed system. Their main disadvantage is the sometimes high operational costs associated with the nature of the manufacturing process. This will continue to be the case until a procedure is developed that makes it unnecessary to remove the water. Another small disadvantage is that often the granules have a high bulk volume, which is particularly a problem when the use rate of the formulation is also high.

The future of these granules will largely depend on the attitude of regulatory authorities. If solvent-based formulations are banned, then these granules could represent the only formulation available. Similarly, if plastic containers are suppressed and paper or fibreboard have to be used, water-dispersible granules could be the formulation of choice.

REFERENCES

(1) Catalogue of Pesticide Formulation Types and International Coding System, *Technical Monograph No. 2*, GIFAP-Belgium.

(2) C. E. Capes *Particle Size Enlargement*, (Elsevier).

(3) Suppliers include Dravo Lurgi Corp., McDowell Wellman Corp. and Polysius.

(4) Maschinenfabrik Gustav Eirich.
(5) Niro Atomizer, Denmark.
(6) *Briquetting, Pelleting and Extrusion*, Centre for Professional Advancement.
(7) K. Masters, *Spray Drying Handbook*, Halstead Press.
(8) Alpine. Hosokawa Micron Europe, Holland.
(9) *CIPAC Handbook* Vols. 1, 1A, 1B, 1C, Collaborative International Pesticides Analytical Council, Harpenden, England.

USEFUL READING

Particle Size Measurement, T. Allen (Chapman Hall).

H.M., Collins, E.E. Newsome and K.L. Turpin, "Water dispersible granule formulation techniques," *Pesticide Formulation and Application Systems, Vol. 15*, H.M. Collins, F.R. Hall and M. Hopkinson, eds., pp. 124-131. (American Society for Testing and Materials, 1996).

CHAPTER

10

The Formulation of Biologicals

T. R. Shieh

Sandoz Crop Protection Corporation
Research Division
975 California Avenue
Palo Alto, California 94304, United States

1 INTRODUCTION

In its broad sense, the biological control of pests and plant diseases consists of (a) classical biological control, which entails the release of beneficial arthropod, which are natural enemies of insects and weeds and serve to suppress them, (b) the application of biochemicals derived from natural products such as enzymes and pheromones to control both pests and diseases and (c) microbials, including bacteria, fungus, viruses and protozoa, for the control of insects, weeds and plant pathogens.

The technology associated with preserving or prolonging the longevity of an arthropod species for its use in biological control practices differs considerably from that the formulation technology aimed at facilitating the mixing and application of microbial-based pesticides and improving their efficacy. Thus, the discussion in this chapter will be limited to the formulation of microbials and microbial proteins that have pesticidal activity. However, the same principles and practical approaches used for chemical pesticide formulations do however, apply to small molecule biochemicals of a natural product.

2 MICROBIAL AGENTS AS PESTICIDES

During the last several decades, numerous microbial agents have been studied and evaluated for use as alternatives to chemical pesticides for the control of pests (1, 2). The agents most commonly evaluated for the control of insects, weeds and plant pathogens are classified as shown in Table 10.1.

Table 10.1. Classification of common microbial pesticides

Classification	*Pest*	*Microbial*
Pesticide	Insects	Bacteria
		Baculoviruses
		Protozoa
		Fungi
	Mites and aphids	Fungi
Mycoherbicide	Weeds	Fungi
Microbial antagonist	Plant pathogens	Bacteria
		Fungi

Microbial pesticides that have been made commercially available for the control of various targets around the world are listed in Table 10.2.

Information on the application rates proved to achieve certain efficacy objectives and the production yields of each microbial agent led to determining economic and formulation structure of a product.

Currently, the most widely used microbial agents are various strains and formulations of *Bacillus thuringiensis,* which control various insects. *Bacillus thuringiensis* is an aerobic spore-forming bacterium, commonly known as B.t., that is characterized by the formation of an insecticidal para-sporal protein crystal known as delta-endotoxin. A number of strains varying in their serotype of flagellar antigens were found to produce a specific toxin against various *Lepidoptera, Diptera* and *Coleoptera* species (3).

Other microbial agents have been registered but used in only a limited quantity at one time or other around the world, include nuclear polyhedrosis viruses, protozoa, fungi and gram-negative bacteria for the control or suppression of insects, weeds, aphids, mites and plant pathogens.

Information on new developmental activities by institutions and industry is rather scarce. Table 10.3 lists the new organisms currently under development by various companies.

Table 10.2 Target pests and field application rates of microbial pesticides

Organism	*Target*	*Application rate*
Bacteria		
Bacillus thuringiensis var. *kurstaki*	*Lepidoptera*	$10–100 \times 10^9$ IU/ha [a] $10–100 \times 10^9$ SU/ha [a]
Bacillus thuringiensis var. *israelensis*	Mosquito, blackfly	$0.44–1.70 \times 10^9$ AAU/ha [a] 0.8 –2.0 mg/L/10 min
Bacillus popilliae	Japanese beetle larvae	1×10^7 spore/larvae
Bacillus thuringiensis var. *tenebrionis*	Colorado potato beetle	$10–20 \times 10^9$ TU/A [b]
Viruses		
Nuclear polyhedrosis virus (NPV)		
Heliothis zea NPV	*Heliothis* spp.	5×10^{11} PIB/ha [c] 4×10^6 HZU/ha [a]
Orgyia pseudotsugata NPV	Douglas fir Tussak moth	2.5×10^{11} PIB/ha
Llymantria dispar NPV	Gypsy moth	$2.5 \times 10^{11}–10^{12}$ PIB/ha (62.5×10^6 GMPU/ha) [d]
Granulosis virus (GV)		
Carpocapsa pamonella	Codling moth	$2 - 5 \times 10^{12}$ GV/ha [e]
Protozoa		
Nosema locustae	Grasshopper	$1–2.5 \times 10^9$ spore/ha
Fungi		
Hirsutella thompsonii	Citrus rust, mites	$2–5 \times 10^6$ conidia/tree
Verticillium lecanti	Aphids in glasshouses	10^7 conidia/ml
Collectotrichum gloeosporiodes	Northern joint vetch	2×10^{11} conidia/ha
Metarhizium anisopliae	Spittle bugs, grubs	10^{15} conidia/ha
Beauveria bassicae	Colorado potato beetle, codling moth	4.5×10^{13} conidia/ha

(a) See Table 10.4 for potency unit designations.
(b) TU/A refers to *tenebrionis* unit designation by immunological assay.
(c) PIB = polyhedral inclusion Body.
(d) GMPU = Gypsy Moth Potency Units designated by bioassay.
(e) GV = granulosis virus.

Table 10.3 New microbials under development by companies

Company (Product name)	*Organism*	*Formulation*	*Target*
Monsanto N.Z. (Invade)	*Serratia entomophilia*	Liquid cell suspension	Grass grubs
W.G. Grace	*Paecilomyces fumasoreseus*	Conidia powder	Whitefly, citrus root weevils, mole cricket
Bayer A.G. (B10-1020)	*Metarhizium anisopliae*	Mycelia granule	Mole circket, termites, sugar cane weevils, *Lepidoptera* spp., *Coleoptera* spp., *Hymerptera* spp., *Diptera* spp., *Orthoptera* spp.
EcoScience	*M. anisopliae*	Conidia trap	Cockroach
Crop Genetics International (Incide)	B.t. gene in claribacteria	Pressure impregnation	European corn borer

3 CONSIDERATIONS FOR FORMULATING MICROBIAL PESTICIDES

The principles and types of formulations applied to microbial agents are quite similar to those of chemical pesticides. The purpose is to produce a suitable product that optimizes the effective and economical use of the product under the particular environment and circumstances. In general, formulation processes and specifications of microbial pesticides closely mimic those of chemical pesticides. In fact, chemical pesticides are often used as a standard for establishing specifications for microbial agents. Microbial agents, however, differ from chemical pesticides in several aspects, as outlined below.

3.1 Microbials Are Particulate Matter

Unlike chemical molecules, insect pathogens are insoluble living entities with an organized cellular structure entrapped in a generally hydrophobic cellular membrane. The particle size can vary from 0.2-0.8 μm, up to 7-8 mm (Table 10.4). The hydrophobicity and particulate nature of microbials affects the wettability, dispersibility and suspensibility of formulated products.

Table 10.4 Sizes of micro-organisms used for pesticides

Organism type	*Active form*	*Size*
Bacillus thuringiensis	Spore	0.5–0.8 × 1.5–1.8 μm
	Delta-endotoxin crystals	0.1–0.3 × 0.1–1.5 μm
Bacillus popilliae	Spore	1.8 × 0.9 μm
Baculoviruses	Nuclear polyhedrosis viruses	1–1.5 μm
	Granulosis viruses	0.2–0.5 μm
Protozoa	Spore	1.5–3.5 × 3.5–5.5 μm
Fungi	Conidia	2–4 × 2–8 μm
	Mycelium pellet	7–8 mm

3.2 Production and Recovery Processes Affect the Nature of Microbial Products

Chemicals, whether synthesized, extracted from fermentation broth or natural products, may be purified to a relatively high-purity technical-grade material that meets the standard specification requirements for formulation into a product.

Microbial agents, on the other hand, are produced either by submerged or solid surface culture fermentation or by multiplication of pathogens in a host insect population. Such biological processes require multiple nutrient systems to support the complicated growth cycle of the pathogen as well as the host organisms. The production yield, measured in terms of the potency or activity, varies widely depending on the strain of the organism and the conditions used for its production.

Variations in growth conditions applied for a particular strain of organism also affect the downstream recovery processes associated with the concentration of active ingredient and thus affect the quality of the technical material produced for the final formulation.

3.3 Microbials Must Be Ingested by Target Pests

With the exception of fungal pathogens, microbial agents such as bacteria, viruses and protozoans must be ingested by target pests in order to be infectious. Thus, microbial formulations containing various feeding stimulants or baits have been demonstrated to increase the effectiveness of pathogens for the control of pests (4-7). The addition of an antifeedant in the spray mix, on the other hand, would reduce the intake, by feeding, of the pathogen, thus reducing the efficacy.

Fungus preparations, however, can grow on the outer structure of plant or pest surfaces and spread into their inner structure under certain favourable environmental conditions. Compounds that inhibit the germination or growth of fungi should not be incorporated into the formulation or into the tank mix.

3.4 Stability Characteristics

Since microbials are generally more sensitive to the surrounding environments than most chemicals, a microbial formulation requires special attention to the preservation of its biological activity.

The factors that affect the stability and bioactivity of biological formulations are as follows:

(a) *Maturity of organisms*. Premature harvesting of organisms may result not only in poor activity yields but also in the death of the organisms;

(b) *Presence of degradative enzymes*. Residual destructive enzymes from host cells or contaminating organisms can destroy cellular structures and inactivate the micro-organisms. Changes in physical properties of formulation may also occur;

(c) *pH*. Neutral pH is often preferred for the preservation of activity. Higher pH levels often result in foul odours and bacterial growth. Low pH (below 4.0) tends to reduce the shelf-life of microbial products;

(d) *Residual nutrient availability*. Carryover nutrients, particularly in a liquid formulation, can facilitate the regermination of spores and mycelia and the growth of other microbials during storage;

(e) *Ionic strength and osmotic pressure*. Various organisms respond differently to the ionic strength and osmotic pressure of the medium. High osmotic pressure may be used for the suppression of microbial growth but it can also destroy certain types of organisms;

(f) *Type of preservative used.* Preservatives of a proton donor type are often effective in the lower pH ranges for suppression of contaminant growth. However, they can also be deleterious to the bioactivity of the microbial formulation;

(g) *Use of surfactants.* Surfactants are often used for improving suspensibility, dispersibility, and the spreading characteristics of chemical and microbial formulations. However, cationic and anionic surfactants often denature the protein structure and affect the thermal death time of a bacterium (8);

(h) *Temperature.* Microbial pesticides are inherently less stable in pesticidal activity than most chemical pesticides. Loss of infectivity is often temperature-dependent between 10°C and 60°C. Therefore prolonged exposure to high temperature should be avoided during the production processes. In general, a shelf-life of at least 18 months at ambient temperature is critical for commercialization of a formulated product (9, 10).

3.5 Selection of Inert Ingredients

To make microbial agents suitable for use in environmentally sensitive areas, the formulation ingredients and inert carriers used for biological formulations may not contain certain hazardous chemicals specified by regulatory authorities. This is also necessary to qualify for exemption of tolerance from residue.

3.6 Bioassay to Measure Active Ingredients

A formulation active ingredient is measured by bioassay against representative species of pests and in some cases by the viability of spore (conidia counts). Alternative chemical methods, such as direct measurement of protein or immunoassay of delta-endotoxin in Bt formulations, are valid only if extensive correlation studies are established between the bioactivity and chemical quantifications.

4 TYPES AND CODING SYSTEMS FOR COMMONLY USED MICROBIAL PESTICIDE FORMULATIONS

Most microbial pesticide formulations utilize the international two-letter code produced by GIFAP (11) and adopted by FAO specification and by CIPAC for methods of analysis (Table 10.5).

Table 10.5 GIFAP codes for pesticide formulations

Code	*Formulation*	*Definition*
SC	Suspension concentrate (Flowable concentration)	A stable suspension of active ingredients in a fluid intended for dilution with water before use
UL	Ultra-low volume liquid	A homogenous liquid ready for use with UL equipment
OF	Oil-miscible flowable	A suspension of active ingredients in an oil intended for dilution in water before use
WP	Wettable powder	A powder formulation to be applied as a suspension after dispersion in water
WG	Water-dispersible granule	A formulation consisting of granules to be applied after disintegration and dispersion in water
TK	Technical concentrate	A technical material, either in solution or diluted with solid adjuvants, for use only in the preparation of formulations
GR	Granule	A free-flowing solid product of a defined granule-size range ready for use
DP	Dustable powder	A free-flowing powder suitable for dusting
BR	Briquette	Solid block designed for controlled release of active ingredient

5 ADVANTAGES AND DISADVANTAGES OF VARIOUS TYPES OF MICROBIAL PESTICIDE FORMULATIONS

5.1 Suspension Concentrates

SC formulations include emulsifiable and aqueous concentrates, which can contain as much as 50% solids, which often results in a high viscosity profile.

The advantages of SC formulations of microbial agents are as follows: (a) relatively simple recovery and formulation and (b) convenience of application by growers. Their disadvantages include

(a) relatively low potency per unit weight of product due to liquid diluent; (b) the high costs for transportation, storage and container disposal due to the bulkiness of the liquid formulations and (c) often shorter shelf-life, presumably due to a higher rate of proton transfer across the microbial cellular membrane in an aqueous media.

5.2 Ultra-Low Volume

A UL formulation is more potent per unit volume and an undiluted formulation is frequently used for a large scale aerial spray on forest or river terrain. The neat application of UL formulations by aerial spray on forest can be carried out if (a) potency is sufficient to allow an aerial application rate of 7-9 liters or less per hectare; (b) the viscosity of the formulation is sufficiently low to permit mechanical pumping of the fluid to an aircraft spray tank and its passage through spray nozzles to provide optimal spray particle size and swath coverage; (c) the formulation contains sufficient anti-evaporant to prevent the evaporation of spray droplets during their free fall to minimize drift and (d) the formulation must not be corrosive to aircraft and spray equipment.

In addition to the above conditions, the aerial treatment of rivers for blackfly control requires that the liquid drops spread quickly upon impacting the water surface without causing sedimentation. This will allow the microbials to be carried along the upper surface of the river water.

5.3 Flowables

OF formulations are often used in UL applications on forests since they are non-evaporative and resistant to wash-off by rain. They may also be used as SC formulations when emulsified into water before spray. For microbials oil formulations have the disadvantage of higher cost and relatively poor shelf-life. For agricultural crops, users should be aware of the possibility of phytotoxicity on sensitive crops under certain conditions.

5.4 Wettable Powders

Wettable powders are formulated for dilution with water into a spray. Due to the hydrophobicity and large particle size of microbial entities, the surfactants and diluents to improve the wettability and suspendability of a formulation must be carefully selected.

Wettable powder formulations have advantages over SC formulations because they are more potent per unit weight and more importantly, they are more stable than that of liquid formulations. Disadvantages include (a) the greater capital investment for equipment; (b) the great amount of energy needed for the removal of water and (c) the added costs for diluents, surfactants and processing to produce a formulation suitable for field application.

5.5 Water-Dispersible Granule

WG formulations are made from a finely powdered technical material that is agglomerated with binders, or coalesced, to a larger particle size. A WG formulation must be non-caking, have good wetting and dispersibility characteristics and must break up instantaneously into particles that will pass screens and nozzles during spraying.

The advantages of a WG formulation include (a) high potency unit per weight and (b) uniform, free-flowing, dustless granules. Its production, however, requires a large initial capital investment for a fluid-bed granulator (12).

5.6 Technical Concentrate

A TK, produced from a technical material in bulk form, is used in the manufacture of more dilute formulations. TK product may be used for the formulation of various liquid or dry formulations and has the advantage of reducing transportation and import tariff costs. Local manufacture utilizing a granular form of a B.t. technical concentrate gives a product for the control of mosquitoes in an aquatic habitat and European corn borer on corn plants.

5.7 Granular Formulation

GR formulations of microbial pesticides frequently use an inert granule of a particular size as a carrier to which a microbial agent is bonded with various sticking agents.

Granule applications are for penetrating dense foliage for aquatic mosquito larvae control and also for whirl deposition on corn plants for corn borer larvae control. To be useful, granular formulations must be non-caking during storage and free-flowing.

Due to the large surface area of GR formulations, a microbial agent is exposed to higher oxidative potentials so the shelf-life is often short. In addition, due to bulkiness of a GR formulation, which increases transportation and storage costs, a granule formulation is precluded from large-scale commercial usage. Thus, microbial granule formulations are, for the most part, limited to on-site preparation from TK materials for seasonal usage based on previous good usage patterns.

5.8 Dustable Powders

DPs are dry formulations diluted to field strength with various clay diluents such as calcium carbonate, silica and talc. Dusts must be free-flowing with a particle size usually under 200 mesh. The formulations provide good deposition on certain crop leaves and, owing to their massive inert particles, better shielding against sunlight inactivation of microbials.

Aerial dusting practices on agricultural crops has been reduced in recent years due to the greater chance of drifting of the dusts. Use of dust formulations for home and garden applications, however, seems to gain in popularity for reasons of both convenience and efficacy.

5.9 Briquette

A BR is a solid block formulation with a diameter ranging from 2 to 6 cm. Briquettes are generally formed by mixing the active ingredients with low-density, inert granules and binding agents. The formulations allow timely release of individual granules and are useful for spreading microbials on water. Briquettes are convenient for the manual application of pesticides into an aquatic environment where spray applications are not available or functional.

Briquette formulation with *Bacillus thuringiensis* var. *israelensis* has been used to prolong the field efficacy up to 2-6 weeks.

6 COMMONLY USED FORMULATION INGREDIENTS FOR MICROBIAL PESTICIDES

Most inert carrier ingredients and surface-active agents used for chemical pesticides may also be used for formulating microbial pesticides. However, the choice of the formulation ingredients for microbial pesticides is subject to the general provisions that it must not be harmful to the environment and must be non-toxic to the pathogen being used. A wide range of compounds are available; those of the type permitted in foods and cosmetics have been used with success.

6.1 Diluents and Carriers for Dry Formulations (WP, WG, DP)

Plant materials

- Soybean flour
- Walnut-shell flour
- Sucrose
- Lactose

Minerals

- Silicon oxides and silicates
- Celites
- Celatom
- Mica
- Talc
- Clays
  - Attapulgite clay
  - Olancha clay
  - Kaolin clay
  - Bentonite

6.2 Liquid Diluents for Liquid Formulations (SC, UL, OF)

- Water
- Vegetable natural oils
- Spray oils

High molecular weight polyglycols
Molasses
50% sorbitol or sucrose in water solution

6.3 Dispersing and Suspending Agents

Bentonite
Attagel
Culigel (super-absorbent polymers)
Soloid
Marseperse
Darvon
Daxad

6.4 Emulsifiers

Non-ionic surface-active emulsifying agents
Span
Tween
Atlas
Triton

6.5 Other Ingredients

Adhesives and stickers

Latex polymers
Plyac
Bond
Lovo
Casein
Dried blood

Granular carriers

Attaclay granular
Sand
Corncob granules
Cornmeal
Soybean oil meal
Vermiculite

Anti-caking agents

Hi-sil
Aero-sil

Buffering agents

Fermentation residue
Protein hydrolysate
Phosphates

Preservatives for aqueous formulations

Sorbic acid, sorbate
Methylparaben, propylparaben
Polyphosphates
Ethylene diamine tetraacetic acid (EDTA)
Ascorbic acid
Propyl gallate

7 MECHANISM OF ACTIONS

The principle mode of action of microbial pesticides lies in their infection by and multiplication of said organism in the target organism. With the exception of microbial antagonists and fungal preparations, bacterial, viral and protozoan based insecticides require ingestion by target insects in order to be effective. Infectious processes may involve direct penetration, pinocytosis or phagocytosis through a cellular membrane. Such infection and multiplication processes may involve the secretion of enzymes or toxins during the uptake processes of the target insects, leading to their destruction.

The mode of action of *Bacillus thuringiensis* against various types of insects is unique. After a larva has ingested the crystal protein and spore, the crystal protoxin is solubilized by a mid-gut juice of high pH and then digested by a proteolytic gut enzyme to form active toxin. Since the active toxin is formed only in the mid-gut of insects, the crystal itself is not toxic to non-target organisms and humans.

The solubilization and enzyme digestion processes for the conversion of protoxin to toxin in the mid-gut are specific to the target insect. Tests show that human and non-target organisms do not have the same system. The active toxin molecule produced in the mid-gut then binds specifically to receptor sites of the epithelial cells' brush border membrane vesicle, forming a cytolytic n-terminal trans-membrane pore of 20°A (13, 14). The formation of pores in mid-gut disrupts the ionic balance across the hemolymph membrane and causes disintegration of the epithelial cells' brush border membrane. These events inhibit feeding within a few minutes and eventually cause disintegration of the gut walls. At the same time, the germinated cells penetrate through the disintegrated gut walls entering, hemolymph and causing septicemia and eventual larval death (15).

8 DETERMINATION OF BIOLOGICAL ACTIVITY

Active ingredients used in the formulation are measured by bioassay against representative species of pests and in some cases by viability of spore or conidia counts. Alternative chemical methods such as direct measurement of protein or immunoassay of delta endotoxin in B.t. formulations are valid only if extensive correlation studies have been established between the bioactivity and chemical quantifications. For *B. thuringiensis* preparations, standardized bioassay methods have been established and the potency of the product is calculated by the following formula (16):

$$\text{Potency of sample} = \frac{LD_{50}\ \text{(reference)}}{LD_{50}\ \text{(sample)}} \times \text{(Designated unit activity/mg)}$$

Selection of a reference standard for a particular product may depend on the type of product and insect used for the registration of product. Examples of bioassay potency units used for commercial products are shown in Table 10.6.

Viable conidia or mycelial counts of fungus preparation may be used as an indicator for biological activity of a product if viable cell counts and infectivity relationships are established for the product. However, since changes in the strain or production processes could affect infectivity of the product, if such changes are contemplated or suspected cell count-infectivity relationships will need to be reestablished to assure biological activity of the product.

Table 10.6 Bioassay potency units designations of commercial microbial insecticides

Product	*Strain variety*	*Test insect*	*Potency units designation*
Dipel	*kurstaki* HD-1	*T. ni*	IU/mg[a]
Thuricide	*kurstaki* HD-1	*T. ni*	IU/mg[a]
Javelin	*kurstaki* SA-11	*S. exigua*	Spodoptera unit (SU)[c]
Vectobac	*israelensis* HD-507	*A. aegypti*	International toxic unit (ITU)[b]
Teknar HPD	*israelensis* HD-567 Asporogenic	*A. aegypti*	Aedes aegypti unit (AAU)[c]
Certan	*Aizawai* SA-2	*Galeria mellonella*	Galeria mellonella unit (GMU)[c]
Elcar	*H. Zea* NPV	*H. Zea*	H. Zea unit (HZU)[c]

Reference standards are:

(a) International standard E-61 or U.S. standard HD-1-5-1980;

(b) International standard IPS-78;

(c) Internal reference (respective products).

REFERENCES

(1) Beltsville Symposium in Agricultural Research, No. 5, *Biological Control in Crop Protection*, 18 May, 1980.

(2) *Microbial Control of Pests and Plant Disease, 1970-1980*, H.D. Burges ed., (Academic Press, 1981).

(3) T.R. Shieh, "*Bacillus thuringiensis* biological insecticide and biotechnology", *ACS* Symposium Series No. 362, 1988, p. 207.

(4) R.E. McLaughlin and others, "Development of the bait principle for boll weevil control. IV. Field tests with a bait containing a feeding stimulant and the sprorozoans *Glugea gasti* and *Mattesia grandis*", *J. Invertebrâte Path.* 13, 1969, p.429.

(5) G.L. Andrews and others, "Evaluation of heliothis nuclear polyhedrosis virus in a cottonseed oil bait for control of *Heliothis virescens* and *H. zea* on cotton", *J. Economic Entomology*, 68 (1), 1974, p.87.

(6) M.R. Bell and C.L. Romine, "Tobacco budworm field evaluation of microbial control in cotton using *Bacillus thuringiensis* and a nuclear polyhedrosis virus with a feeding adjuvant", *J. Econ. Entomol.* 73, 1980, p. 427.

(7) T.R. Shieh and G.T. Bohmfalk. *Production and Efficacy of Baculoviruses Biotechnology and Bioengineering*, vol. XXII. 7, 1980, p. 1357.

(8) L.J. Rode and J.W. Foster, *Agric. Microbiology*. 36, 1960, p. 67.

(9). T.L. Couch and C.M. Iganffo, "Formulation of insect pathogens," *Microbial Control of Pests and Plant Diseases*, H.D. Burgess, ed. (Academic Press, 1981), p. 621.

(10) T.R. Shieh, "Industrial production of viral pesticides," *Advances in Virus Research*, 36, 1989, p. 315.

(11) GIF, *Catalogue of Pesticide Formulation Types and International Coding System,* Technical monograph No. 2, 1984.

(12) K.C. Lin, "Development of solid pesticide formulations by fluidized-bed technology," *Pesticide Formulations, Innovations, and Developments,* Barrington Cross and Herbert B. Scher, eds., ACS Symposium Series 371, 1988, p. 251.

(13) B. Knowles, "Molecular interactions between insecticidal bacterial proteins and the insect midgut epithelium," *Proceedings of the Entomological Society of America,* New Orleans, Lowsiana, 4, December. 1990.

(14) J. Van Rie, "Role of midgut receptor proteins in determining the susceptibility of insects insecticidal bacterial toxins,". *Proceedings of the Entomological Society of America,* New Orleans, Lowsiana, 4 December 1990.

(15) P. Luthy and R. Ebersold, *Pathogenesis of Invertebrate Microbial Diseases,* R. Davidson, ed. (Allanheld, Osmur Publishers, 1981), p. 260.

(16) H.D. Burges, "The standardization of products based on *Bacillus thuringiensis,*" *Proceedings of the International Colloquium on Insect Pathology and Microbial Control,* P.A. van der Laan, ed. (Amsterdam, North Holland Publishing Company, 1967) p. 306.

CHAPTER

11

Pesticide Activity Modification

J. Wade Van Valkenburg

Accessible Technologies Inc.
494 Curfew Street
St. Paul, Minnesota 55104, United States

1 INTRODUCTION

The synthesis chemist has an easy job compared to the pesticide formulator. The chemist designs molecules that are initially examined for biological activity under ideal conditions that yield the maximum level of activity inherent in the molecule. The formulator chemist, however, will never achieve that level of biological activity because of the obstacles that nature, chemistry and physics put in his way. This chapter discusses some of the methods pesticide formulators use to overcome these obstacles so that the observed pesticidal activity approaches the level originally observed in initial screening in the greenhouse.

Some of the obstacles and requirements that must be met by the formulator are as follows:

(a) The pesticide must reach the site of action. This may involve passing through membrane barriers that inhibit passage to an active site;

(b) The pesticide must be distributed evenly and uniformly over a target area;

(c) The physical form of the pesticide must optimize biological activity;

(d) When used, the pesticide must have suffered minimal degradation by containers, carriers and the storage environment;

(e) During the application and residence time on the target, adverse environmental factors of heat, light and rain must be minimized.

Thus, there are all kinds of forces and conditions that modify the observed biological activity of a pesticide. It is desirable that the pesticide formulator take measures to minimize the adverse factors, surmount the obstacles to performance of the pesticide and judiciously use additives that increase the observed biological activity of the pesticide.

2 MINIMIZING ADVERSE FACTORS

2.1 Environmental Factors

Water sprayed in 50 μm drops at 20°C and 50% relative humidity will have a life expectancy, in air, of four seconds (1). On the other hand, in four seconds, a 200 μm droplet will only decrease in size to 193 μm. In the same amount of time a 100 μm drop will have decreased to 87 μm. Consequently, a water-based low-volume mist spray will yield very poor results if the time between the spray nozzle and leaf impingement is long. Improved results will come from a higher volume spray, a shorter delivery time and a slightly coarser spray. Improved results may also be obtained if the formulator adds to the application system a small amount of water-soluble polymer that will inhibit the evaporation process.

The above discussion relates to water-based sprays. Emulsions of oil in water will maintain droplet integrity for a longer period of time. Oils and emulsifiers in an emulsifiable concentrate will reduce the amount of water available to the surface of the spray droplet, reducing the rate of evaporation.

The pesticide formulator might go to the ultimate extreme and use an all-oil carrier for his spray. This would reduce evaporation but raise other obstacles. First, it could increase phytotoxity to desirable plants. Secondly, it could cause drift: oil sprays tend to promote very small spray droplets that do not drop to the ground nearly as rapidly. Instead, they remain suspended for longer periods of time and in both calm and windy conditions, will drift hundreds of yards from the target.

2.2 Deposit

Although it is desirable to reduce volatilization of a liquid carrier by increasing spray droplet size, one must reach a balance between droplet size, spray volume per acre and the wetting characteristics of the spray. High-volume, large droplets and good wetting of foliage will decrease the amount of spray that deposits on a target site (2). Interestingly, it has been shown that large droplets that wet a plant poorly will deposit a larger amount of spray on a plant than a good wetting formulation. However, the deposit will not be uniform. Since it is better to deposit small droplets uniformly over an entire surface, tank additives of small amounts of "spreader-stickers" are desirable. Many such products are commercially available.

2.3 Ambient Moisture

Water, dew, rain, indeed moisture in any form, can affect biological activity in many ways. If it rains shortly after application, a pesticide can be washed off a target. The presence of moisture on a plant surface can cause rapid hydrolysis of organophosphate esters and carbamates, particularly if there is a thin coating of an alkaline soil on the target. Alkalinity catalyzes the decomposition of organophosphate esters and carbamates. Since the rate of hydrolysis is pH-sensitive, if this is a problem in a given area, hydrolysis on a plant surface can be reduced by the addition of stabilizing buffers to the pesticide formulation.

On the other hand, if biological activity depends upon penetration through some barrier membrane, the presence of moisture may be advantageous. Penetration through a barrier membrane is enhanced as long as the spray deposit is in a liquid form. In a dried or highly viscous form, penetration is reduced. Environmental moisture will reduce the rate of evaporation of a spray droplet on a leaf surface.

As mentioned previously, if one adds a small amount of water-soluble polymer to the spray solution, evaporation can be retarded and one can increase the residence time during which a formulation will penetrate through barrier membranes.

2.4 Storage Stability

All pesticide formulation concentrates or ready-to-use compositions should be checked for shelf-life by analysis after storage at both high and low temperatures. With liquid formulations it is desirable to cycle the product through both high and low temperatures as well as to add seed crystals at low temperatures to determine the possibility of phase separation upon storage. If separation does occur, resolubilization may not occur under normally ambient conditions and an unsuspecting user may apply too little pesticide to achieve the desired results. It is a good idea to add a seed crystal of the pesticide to see if this causes separation. Many pesticide formulations have a tendency to supercool. Of course, besides the phase separation studies one should analyze the composition to determine that the assay of active ingredient does not fall below specifications during a normal storage period.

It would be fortunate if the so-called inerts in a pesticide formulation were really inert and had no effect upon the active ingredient incorporated therein. Such is not the case. In wettable powders, clay diluents contain acid or basic sites that cause the decomposition of pesticides (3). Acid sites cause decomposition of chlorinated hydrocarbons, while basic sites cause hydrolysis of organophosphate esters and carbamates. Mildly basic materials, such as urea, can be incorporated into a formulation to neutralize acid sites. Materials such as precipitated calcium carbonate and some silicates should be avoided as carriers to phosphate esters.

As one evaluates a number of potential mineral or organic carriers for a pesticide, it may be well to correlate stability of the active ingredient with the acidity or basicity of the carrier. A useful tool for measuring this parameter is through the use of dyes called Hammett indicators. A small sample (0.5 g) of the carrier is added to 2 ml of indicator solution, the appearance of the acid conjugate colour indicates a surface acidity higher than the ionization constant (pK_a) of the indicator (2) Hammett indicators are characterized in Table 11.1.

Table 11.1 Properties of Hammett indicators

Eastman cat. no.	*Indicators*[a]	pK_a	*Basic colour*	*Acid colour*
304	Anthraquinone	-8.2	Colourless	Yellow
1254	Chalcone	-5.6	Colourless	Yellow
2926	1, 9-Dipheny1-1, 3, 6, 8-nonatetraen-5-one	-3.0	Yellow	Red
3651	4, 4', 4"-Methylidyne-tris-(N, N-dimethylaniline)	+0.8	Yellow	Blue
2494	N, N-Dimethyl-p-1-napthyl-azoaniline	+1.2	Yellow	Blue
1714	4-Phenylazodiphenylamine	+1.5	Yellow	Purple
1210	4-o-tolyazo-o-toluidine	+2.0	Yellow	Red
1375	p-Phenylazoaniline	+2.8	Yellow	Orange
338	N, N-Dimethyl-p-phenyl-azoaniline	+3.3	Yellow	Red
5548	4-Phenylazo-l-napthylamine	+4.3	Yellow	Red

(a) Indicator solutions are made up with 10 mg of indicator/100 ml water-free benzene (or equivalent).

Cycling temperature studies from hot to cold is an appropriate practice with some dry formulations. This is important if there is some moisture content in the composition and some ingredients have a degree of water solubility. As the temperature cycles, the water-soluble ingredient solubilizes at elevated temperatures. Then it crystallizes out upon cooling. This process can bind solid particles together, transforming the formulation into a hard non- flowable mass which may be unusable by the consumer. If the resulting product does not then disperse well in water, the result can be reduced biological activity, clogged nozzles and all kinds of things that make a customer very unhappy.

2.5 Packaging

The materials of construction of the packaging could have an adverse effect on activity. In liquid formulations of esters, acids may be present that can corrode tin and iron containers. If iron drums are the container of choice, the formulator should check the corrosiveness of the formulation against the exact materials of construction of the container. For example, a container may look like black iron but may be phosphatized. Black iron is more susceptible to acid attack than is phosphatized iron. If iron containers are corroded by a candidate formulation, an epoxy lining on the drum may be necessary.

3 SURMOUNTING OBSTACLES

Knowledge of the mechanism of action of a pesticide can be very beneficial to the pesticide formulator. In many instances the pesticide must interact with an enzyme, a cellular component or some factor that affects a metabolic process of the target plant or insect. This component is frequently referred to as a receptor site. But, since biological systems are complex and a pesticide may be remote from a given receptor site (4), it will have to migrate through lipid and waxy barriers, pass through phase boundaries and undergo sorption and desorption processes as it passes through cellular membranes.

In view of the above, the mechanism of action of a pesticide is viewed as a two-step process involving (a) transport and (b) reaction at a receptor site.

3.1 Biological Correlation Studies

Biological correlation studies are generally designed to study an homologous series of compounds to optimize the physical and chemical properties of a candidate pesticide. Indeed, the Bayer organization used this approach in the development of α-cyanocarbonylphenylhydrazones as acaricides and insecticides (5).

Biological correlation studies can also be used to adjust the physical properties of a pesticide application system to improve transport properties, that is, to enhance the movement of a pesticide to a receptor site, thereby improving the biological response to the pesticide. Van Valkenburg and Yapel (6) studied the effects of 0.1% surfactant on the pre-emergent herbicidal activity of meta- and para-monosubstituted trifluoromethane-sulfonanilides. The physical characteristic that most affects the transport property of a pesticide is the partition coefficient, which relates the solubility of the pesticide in octanol to its solubility in water. The addition of a non-ionic surfactant caused an apparent increase in water solubility, thereby decreasing the partition coefficient.

It is believed that there is an optimum partition coefficient for maximum performance of a pesticide against a given plant or insect. A given pesticide may not specifically match this ideal partition coefficient but may be either slightly too lipophilic or slightly too hydrophilic.

The effect of the surfactant was to increase the apparent hydrophilicity of the pesticide. Thus, if a given compound already was slightly too hydrophilic to the optimum partition coefficient, the surfactant decreased the observed biological activity. If, however, the candidate pesticide was slightly too lipophilic, the decrease in the apparent partition coefficient, as affected by the surfactant, caused an improvement in biological response. Thus, a given surfactant can either increase or decrease apparent biological activity, depending upon the relationship between the partition coefficient of a given pesticide and an idealized optimum partition coefficient.

3.2 Hard and Soft Acids and Bases

A carrier solvent in an emulsifiable concentrate or an emulsion concentrate can be used to increase the transport properties of an active ingredient. When the pesticide formulator selects a solvent for a given system, his primary concern seems to involve solubility characteristics, emulsifiability, phytotoxicity and packaging. It is all to easy to select a solvent for an emulsifiable concentrate on a 100°C flash xylene solvent because it has worked well in the past. Little thought has been given to blending in either aliphatic solvents or polar compounds that will shift a partition coefficient in the direction that would increase the transport aspect of a pesticide and thereby increase biological activity.

An interesting concept that can be applied to the selection of solvents is the concept of hard and soft acids and bases (7-8).

Chemical compounds can be considered either acids or bases. The modern theory of acids and bases is that acids are electron acceptors and bases are electron donors. Strong or hard acids are ones that are highly polarizable, while weak or soft acids are less so. As a general rule, hard acids are soluble in hard bases and soft acids are soluble in soft bases. It is interesting to note that the concept of "like dissolves like" follows here in that weak bases are more soluble in soft acids than they are in hard acids.

Compounds that contain oxygen, nitrogen, sulphur and related elements tend to be basic. The π electrons of olefins and aromatics, and all nucleophilic groups, are basic sites in molecules, and anions tend to be bases.

Thus, if a barrier layer of an insect is very lipophilic and slightly basic in characteristic, a combination of an aliphatic solvent and a small amount of a relatively polar ether or ketone solvent may be more effective than a straight aromatic solvent. The idea here is that the aliphatic solvent may be more compatible with the lipid layer and hence do a better job of transporting the pesticide through it to an active site. Adding polar ketones or ethers may detract from this but will assist in solubilization of the active ingredient. The critical suggestion here is that the properties of a solvent carrier system should be matched to the properties of the target in addition to optimizing their solubility and emulsification properties.

3.3 Antagonism

Acid-base interactions came into play in a recent study on bentazon antagonism on sethoxydim absorption and activity (9). It was found that the diffusion of sethoxydim was inhibited by the presence of the sodium salt of bentazon. The results were interpreted to mean that some of the sodium ions in the bentazon formulation were exchanging with the H^+ of the sethoxydim hydroxyl group to form a more polar sodium salt of sethoxydim. Following up on this hypothesis, the investigators added weak acid salts of lithium, potassium, caesium, calcium and magnesium and found that all reduced sethoxydim absorption. Having discovered this, the investigators failed to take a step very valuable to pesticide formulators, wherein they could have added a weak organic base to the sethoxydim formulation to see if metallic salt formation was inhibited.

3.4 Soil Degradation

Acid-base relationships may also be critical to the activity of herbicides in soil. A sulfonylurea herbicide, identified as DPX-A 7881, was found to be more residually active in slightly alkaline soil as opposed to acid soil (10). Actually the maintenance of biological activity in various soils is analogous to the well-known hydrolysis of chlorinated hydrocarbons when mixed with an acid mineral diluent. Where a material such as this sulfonylurea is going to be used extensively on acid soils, it may be desirable to add an alkaline buffer agent.

3.5 Absorption

For soil-applied pesticides, one should be concerned not only with the stability of the pesticide but also with the possibility of its absorption on a material that will bind it and not release it (11). In some areas, straw and stubble harvest may be burned at the conclusion of a season. The resultant carbon becomes incorporated into the surface layer of the soil by winter rains. With subsequent treatment of the soil by chlorpyrifos for control of crickets, reduction in activity of the insecticide was attributed to the absorption of chlorpyrifos on the residual carbon. Highly polar formulations will have less of a tendency to absorb but will have the drawback of leaching out of the soil more rapidly.

4 ADDITIVES

4.1 Resistance

The resistance of insects to a number of pesticides is well recognized. As in many cases, the resistance of the soybean looper to permethrin is attributed to the ability of the insect to metabolize enzymes at a more rapid rate (12). Appropriate additives for the formulator to consider are materials that inhibit enzymatic activity, such as glutathione transferase, monooxygenases and hydrolases.

Another additive approach for permethrin is a combination of pesticides. In India, extensive use of synthetic pyrethroids has resulted in a considerable increase in the whitefly population, which the synthetic pyrethroids do not control. Entomologists in India have suggested, informally, that this problem can be solved by adding natural pyrethrins to the synthetic insecticides.

Perng et al. (13) recommend the addition of piperonyl butoxide, an inhibitor of microsomal oxidases, as a means of maintaining the activity of teflubenzuron against the diamondback moth when resistance becomes a factor in the control of these insects.

4.2 Antidotes

There are times when an herbicide is effective against an unwanted weed species but too active against a desirable crop. "Antidote" additives are designed to reduce the phytotoxicity of an herbicide to an agronomic crop. For example, treating sorghum seedlings with antidotes such as flurazole, oxabetrinil and naphthalic anhydride resulted in the appearance of four to five species of glutathione s-transferase isozyme which exhibited activity towards metolachlor (14). In other words the antidotes safened metolachlor when applied to sorghum seedlings.

In another example, the antidote naphthalic anhydride increased the metabolism of imazaquin to less toxic metabolites, resulting in a reduction of imazaquin injury to corn and sorghum.

4.3 Surfactants

In 1958 Freed and Montgomery (15) postulated three critical variables necessary for the enhancement of herbicidal activity by surfactants:

- (a) Appropriate wetting characteristics of the herbicide formulation;
- (b) Low viscosity;
- (c) Positive interaction with a plant cuticle.

With regard to wetting characteristics, the spray composition must not wet the foliage with a totally spreading spray (thin film) but must form a positive contact angle with the solid surface. There are many reasons for this. By having a finite droplet, the volume of liquid is larger than in a thin film and the spray will remain in a liquid state for a longer period, because it will take a longer time for the larger volume to evaporate. Penetration of foliage is favoured by a liquid state as opposed to a viscous liquid or solid residual state after evaporation.

Droplet size and spray volume must be balanced to have the highest retention of deposits on foliage. If the drops are too large, gravity takes over and drops will either bounce off or roll off foliage.

Volume is critical because if the droplets completely cover a surface and have a chance to coalesce, the deposit will run off the foliage and not be retained. For further information on the spreading and retention of agricultural sprays on foliage, the reader is referred to an article by D.R. Johnstone (16).

Freed's third requirement for penetration calls for a positive interaction between a surfactant and the plant cuticle. He called for a favourable hydrophile-lipophile balance (HLB) for either solubilizing some of the plant waxes or swelling the cuticle to allow penetration. HLB is an arbitrary numerical classification system for surface-active agents. Surfactant molecules are usually constructed of a segment that is lipophilic, or oil-loving, and a hydrophilic, or water-loving, segment. Molecules with a high HLB are quite hydrophilic and tend to be wetting agents. Molecules with a low HLB tend to be lipophilic and tend to be useful for making invert or water-in-oil emulsions.

A surfactant must have a proper solubility parameter or solubility relationship to the material that a composition must penetrate to reach an active site, and HLB is one way of specifying this.

The barrier of a leaf cuticle and its associated penetration is better understood from Figure 11.1, which is reproduced from Hull et al. (17).

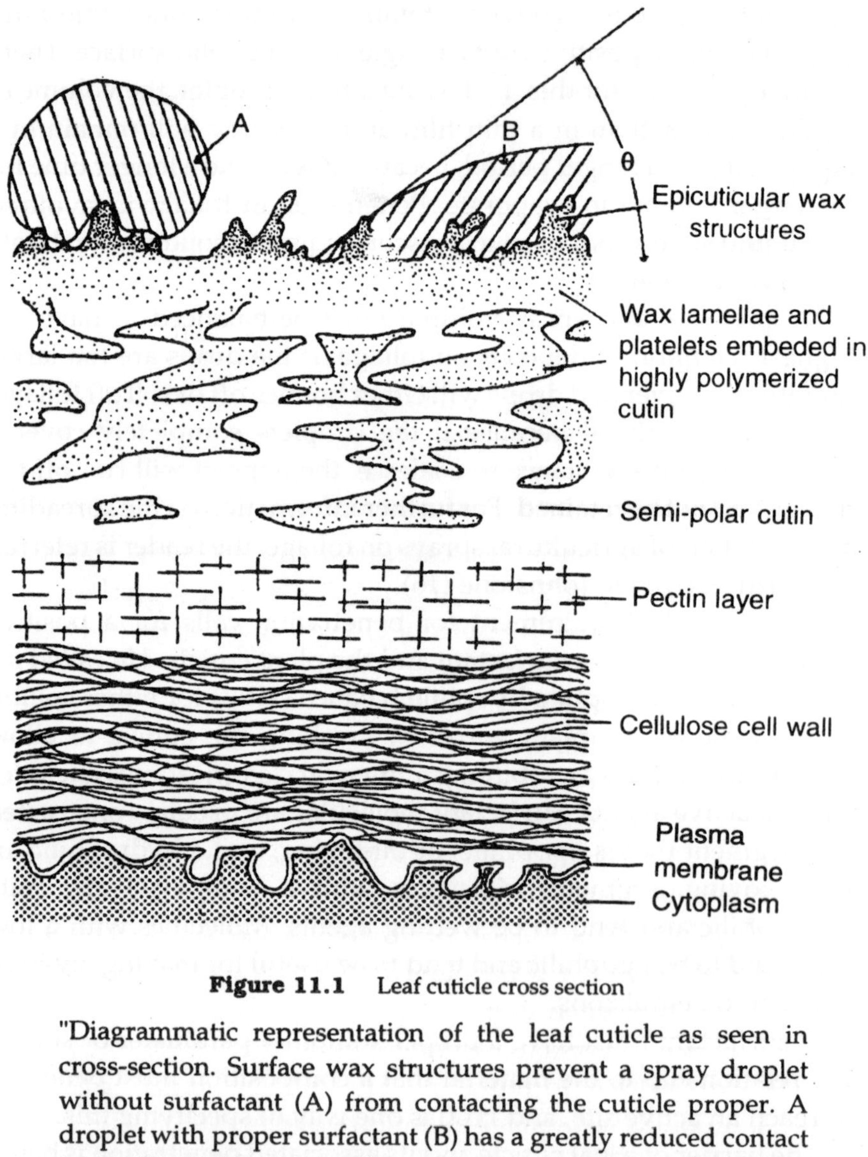

Figure 11.1 Leaf cuticle cross section

"Diagrammatic representation of the leaf cuticle as seen in cross-section. Surface wax structures prevent a spray droplet without surfactant (A) from contacting the cuticle proper. A droplet with proper surfactant (B) has a greatly reduced contact angle (θ), enabling it to effectively contact the leaf surface. Relative thickness of the cell wall and the various cuticular components depicted can vary markedly among different plant species and with extent of foliar development."

4.4 The Drying Process

Once an aqueous spray solution is on foliage, the water begins to evaporate and the composition of the remaining solution is constantly changing. In this residual solution, which may contain a surfactant or wetting agent, a solvent and the herbicide, interactions between the components occur during the drying process. The surfactants tend to be non-volatile so that during evaporation their concentration tends to increase. The resultant mixture forms a new solvent solution in which some of the pesticide may dissolve. Now you have a new solution and new physical properties that will dictate whether or not a pesticide is transported to an active site. Once again there must be a favourable solubility relationship for the residual pesticide to penetrate and be translocated to an active site.

4.5 Penetration and Translocation

Earlier it was mentioned that a surfactant might either enhance or decrease the apparent activity of a pesticide and that the partition coefficient of the pesticide could be optimal, slightly hydrophilic or slightly lipophilic with respect to maximizing transport. Surfactants cause an apparent shift in this partition coefficient that can enhance or decrease penetration.

The following two examples will illustrate the point. Dow Chemical discovered that the sodium salt of dichloropropionic acid (Dowpon) is a good grass killer. As the sodium salt, the pesticide is very water-soluble and consequently very hydrophilic and very polar. A relatively small proportion of the applied grass killer actually penetrates the waxy outer layers of grass and translocates. Dow Chemical wanted to increase this penetration and translocation. Originally an anionic wetting agent was added to the commercial spray-dried formulation. The wetting agent caused the spray solution to wet foliage better but, quite surprisingly, did not significantly increase penetration. In fact, with this now obsolete formulation, the surfactant may have actually decreased the penetration slightly.

There is a logical explanation for this. An aqueous solution of dalapon contains a very hydrophilic polar salt. Adding a hydrophilic surfactant does not move the partition coefficient of the system towards more lipophilic properties that are more compatible with the lipophilic waxy layers of the cuticle.

However, when Dow added a non-polar, non-ionic surfactant to the dalapon, the solution became more compatible with the waxy cuticle, and penetration and translocation were increased.

In later years, 3M developed a slowly translocatable herbicide, 2,4–dimethyl–5–acetamidotrifluoro–methanesulfonanilide (Embark).The product had limited commercial success but was found much more effective if combined with a contact herbicide. The problem was that when the contact herbicide was used in sufficient quantities to be effective, it caused so much damage to plants that the Embark did not translocate and was rendered ineffective. The solution seemed to be to use an effective amount of the contact herbicide but to slow down its activity so that it would be less disruptive of the growth of the plant and the two herbicides could work in concert.

The very effective solution to the problem was to add a very hydrophilic surfactant that would form a viscous hydrophilic deposit as the spray solution dried. Amazing results were obtained by adding a 40 mole ethylene oxide adduct of a fatty acid. This diminished the apparent activity of the contact herbicide by decreasing its ability to penetrate foliage.

5 CONCLUSION

The pesticide formulator must be totally knowledgeable about all factors that may affect the activity of his product. This entails a complete knowledge of all the chemistry and physics relating to the manufacture, storage and use of the product. He must also be knowledgeable as an entomologist, plant physiologist and microbiologist to ensure that his product does the job intended.

REFERENCES

(1) G.S. Hartley and I. J. Graham-Bryce, *Physical Principles of Pesticidal Behavior* (Academic Press, 1980).

(2) W. Van Valkenburg, *Pesticide Formulations* (Marcel Dekker, 1973) p. 160.

(3) C. J. Walling, *J. Am. Chem. Soc.*, 72, 1950, p. 1164.
(4) T. Fujita, *Biological Correlations*, Advances in Chemistry Series No. 114 (American Chemical Society, 1972).
(5) K. H. Buchel, ibid. p. 141.
(6) W. Van Valkenburg, ibid., p 252.
(7) R. S. Drago, G. C. Vogel and T. E. Needham, *J. Am. Chem. Soc.*, 93, 1971, p. 6014.
(8) F. M. Fowkes, *Surface and Interfacial Aspects of Biomedical Polymers* (Plenum, 1985).
(9) G. Wanamarta, D. Penner and J. J. Kells, *Weed Science*, 37, No. 3, 1989, p. 400.
(10) H. J. Beckie and A. B. McKercher, *Weed Science*, 37, No. 3 1989, p. 412.
(11) J. A. Kamm and M. L. Montgomery, *Journal of Economic Entomology*, 83, No. 1, 1990, p. 55.
(12) R. L. Rose and others, *Journal of Economic Entomology*, 83, No. 3, 1990, p. 672.
(13) F. S. Perng and others, *Journal of Economic Entomology*, 81 No. 5, 1988 p. 1277.
(14) J. V. Dean, J.W. Gronwald and C.V. Eberlein, *Plant Physiology*, 92, No. 2, 1990, p.467.
(15) V. H. Freed and H. Montgomery, *Weeds*, 6, 1958, p. 386.
(16) D. R. Johnstone, *Pesticide Formulations*, W. Van Valkenburg, ed. (Marcel Dekker, 1973).
(17) H. M. Hull, D. G. David and G.E. Stolzenberg, *Adjuvants for Herbicides* (Weed Society of America, 1980).

CHAPTER

12

Aerosol Formulations

Stan J. Flashinski

S. C. Johnson Wax

Racine, Wisconsin 53403, United States

1 INTRODUCTION

Approximately eight billion aerosol units were produced worldwide in 1990, including almost three billion in the United States, three billion in Western Europe and two billion in the remainder of the world (1). Aerosol insect sprays account for over 190 million units, or 6.5% of the total in the United States. The United Kingdom contributes an additional 50 million insecticide aerosol units and Australia produces a substantial 20 million units. Based on per capita consumption in Argentina, Brazil and Mexico, over 150 million aerosol insect sprays are produced in those three countries (2). Worldwide, insecticide aerosols account for over 600 million units, and it appears that the advantages offered by aerosol packaging will maintain this production trend well into the future.

1.1 Advantages and Disadvantages

Aerosol insect sprays have many advantages over other types of delivery systems. The aerosol package is safe and convenient for the consumer. The sealed package minimizes deterioration of the contents and shelf-lives of 5-10 years are not uncommon. The contents of the aerosol units are unavailable unless activated by the dispensing system, which controls the delivery rate. In actual usage, only a few seconds of spray are needed to provide the performance desired. Aerosols also provide for a variety of delivery systems, which allows dispensing the product in a form that is optimum. The forms includes a fine particle spray for spraying a space, a coarse wet spray for surface treatment and any degree of foaming for either a crack and crevice treatment or a foam tracer. All these parameters combined result in a product form that is most efficient and effective for insect control.

There are very few disadvantages of aerosol insect sprays from a technical standpoint. However, price, lack of visibility of the product, little differentiation between products and confusion concerning use of the actuator function are some disadvantages that inhibit market growth. The major technical disadvantages of aerosols are their flammability (for hydrocarbon-propelled systems) and ozone depletion (for fluorocarbon-propelled systems). Flammability concerns are more important in shipping and storage since very few consumer problems have occurred in relation to the number of units used. The ozone depletion problem is a worldwide concern. Fluorocarbons are known to deplete the ozone layer, which protects the earth from ultraviolet radiation. Depletion of the ozone layer results in more exposure to UV rays and a correspondingly increased risk of skin cancer. This potential problem has resulted in a ban on the use of fluorocarbons in aerosol products in the United States as well as in many other countries. The fluorocarbon propellants have been replaced by hydrocarbons, which are the propellants of choice at present time.

The removal of the fluorocarbons did not, however, ease consumer concerns regarding aerosols. The Roper Organization was commissioned to do a study in the United States on consumer perception of aerosols (3). A majority of the people who knew about ozone thought that aerosols were the main contributor to the problem and 78.0% would support regulations to ensure the use of safer propellants. They were unaware that regulations had been in place since 1978 in the United States and that present-day aerosols did not contribute to the ozone depletion problem. Similar regulations will phase out fluorocarbon propellants by the year 2000 in developed nations and by 2010 in developing nations. This negative consumer perception about aerosols is the major disadvantage for the aerosol industry and needs to be addressed by a programme now in place that will provide accurate information to environmental groups, lawmakers, the media and consumers.

1.2 Definition

The word "aerosol" was first used in colloid chemistry to describe a suspension, in air or gas, of solid or liquid particles having a diameter less than 50 μm. Using this definition as a baseline, the United States Department of Agriculture defined a pressurized aerosol insecticide product as a system where 80% of the particles, by weight, must be less than 30 μm in diameter and all particles must be less than 50 μm (4).

In 1955, the Chemical Specialties Manufacturers Association (CSMA) adopted the following definition for aerosol products (5): "a self-contained sprayable product in which the propellant force is supplied by a liquefied gas." This definition includes space, residual, surface coating, foam and various other types of products but does not include gas-pressurized products such as whipping cream.

At the present time, the word aerosol is understood by the general public to mean any product that is dispensed with the aid of a propellant.

1.3 Aerosol Operation and Theory

The aerosol package consists of a sealable container equipped with a valve and containing a liquefied gas propellant, solvents, active ingredients and other miscellaneous materials packaged under pressure.

The aerosol container is never completely filled so that the propellant can exist in a gaseous form in the head space, thereby maintaining a constant pressure in the container.

When the valve is opened, the internal pressure forces the product up the dip tube and out the valve. As soon as the product leaves the container, some of the propellant molecules vapourize and fly off into space. In doing so, they collide with solvent molecules and cause collisions between the solvent molecules themselves thereby disrupting the attractive forces between them. As the propellant vapourizes in air, the solvent molecules are left almost isolated in relatively small groups of particles. The liquefied gas propellant, therefore, not only provides the pressure to force the product out of the container when the valve is opened, but also breaks up the concentrate into small droplets or particles once the aerosol has reached the atmosphere.

2 TYPES OF AEROSOL FORMULATIONS

Aerosol formulations can be classified as homogeneous or heterogeneous systems, based on the solubility of the components in the system.

2.1 Homogeneous Systems

All components of homogeneous systems are mutually soluble or miscible, forming a two-phase system consisting of a single liquid phase and a vapour phase. These systems do not require shaking before use and are used for some insecticide as well as insect repellent products.

The liquid phase of these solution-type systems can be liquid solvents or a combination of a liquid solvent and a liquid propellant, with all components soluble in the liquid phase. The liquid phase can consist of only liquid solvents (mineral spirits, methylene chloride, alcohol), alcohol-water solutions or water-dimethyl ether propellant solutions. The vapour-propellant phase can consist of various propellants, including hydrocarbons, fluorocarbons or dimethyl ether.

2.2 Heterogeneous Systems

Heterogeneous systems are not mutually soluble and result in a three-phase system consisting of two liquid phases or a liquid and a solid phase in combination with a vapour phase. These systems require shaking before use and are used for some insecticide products as well as numerous personal care products.

Heterogeneous systems can be further classified as emulsion or suspension systems. Emulsion systems are aqueous systems using emulsifiers to incorporate the non-soluble components. Suspension systems contain material that is insoluble in the mixture of the propellant and solvent or insoluble in an emulsion system. These systems usually contain solid material (talcs, powders, clays) suspended in the liquid phase. Since the solid material settles, shaking is necessary before use.

Insecticide aerosol formulations can also be classified as solvent-based systems, oil-out emulsion systems or water-out emulsion systems.

2.3 Solvent-Based Systems

Solvent-based systems are non-aqueous systems consisting of an organic solvent as a carrier and any one of several propellants to expel the product from the container. These are usually two-phase homogeneous systems but can also be heterogeneous systems if powder is added to serve as an insecticide carrier. The powder will settle upon standing and the container will need shaking before using the product. Non-powder solvent systems do not require shaking. Mineral spirits, or non-deodorized kerosene blends, are commonly used solvents due to availability and economics. Most insecticides are readily soluble in kerosene, but a co-solvent (methylene chloride, trichloroethane, isopropanol) may be needed for some compounds. Other components, such as fragrances, synergists and inhibitors, can also be added to solvent systems. Most solvents are quite compatible with most of these materials and the propellants now used in insect control products.

Solvent-based insect control aerosol products are marketed throughout the world as surface, space and multi-purpose sprays.

2.4 Oil-out Emulsion Systems

Oil-out emulsions contain a fairly large amount of propellant miscible with the oil of the formulation, which forms the external phase of the emulsion. This external phase encloses the internal aqueous phase. These systems are quite stable, but may also separate over time and require shaking before use. Many of the space spray insecticide products are formulated as oil-out emulsions. Performance comparable to that of all solvent-based systems against flies and mosquitoes can be obtained with properly formulated oil-out emulsions. Against larger insects such as cockroaches, oil-out emulsions are also effective. However, there is a longer time period between direct spray treatment and mortality than with solvent-based systems. Formulated as a residual preventative treatment, aqueous-based systems can be just as effective as solvent-based systems, and they offer the advantage of using less organic solvent.

2.5 Water-Out Emulsions

Water-out emulsion systems, or foam emulsions, have a reduced level of propellant. The propellant and solvent of the system creates the internal phase of the emulsion, while the aqueous intermediate creates the external phase. The propellant travels through the emulsion when exposed to the atmosphere and creates a foam as it expands the external phase. Various formulation techniques can be used to create stable or quick-breaking foams for various applications in insect control. However, since the aqueous external phase provides performance characteristics that are insufficient for direct spray application against insects, the systems have been little used worldwide. Some water-out emulsion products are available as crack and crevice foam treatments and provide excellent performance in this application.

3 COMPONENTS OF AEROSOL SYSTEMS

The complete aerosol system consists not only of the chemical formulation, which includes the intermediate and the propellant, but also the valve and actuator, which influences the performance obtained from the formulation. These three important components of all aerosol insecticide systems will be discussed here.

3.1 Aerosol Intermediates

The intermediate of an aerosol system consists of all the chemical components in the formula with the exception of the propellant. The intermediate can contain various chemicals necessary or desired for specific functions in the formulation. Depending on the type of insecticide formulation, the intermediate may include:

(a) Active ingredients;
(b) Solvents;
(c) Water;
(d) Emulsifiers;
(e) Inhibitors;
(f) Fragrances.

3.1.1 *Active ingredients*

For our discussion, active ingredients will include conventional insecticides as well as synergists, insect growth regulators, repellents and attractants, as all these chemicals must be registered with EPA.

Table 12.1 provides a classification of active ingredients by function as well as by aerosol product application. It provides a good guideline for selecting active ingredients for a specific purpose. For each product type, active ingredients were ranked to indicate preference based on present usage as well as consumer acceptance and worldwide availability.

3.1.2 *Solvents*

Hundreds of solvents are available but only a few are commonly used in aerosol insecticide formulations. The choice of a solvent is to a degree based on what it must dissolve, using "like dissolves like" as a guideline. For instance, if an organic chemical must be dissolved, an aromatic solvent would be the best choice. However, this will depend on the total properties necessary for developing an acceptable product.

3.1.2(a) *Solvent properties:* In general, the following solvent properties should be reviewed when developing an aerosol product: solvent power, flash point, toxicity, speed of evaporation, cost, aniline point, distillation range, specific gravity, weight per gallon and solubility parameter.

Solvent power may be defined as a liquid's ability to dissolve a given substance. It is usually expressed as so many grams of a particular substance in 100 cc of solvent. For instance, 100 cc water dissolves 36 g salt, while 100 cc alcohol dissolves 3 g. This solvent

Table 12.1 Classification of active ingredients for use in aerosol formulations

Product type	*Function of active ingredients*					
	Knockdown agent	*Synergist*	*Kill agent*	*Residual kill agent*	*Insect growth regulator*	*Repellent*
Space spray aerosol	Pyrethrin Tetramethrin Pynamin forte Bioallethrin Neopynamin forte	PBO	Sumithrin Resmethrin Permethrin DDVP			
Surface aerosol	Pyrethrin Tetramethrin Neopynamin forte DDVP	PBO MGK 264		Permethrin Cypermethrin Deltamethrin Cyfluthrin Propoxur Dursban Malathion Sumithion	Hydroprene Fenoxycarb Dimilin	MGK 874
Multi-purpose aerosol	Pyrethrin Tetramethrin Neopynamin forte Pynamin forte Bioallethrin Esbiothrin Prallethrin		DDVP	Propoxur Cyphenothrin Permethrin		
Aerosol fogger	Pyrethrin Tetramethrin		DDVP	Permethrin Cyphenothrin	Hydroprene Fenoxycarb	
Aerosol repellent					Deet	
Aerosol mothproofer			Permethrin Vaporthrin			

power is usually expressed as the kauri-butanol value (k.b. value) for the solvent. This value is the millilitres of that solvent that will produce turbidity when added to 20 g of a solution of kauri gum resin in butyl alcohol at 25°C. For toluene, 105 ml are required, showing toluene to have excellent solvent power, which is used as the standard. The k.b. value is only used in describing hydrocarbon solvents since alcohols are better solvents than toluene and a measure of relative solvent power would be less meaningful when using k.b. values.

The flash point of a liquid is the temperature at which the saturated vapour will ignite in the presence of a spark or flame, or the temperature that must be reached before the material will give off sufficient vapour to form an explosive mixture with air when exposed to a test flame. The standard method of measurement is called the Tag Closed Cup as defined by ASTM (6). Flash points determined in a closed cup are always lower than determined in an open cup.

Toxicity of the solvent is important since it makes a significant contribution to the toxicity of the finished product. Methylene chloride is a commonly used solvent in aerosol systems but is now under question and should not be considered for aerosol formulas. The threshold limit value (TLV) is the maximum airborne concentration of the solvent that a person could be repeatedly exposed to day after day without adverse effects. The flammability rating is a relative number, i.e. the lower the number, the less flammable the solvent. Carbon tetrachloride, which is non-flammable, is rated 0 and benzene rated 100.

Speed of Evaporation is expressed in relation to the evaporation rate of toluene, which is standardized at 1. All products with evaporation rates greater than 1 evaporate faster than toluene and conversely, numbers lower than 1 indicate a slower evaporation rate.

Cost is always a consideration in attempting to develop a cost-effective formulation.

Aniline point is the temperature at which a volume of hydrocarbon will dissolve an equal volume of aniline oil. 5 ml of the solvent and aniline oil are heated in a test tube until the mixture becomes homogeneous.

Distillation range is a series of temperatures recorded while boiling a solvent sample. The initial boiling point (IBP) occurs when the boiling starts and the dry end point occurs when the sample has just boiled away. The IBP gives an indication of flash point, i.e. a solvent with a higher IBP will have a higher flash point and the dry end point will show which solvent will be completely evaporated first. The distillation range also indicates chemical impurities: the shorter the range, the fewer constituent products.

Specific gravity is the ratio of the weight of a given volume of liquid to the weight of an equal volume of distilled water at a specific temperature. This is important when comparing costs of different liquids sold by weight and if the final product is sold by volume.

The weight per gallon of any liquid is determined by multiplying the weight of a gallon of distilled water (3.78 kg) by the specific gravity of the liquid.

The solubility parameter is a physical constant for a compound and provides a means of predicting (a) whether mixtures of liquids will be miscible, (b) whether an amorphous solid will dissolve in a solvent or a mixture of solvents and (c) whether a valve gasket will be affected by an aerosol propellant, solvent or co-solvent of the formulation. If two compounds have similar solubility parameters, the chances are that they will be miscible.

3.1.2(b) *Solvent Types:* Insecticide aerosol solvents fall into three chemically distinct groups:

(a) Hydrocarbons;
(b) Chlorinated solvents;
(c) Alcohols.

Hydrocarbon solvents consist of only carbon and hydrogen and include petroleum naphthas, benzene, toluene, xylene, aromatic petroleum solvents and turpentine. The petroleum naphthas, aromatic petroleum solvents and turpentine are mixtures of several chemical hydrocarbons (7). Aliphatic or straight-chain hydrocarbons, also commonly known as paraffins, make up over 85% of the petroleum naphthas. Aromatic hydrocarbons have stronger solvent power than the aliphatics and include benzene, toluene and xylene. Cyclic hydrocarbon solvents are somewhat of a compromise between the aliphatics and aromatics and include the naphthenes, which are formed to a small extent in petroleum naphthas and to a larger extent in aromatic petroleum naphthas. Terpenes are also included in this group and include pinene, the main constituent of turpentine.

Chlorinated solvents consist of carbon, hydrogen and chlorine or carbon and chlorine. Most chlorinated solvents are quite pure and therefore have a narrow boiling range. They also have excellent solvent properties and are commonly used for dissolving oils and fats. Methylene chloride is a fast-evaporating chlorinated solvent with excellent lifting power suitable for paint stripping applications. It is also used in insecticide products but is now banned in many countries. 1,1,1-trichloroethane is also widely used in aerosol formulations, including insecticides, and is sold under various trade names. Chlorothene NU contains a corrosion inhibitor. Chlorothene VG is 1,1,1-trichloroethane proprietary grade to replace trichloroethylene where restrictions on solvent vapour emissions have been legislated.

Alcohols vary greatly in their solvent properties when compared to hydrocarbons. The low-boiling alcohols, methyl, ethyl and isopropyl, are all miscible with water, while the hydrocarbons are not. Alcohols dissolve various compounds that cannot be dissolved by hydrocarbons, including shellac, vanilla gum, natural kauri resin and synthetic resins. The alcohols also differ from hydrocarbons in having oxygen in addition to carbon and hydrogen. Primary and secondary alcohols of the same name have the same formula but a different isomeric structure and, therefore, different chemical properties. Ethanol, or ethyl alcohol, can be produced by fermentation, but most industrial grades are produced by less expensive synthetic routes using petroleum gases. Isopropyl alcohol is produced from natural gas and is widely used because it is non-poisonous. It can be marketed as anhydrous or as an over 90% grade, the remaining portion being water.

3.1.3 Water

Water has many properties that make it desirable for use in aerosol systems. It is an excellent solvent, non-flammable, readily available, inexpensive and low in toxicity. Its major disadvantage is its immiscibility with liquefied gas propellants. This immiscibility can be overcome using coupling solvents such as ethyl alcohol or using a dimethyl ether propellant system. The most common method is to use surface-active agents to produce stable emulsion systems that can be dispensed as a foam or a spray.

3.1.4 Emulsifiers

Surface-active agents are commonly used in water-based aerosol systems to produce stable water-out (foam product) or oil-out (spray product) emulsions, as described previously.

3.1.5 Inhibitors

Inhibitors can be used in aerosol products as stabilizers for the active ingredients or to prevent corrosion of metal containers used for aqueous-based formulations.

3.1.6 Fragrances

Fragrances are commonly chosen to provide a preferred odour for the product or to mask the odour of the solvent used in the formulation.

3.2 Aerosol Propellants

An aerosol propellant can be defined as a material that is capable of expelling the contents of an aerosol container. They are usually classified as follows (8):

(a) Fluorocarbons, either true fluorocarbons containing carbon, hydrogen and fluorine (FC-152a) or chlorofluorocarbons containing carbon, hydrogen, fluorine and chlorine (FC-11, FC-12, FC-114, FC-22, FC-142b);

(b) Hydrocarbons containing carbon and hydrogen;

(c) Dimethyl ether containing carbon, hydrogen and oxygen;

(d) Compressed gases consisting of carbon dioxide, nitrogen and nitrous oxide.

3.2.1 General properties

The general properties of the more commonly used propellants in insect control products will be emphasized in the discussion.

3.2.1 (a) *Fluorocarbons* The term fluorocarbon is used for all organo-fluorine compounds, while the non-flammable fluorocarbons bear the Freon trade mark of DuPont, including Freon 11, 12, 22 and 114. Fluorocarbons 11 and 12 were once the most widely used materials in aerosol insect control products due to their non-flammable nature. Their present use in insect control is diminishing, however, due to their contribution to the depletion of the protective ozone layer in the upper atmosphere. This has led to the widespread use of hydrocarbon propellants.

3.2.1 (b) *Hydrocarbons:* The major hydrocarbon propellants are propane and isobutane, with n-butane being used to some extent. The hydrocarbons are stable, noncorrosive and essentially odourless liquefied gases with low toxicity and good environmental acceptance. They are also considerably cheaper than fluorocarbons. Their major disadvantage is their flammability.

Aerosol-grade hydrocarbons are designated by the letter "A" followed by a number that indicates the propellant vapour pressure in psi at 70°F. Cosmetic-grade hydrocarbons use the letter "C" in the same manner.

Specific properties of the most commonly used hydrocarbons are described below:

(a) Propane is a high-pressure propellant that can be used alone or in blends with other hydrocarbons or fluorocarbons. It is non-polar and has poor overall solubility and is essentially insoluble in water. Water can be used in a propane system if alcohols are used as a co-solvent to form a homogeneous solution system;

(b) Isobutane has many properties similar to those of propane as regards solubility, but it has a lower vapour pressure and can easily be used in many applications in blended form. It is slightly heavier than propane and also slightly more flammable. It is commonly blended with propane when higher pressure propellants are needed;

(c) N-butane is rarely used alone due to its low vapour pressure and is usually blended with propane or isobutane;

(d) Dimethyl ether (DME) is not widely used in the United States but is commonly used in Japan and Europe. The propellant has a slight ethereal odour that is sometimes considered disagreeable. It is unique among propellants in having a high solubility in water and also functions as an excellent solvent for many substances. DME is highly flammable but produces less flame extension than to hydrocarbons. With DME, higher air concentrations are tolerated, compared to hydrocarbons, before a mixture becomes flammable;

3.2.1 (c) *Compressed gases* A compressed gas is generally defined as a high-pressure propellant that must be injected into containers in the gaseous form instead of as a liquid under pressure. These gases comprise of carbon dioxide, nitrous oxide and nitrogen. The solubility of the compressed gas in the liquid concentrate is one of its most important properties since a higher solubility will result in (a) a better initial spray and (b) during use, fewer changes in pressure and spray characteristics. Nitrogen has extremely low solubility and for that reason is not used to any significant degree as a propellant. Carbon dioxide, on the other hand, is considerably more soluble and is more widely used in insecticide aerosol products.

Compressed gases offer many advantages since they are basically odourless, colourless, non-toxic, non-flammable, inexpensive and environmentally acceptable. Temperature variations do not result in pressure changes to any significant degree when compared to liquefied gas propellants.

Their disadvantages include an inferior spray (wetter and less forceful). In addition, the spray becomes coarser during product use as the pressure is decreased. There is also more potential that the product will no longer discharge if it is misused. The presence of moisture can result in the formation of carbonic acid, which may lead to corrosion. Finally, filling equipment is usually slower with carbon dioxide than with other propellants.

3.3 Valves and Actuators

As the valve and actuator of an aerosol system play a significant role in the performance and quality of the final product they are included in our discussion. The variety of valves available from numerous manufacturers worldwide allows for the flexibility to use an appropriate valve for any type of aerosol product in steel, aluminium, glass or plastic containers. The purpose of the valve is to control the flow of product from the container and allow dispensing the contents in desired amounts as needed.

The valve also greatly influences the characteristics of the product dispensed, which may be a foam, a coarse spray or a fine spray. All valves work in a similar fashion and consist of basically the same components.

3.3.1 Mounting Cups and Corresponding Gaskets

Mounting cups and corresponding gaskets are used to clamp the valve stem, stem gasket, spring and body together and to provide a seal to the opening of the container.

3.3.2 Valve Stem Gaskets

Valve stem gaskets are crucial for the proper operation of the aerosol product. The stem gasket provides the seal to prevent leakage of product and unacceptable weight loss. A proper selection of gasket material is also necessary (a) to deter interaction with the product and (b) to provide proper gasket swell to close the opening in the stem when the valve is not in operation. Optimum swell is between 2 and 9% depending on the product and valve system used. Gaskets can be checked for compatibility with the product by immersing them gasket in the intermediate for 48 hours. Measurements of diameter, hardening or softening of gasket material and deterioration can be compared before and after immersion in the product.

3.3.3 Valve stems

Valve stems contain the metering orifices that determine the spray rate. The stem also contains an expansion chamber to provide a smaller particle size in the spray.

3.3.4 Springs

Springs control the upward and downward movement of the valve stem and are usually made from stainless steel.

3.3.5 The Valve body (housing)

The valve body provides a base for the spring and also serves to make a connection with the dip tube. When the body orifice is less than the stem orifice, the valve body orifice becomes the metering orifice. The valve body can also contain a vapour tap orifice, which allows additional vapour phase propellant to mix with the product and provide a finer particle size. Vapour taps are commonly used for space spray products.

3.3.6 Dip tubes

Dip tubes function (a) to transfer the product to the valve body, (b) to serve as a metering device in some systems and (c) to be used as a product reservoir for the inverted spraying of products. Inverted spraying is common with insect repellent sprays.

3.3.7 Valve actuators

Valve actuators may be simple spray buttons or an integral part of various spouts or spray domes. The actuator not only allows operation of the valve but can also be important in determining whether the product will be dispensed as a foam, spray or stream pattern. It will also influence the spray rate and the particle size of the spray.

Actuators come in tremendous variety of sizes and shapes. The actuator orifice is usually chosen larger than the metering orifice to obtain a more uniform particle size in the spray. The terminal orifice may be straight or tapered to produce moderate width sprays or wider sprays, respectively.

Actuators can also contain a swirl chamber and are referred to as mechanical break-up actuators. They function by imparting a swirling motion to the discharged product, resulting in an improved spray pattern.

4 PROPERTIES OF AEROSOL FORMULATIONS

In addition to understanding the ingredients and proper valve selection to provide desired attributes, formulated products should also be evaluated for specific characteristics that influence product performance.

4.1 Spray Properties

One of the most important characteristics of an aerosol product is a satisfactory spray. This is especially true of space spray insecticides, where a correlation exists between particle size and biological activity.

The spray is formed when an aerosol is discharged and the propellant changes from a liquid to a gas at atmospheric pressure. Some flash evaporation occurs in the expansion chamber of the valve, which is at lower pressure than the container. Additional flash evaporation occurs when the vapour-liquid phase exits the outer orifice. After these initial flash evaporations, additional evaporation takes place depending on various factors (9).

The flash evaporizations are primarily responsible for the initial breakup of the liquid product and the resulting spray characteristics, which will also depend on the type of propellant, valve, actuator or intermediate.

There are many ways to evaluate the spray pattern, which can then be classified as follows:

(a) *Fine spray*. Spray disappears at a distance of approximately three to four feet after it leaves the actuator and gives only slight wetting of paper sprayed from a distance of 18 in.

(b) *Medium*. The spray pathway is more horizontal and wetting of paper occurs at a distance of 18 in. The particle size is noticeably larger than a fine spray.

(c) *Coarse*. Particle fallout is very evident and significant wetting of paper occurs at a distance of 18 in.

The fine spray would be appropriate for a space spray application, the coarse spray for a surface application and the medium spray for a multi-purpose application.

The two most important propellant variables that affect the spray are vapour pressure and the concentration of the propellant in the product. An increase in propellant concentration decreases particle size. Also, a high-vapour-pressure propellant provides a finer spray than a low-vapour-pressure propellant at the same concentration. This is a result of the increased flash vapourization of a propellant with a high vapour pressure. More important is the vapour pressure of the total formulation. The use of a high-vapour-pressure propellant may produce a coarser spray than a low-vapour-pressure propellant if a pressure-depressing solvent is included in the formulation.

Valving parameters also have a significant effect on spray characteristics. This is especially important when using a vapour tap valve in combination with a propellant that has a low vapour pressure. Using a vapour tap, the vapour pressure and discharge rate are decreased, resulting in a change in spray during the discharge of the container. These results emphasize the importance of testing spray characteristics initially and after discharging 70-90% of the container.

The actuator, as well as the valving, can have a big influence on spray characteristics. This is especially so when comparing mechanical and non-mechanical breakup actuators.

Depending on the construction of the actuator orifice this same shift in spray pattern can occur with other types of actuators. A study was conducted comparing a twenty-thousandths mechanical breakup button actuator to five alternate actuators. These six actuators were tested with an aqueous oil-out emulsion system with three valve variables. The valving variables included a standard, the standard with a thirteen-thousandths vapour tap and a high discharge valve. Overall, the use of three valving variables with any one actuator hardly changed the spray pattern, but a significant difference in spray pattern could be distinguished when the various actuators with the same valving were compared.

4.2 Particle Size Characteristics

Particle size can be an important characteristic for an aerosol product, especially in insecticidal aerosols. As described earlier, an aerosol was technically defined as having no particles greater than 50 μm and 80%, by weight, smaller than 30 μm. However, many products may not fit this particle size definition but still be classified as aerosols for the purpose of our discussion (10). In general, particles under 50 μm should be effective in a space spray, with 20-30 μm diameter particles being optimum.

The relationship between particle size and biological activity was studied with various propellant levels, as shown in Table 12.2. As the propellant level is increased the particle size is decreased and the number of particles per gram of product is increased. However, biological performance reaches an optimum at approximately 60% propellant, which provides particles having a mean diameter of 20 μm.

Table 12.2 Effect of propellant levels on particle size and biological activity

	Sample No.				
Product composition(%)	1	2	3	4	5
Synergized pyrethrin	1.8	1.8	1.8	1.8	1.8
Mineral spirits	78.2	58.2	38.2	18.2	8.2
Propellant (46 PSI)	20.0	40.0	60.0	80.0	90.0
Total	100.0	100.0	100.0	100.0	100.0
Mean Particle Size (μm)	65.0	38.5	21.5	12.0	9.0
Fly knockdown KD 75 (min)	40.8	23.3	14.1	12.0	14.8
Fly mortality (%)	4.2	32.2	67.1	63.9	43.7

The effect of valving on particle size was also determined and it was again confirmed that the most important contributing factor is the actuator. When the actuator was kept the same but the valve stem orifice was increased, the particle size remained the same. This was most evident with a large orifice actuator. These results correlate with those of our previous discussion, indicating that the stem orifice is the influencing orifice unless the actuator orifice is larger.

To compare particle size and spray characteristics, a standard set of samples can be prepared for comparison. The test sample spray can then be compared to the standard samples. This will provide a good basis for making a judgement on a valve change that may or may not affect performance.

4.3 Discharge Rate

The discharge rate or delivery rate of an aerosol product is the quantity of material '*q*' expelled through the valve in a given period of time at a specific temperature. The typical aerosol valve is made up of a series of channels, orifices and expansion chambers, all of which may influence delivery rate. The prime metering points available for manipulation of the discharge rate, in order of flow, are as follows:

(a) *Dip tube*. The use of a capillary dip tube can be important in metering and influencing the discharge rate;

(b) *Valve body orifice*. This can become the metering orifice when the diameter of the orifice is less than the diameter of the valve stem orifice;

(c) *Vapour tap orifice*. The use of a vapour tap in the valve body will usually produce a softer, warmer and drier spray with a fine particle size and a reduced delivery rate;

(d) *Stem orifice*. This orifice usually determines the spray rate and is thus called the metering orifice;

(e) *Actuator orifice*. Actuator orifices have a great effect on many delivery characteristics of the spray. An orifice of somewhat larger diameter than the metering orifice is often chosen for a more uniform spray.

The effect of valving orifices on the discharge rate is presented in Table 12.3. Samples 1-3 illustrate that a decreasing body orifice will decrease the delivery rate when other parameters are constant. A larger actuator (samples 4-6) provides a higher discharge rate, but the same trend in decreased discharge rate is evident. Addition of a vapour tap decreases the discharge rate, as shown for samples 7-9. As illustrated, the vapour tap can significant by decrease the discharge rate of the product. Increasing the stem orifice increases the discharge rate, as shown for samples 10-13. The discharge rate will also decrease as the contents of the containers are dispensed. This decrease can be significant when using compressed gas aerosol systems.

Varying the discharge rate can also affect the biological activity of insecticide aerosols. Three aerosols that differ significantly in discharge rate but that have similar mean particle size ranges were compared. The data show a trend to improved activity using a higher discharge product that required less time to total fly knockdown. These differences are not great but do indicate that some improvement can be made in biological activity without any major changes in the system.

Table 12.3 The effect of valving orifices on discharge rates

Sample number	*Actuator*	*Stem*	*Body*	*Vapour tap*	*Discharge rate*
			(in.)		*(q/sec.)*
1	0.018	0.018	0.080	none	0.88
2	0.018	0.018	0.040	none	0.80
3	0.018	0.018	0.020	none	0.71
4	0.040	0.018	0.080	none	1.34
5	0.040	0.018	0.040	none	1.23
6	0.040	0.080	0.020	none	0.92
7	0.040	0.018	0.080	0.013	1.15
8	0.040	0.018	0.080	0.025	1.07
9	0.040	0.018	0.080	0.040	0.33
10	0.018	0.024	0.080	none	1.17
11	0.018	0.018	0.080	none	1.28
12	0.018	0.040	0.080	none	1.77
13	0.018	0.025	0.080	none	1.85

4.4 Vapour Pressure

The vapour pressure of a liquid is defined as the pressure of the vapour in equilibrium with the liquid and is constant at a given temperature. Vapour pressure is an important characteristic of aerosol products, not only as it relates to product performance but also due to pressure limitations that regulations place on aerosol containers. The vapour pressure of any aerosol formulation depends upon the vapour pressure of the propellant and the effect the other components in the formulation have upon the vapour pressure of the propellant.

Pressures in aerosol containers are commonly expressed as pounds per square inch gauge (psi). Although air pressure is 14.7 psi, the gauges are set to read zero, so 14.7 would need to be added to the gauge pressure to obtain the absolute pressure.

These measurements of vapour pressure consider only the vapour pressure of the solution of propellant and concentrate, but the concentration of air in the container also has an effect on pressure. The amount of air in the container can be determined by gas chromatography to determine its contribution to total pressure. The amount of air can be controlled to some degree depending on the method of aerosol filling.

Using Raoult's law and Dalton's law, the vapour pressure of any solution can be determined if the composition of the solution and the molecular weights and vapour pressure of the pure components are known. This involves determining the partial pressure of each component by multiplying the mole fraction of each component by its vapour pressure. The total pressure of the solution is obtained by adding together the partial pressures of the components.

Conversely, if the vapour pressure of a blend of isobutane and propane is specified, we can calculate the weight % of each propellant in the blend.

These calculations will allow determining the vapour pressure of any known propellant blend or the propellant blend ratio if the vapour pressure of the blend is known.

4.5 Density and Fill Weight

There is often confusion between density and specific gravity, so we must differentiate between the two. Density has the dimensions of mass over volume, usually expressed as g/cc, and is important in determining the weight of an aerosol product that can be packaged in an aerosol container. Specific gravity is a pure number without dimension: it is the ratio of the weight of a given volume of substance at one temperature to the weight of the same volume of a reference substance, usually water, at the same or a different temperature.

The density of propellant blends can be calculated using the known densities of each propellant in the blend. The density of a formulated product can be calculated in a similar manner. However, in many cases, the concentrate consists of many ingredients, and it is difficult to calculate the density of the concentrate itself. In these cases, a weight determination of the concentrate in a certain volume will provide an acceptable experimentally determined density.

In many instances, ingredients contained in the concentrate at low levels can be combined with an ingredient of known density. Knowing the density of specific formulas, we can now calculate the fill weight for any size container.

Using a 202 x 509 container (290 ml brimful capacity) we can calculate the amount of product for fill weight. The brimful capacity is reduced by 5 ml for the valve, resulting in a 285 ml capacity container. Using 85% fill volume, the fill capacity can be calculated (285 ml x 85%) as (242.25 ml). The density of a theoretical aqueous formula is 0.737 g/cc, so the fill weight would be 242.25 cc x 0.737 g/cc = 178.54 g. Taking the density of a theoretical solvent formula as 0.626 g/cc, the fill weight would be 242.25 cc x 0.626 g/cc = 151.65 g.

It is clear that, more aqueous product is needed to fill the same volume of container. This need for more product may sometimes offset the other advantages of an aqueous-based system. In the theoretical case presented above, the aqueous system required 17% more product to provide a filled unit than the solvent system. If the cost of raw materials for an aqueous formula showed a saving of 15%, the end result would be a slight cost increase per unit fill for the aqueous formula. The above example shows the importance of calculating not only the cost of raw materials used in the formula but also the cost based on fill.

5. APPLICATION OF AEROSOL FORMULATIONS

Aerosol insect control products can be classified in three general categories by type of application: surface sprays, space sprays and multi-purpose sprays. Basic solvent-based formulations for these applications are provided in Table 12.4. The insecticide concentration can vary considerably depending on the active used and the level of performance desired while the solvent and propellant level can be maintained at levels similar to what has been suggested. The valve chosen can also vary considerably but the recommendation provided can serve as an excellent base from which to experiment. Details of these formulations will be addressed individually.

Table 12.4 Basic insect control aerosol formulations

Composition (%)	*Surface spray*	*Space spray*	*Multi-purpose spray*
Insecticides	0.5	0.5	0.5
Solvents	69.5	29.5	49.5
Propellant (46 psi)	30.0	70.0	50.0
	100.0	100.0	100.0
Dimensions of valving orifice (in.)			
Stem	0.024	0.024	0.024
Body	0.080	0.080	0.080
Vapour tap	—	0.018	0.018
Actuator	0.02	0.024	0.024
Spray pattern	Coarse	Fine	Medium

5.1 Surface Sprays

Surface sprays, also known as, crawling insect killers (CIK), are designed to be sprayed onto a surface, providing a residual deposit, or to be sprayed directly on an insect for fast activity. Insecticide actives utilized usually include a knockdown agent in combination with a residual kill agent (Table 12.1). Common knockdown agents for a CIK are pyrethrin, neopynamin or DDVP in combination with one of several residual kill agents, including propoxur, permethrin, cypermethrin or cyfluthrin. Commonly available mineral spirits or kerosene can be used as the solvent carrier for most of the actives.

A co-solvent is necessary for propoxur-based systems. Methylene chloride and trichloroethane have been used, but anhydrous isopropanol is the preferred substitute due to toxicity and environmental concerns with the chlorinated solvents.

The use of a hydrocarbon propellant at 30% will produce a coarse, wet spray suitable for this type of application. Valving parameters are also provided that will deliver a discharge of over 1 g product/sec. Many valve selections could be made for a CIK product since the particle size is less important and valves with delivery rates of 1-3 g producing a coarse wet spray are quite common. Similar levels of fluorocarbon propellants could also be used with similar results but the hydrocarbons are preferred due to the ozone depletion hazard associated with the fluorocarbons. Compressed gas propellants are also compatible with the solvent CIK formulations. Carbon dioxide at a 2-3% level would produce a coarser, wetter spray but one that is still acceptable for controlling crawling insects.

5.2 Space Sprays

A space spray can also be used for direct spraying of insects, although it actually is intended for delivering a spray into the air and thereby controlling insects moving through the space. Therefore, a space spray (or flying insect killer) is designed to deliver particles in a range that will remain airborne for a sufficient period of time to allow flying insects to be exposed to a lethal dose. Target insects usually include various flies and mosquitoes as well as moths. The insecticides used for a space spray should include knockdown and kill agents as described in Table 12.1. Pyrethrin or tetramethrin are used worldwide in combination with resmethrin, sumithrin or a synergist to provide adequate killing power. As with surface sprays, the availability of mineral spirits or kerosene make them the solvent of choice, although a specific solvent can have a significant influence on the performance of the product (11). Hydrocarbon propellants are the most suitable for space sprays at this time. Although they may pose a flammability hazard, they are not responsible for ozone depletion like the fluorocarbons and they can provide the continuous, proper pressure necessary in a space spray but not yet possible with compressed gas technology. The level of propellant and the valve selection are extremely important to achieve a space spray with particles under 50 μm and preferably in the 20-30 μm range. A 60-70% level of propellant with a pressure of 46 psi in combination with a suitable valve will provide this preferred particle size. The valving provided in the basic system (Table 12.4) is appropriate for providing a discharge of over 1 g/sec and a particle size in the range of 20 μm. Depending on the propellant pressure and propellant concentration, other valving options may be required to provide similar properties. Overall, the selection of valves is more limited than for surface sprays, but many appropriate valving systems can be developed with some basic understanding of what is required to produce a good space spray.

5.3 Multi-purpose Spray

Multi-purpose spray is designed to be effective as a space spray for flying insects as well as a surface spray for crawling insects. It requires that the formula be modified to incorporate properties of surface and space sprays and limits the options that are available for these specific formulations. As illustrated in Table 12.4, multi-purpose spray is a hybrid of surface and space sprays. The propellant level is greater than for a surface spray to provide a finer spray to control flying insects but it must also serve similar to that for a surface spray to provide a residual deposit. Valving, a space spray can be used to achieve a break-up of particles. Insecticides similar to those used for surface and space sprays can be used, but the levels of residual agent should be reduced or replaced by kill agents, as described in Table 12.1. This will provide some residual activity when the product is used as a surface spray but minimize residual activity when the product is as a space spray.

There are many parameters to consider in developing an aerosol formulation that will deliver the characteristics desired. The above information can serve as an appropriate starting point in optimizing aerosol formulations using local materials and obtaining the desired results.

REFERENCES

(1) D. Minogue, *Aerosol Age*, May 1989, pp. 18-19.
(2) H. A. Chaluleu, *Aerosol Age*, May 1988, pp. 24-27.
(3) The Roper Organization, *The Environment: Public Attitudes & Individual Behavior*, Report commissioned by S. C. Johnson & Son, 1990.
(4) P. A. Sanders, *Handbook of Aerosol Technology*, 2nd ed. (Litton Educational Publishing, 1979).
(5) CSMA, *Glossary of Terms Used in the Aerosol Industry* (Washington, D.C., Chemical Specialties Manufacturing Association, 1955).
(6) CSMA, *Aerosol Guide*, 7th ed., (Washington, D. C., Chemical Specialties Manufacturing Association, 1981).
(7) Chemcentral, *Organic Solvents* (Chicago, Illinois, Chemcentral Corporation, 1982).
(8) M. A. Johnson, *The Aerosol Handbook*, 2nd ed., (Mendham, N.J., Wayne Dorlund, 1982).
(9) J. J. Sciarra and L. Stoller, *The Science and Technology of Aerosol Packaging* (New York, Wiley, 1974).
(10) G. M. Hidy, *Aerosols an Industrial and Environmental Science*, (Orlando, Florida, Academa Press, 1984).
(11) Sumitomo, "Formulation factors influencing the efficacy of aerosol insecticides," *Pyrethroid World*, Spring 1986.

MANUFACTURING PLANT PERSPECTIVES

CHAPTER

13

Establishment of Pesticide Formulation Plants in Developing Countries

B. Sugavanam

Chemical Industries Branch
United Nations Industrial Development Organization
P.O.Box 300, A-1400 Vienna, Austria

and

R. Teuber Weckersdorf

Kwizda G.m.b.h., Agrarchemie
kwizda-Allee,
A-2100 Leobendorf bei Korneuburg, Austria.

Pesticide formulation has been practised in developing countries for many decades either at the farm level, prior to application, or in a proper formulation factory. An interesting example comes from Malaysia, where in the 1940s farmers used to grow derris roots along vegetable fields, soaked the roots in water and applied the emulsion in the field. Most of the formulations made in the developing countries are the basic types such as emulsifiable concentrates, water-soluble concentrates and dusts. The new era in pesticide formulation started in the 1970s and 1980s in developed countries, when the technology became more and more sophisticated, especially with the introduction of wettable powders (WP), suspension concentrates (SC), capsule suspensions (CS), water-dispersible granules (WG), water-soluble bags, seed dressings (SD), wax baits, pheromone traps, and mixture formulations. In China it is estimated that, on average, one pesticide is formulated in three different types while in the developed countries the average is around six different types.

Developed and developing countries alike are under pressure from their governments and environmentalists to make pesticide formulation safer, more effective, target-specific and friendly to the applicator and the environment. All these factors make the formulations more expensive and either the users will have to pay the price or the formulations will have to be subsidized. While the use of pesticides will increase in the developing countries, the modern technology used in formulation will be elusive for most a majority of these countries.

1 CLASSIFICATION OF DEVELOPING COUNTRIES

During the last 20 years, developing countries have realized the importance of pesticide inputs into agricultural production. More and more countries are looking into local pesticide formulation, importing only active ingredients and other adjuvants that are used in the formulation. In a broad sense, developing countries may be classified into four categories with respect to pesticide use:

(a) *Category 1.* Countries with no manufacture and inadequate use of pesticides;

(b) *Category 2.* Countries with substantial pesticide demand and some distribution system but no local production and only repackaging facilities;

(c) *Category 3.* Countries with sizeable pesticide markets and local formulation plants but no production of basic pesticides;

(d) *Category 4.* Countries with capabilities for the manufacture of pesticides with potential for export and also capable of carrying out R and D work to exploit local raw materials.

2 BASIC ECONOMICS OF PRODUCING ACTIVE INGREDIENTS AND FORMULATIONS

Developing countries always face the dilemma whether to concentrate on formulation only or to take up the manufacture of active ingredients. The manufacture of active ingredients is suitable for only a few developing countries owing to a number of constraints (a) infrastructural requirements, (b) demand in the internal market for the particular active ingredients, (c) their potential for export, (d) the lifetime of the product in agriculture and (e) the patent situation.

In a majority of cases, developing countries opt mainly for patent-free compounds where the profit is low. But still they have to rely on imported know-how, raw materials and, in some cases, annual royalty payments. Moreover, in a free market economy, there will be severe competition from companies that have almost written off the cost of the plant. While the manufacture of active ingredients is within the reach of only a few developing countries, the formulation of pesticides is within the reach of the majority of them. The advantages of local pesticide formulation are as follows:

(a) Investment costs are low;
(b) Different products in terms of both active ingredients types and formulation types can be produced;
(c) Active ingredients and other raw materials can be purchased from different sources;
(d) Products are produced when needed, both for seasonal and emergency needs;
(e) Inventory can be kept to the minimum;
(f) The time between production and application can be kept to the minimum, reducing deterioration during storage;
(g) There is little waste generation;
(h) Possibility exists to carry out R and D that would introduce locally available raw materials and new formulations;
(i) Operation and maintenance of such plants is not complicated;
(j) Getting the know-how for formulations is not difficult;

(k) Unused materials can be recycled subject to quality control requirements;

(l) Product diversification is easy to plan.

If a country imports formulated pesticides in large or small containers they should definitely look into the economics of local production. According to rough estimates, depending on the type of formulation, more than 50% in foreign exchange could be saved because many of the inputs could be provided locally. One example is diazinon 40 EC imported by a developing country at a cost of \$7.03/litre (1985 price). If locally formulated, the cost of formulation would be as shown in Table 13.1.

Table 13.1 Cost of local formulation of diazinon 40EC

	Cost (US\$/litre)
Diazinon 98% (imported)	3.26
Emulsifier (imported)	0.18
Solvent (local)	0.26
Raw Materials (sub. total)	3.70
Packaging (local)	0.70
Labour (local)	0.50
Total cost	4.90

Foreign exchange savings will amount to 51%. The more dilute the formulations are, the greater will be the foreign exchange savings. The detailed techno-economic feasibility studies of various other parameters that should be taken into account prior to any investment decision are beyond the scope of this chapter.

3 REQUIREMENTS FOR ESTABLISHING PESTICIDE PLANTS IN DEVELOPING COUNTRIES

3.1 Market Requirements

Market demand and supply is a key element in establishing pesticide formulation plants. Even a small formulation plant could be viable if it is properly operated and caters to specific requirements. Many developing countries provide a subsidy to farmers, which has a direct bearing on the market demand for pesticides. It is well documented that the consumption of pesticides in developing countries will increase faster than in developed countries. This will increase the operational capacity of the existing plants in developing countries and create opportunities for new plants. However, the proliferation of formulation plants in many developed and developing countries could create problems, in so far as the volume of sales and, consequently, the lower capacity utilization rate in many cases would not allow an economically sound operation. Due to the strong seasonal demand pattern and uncertainties in forecasting market requirements, many plants are operated for only a short period of the year, prior to and during the actual selling season. Consequently, a significant part of the available production capacity remains underutilized. In the developing countries the concept of sharing the regional market has not made headway as it has in European countries. However, it is likely to become more popular in the Asian countries and could eventually take hold in East Africa and Latin America.

3.2 Raw Material Requirements

A majority of developing countries have to import active ingredients, surfactants and other adjuvants. However, the so-called inerts and packaging materials could be locally produced, provided they meet the quality specifications required for use in pesticide formulations. A product might need processing of raw materials, especially particle-size reduction to give optimum physico-chemical properties. It is always advisable to depend on local suppliers to provide suitable material, thereby reducing additional operations in the formulation factory. A team consisting of a procurement officer, a production manager and a quality control chemist should be in charge of raw materials purchase. The raw material requirements for different formulations are described in other chapters.

4 DESIGN REQUIREMENTS

4.1 Site Selection

When choosing a site for a new plant and store, close proximity to residential areas, schools, hospitals and shopping areas should be avoided. Preference should be given to areas already devoted to industrial development. However, even in these circumstances the proximity of premises used to process food and other materials destined for human or animal consumption should be avoided. It is also vital, especially in developing countries, that residential colonies should not be allowed to spread close to the formulation site.

In addition, one should take into account the region in which the finished products are going to be used, so that transport of pesticides can be kept to a minimum. These elements, along with the local availability of inerts (fillers, solvents), may provide sound economic justification for establishing formulation plants close to the market. It is clearly preferable from an economic and logistical point of view to transport the relatively small tonnages of active ingredients and special ingredients (surfactants) a considerable distance (from abroad) to the formulation plant for incorporation into less-concentrated formulations.

4.2 Plant Design

The layout of the plant will vary according to requirements, that is, there is no single plant layout. Figure 13.1 shows a general site plan for an agrochemical formulation plant. Three factors determine the characteristic features of pesticide formulation plants:

(a) Manufacture of an emulsifiable concentrate (EC) needs a single storey building with steel structures for mixing vessels;

(b) Manufacture of suspension concentrates (SC) needs a two storey building. The ground floor is used for filling, while the first floor is used for mills and blenders;

(c) Manufacture of dusts (DP), granules (GR), wettable powders (WP) can be done in two types of buildings: a two storey building with conveying equipment (called horizontal production), which will have cleaning problems, or a four storey building with different blenders and mills (called vertical production), which will be expensive to build.

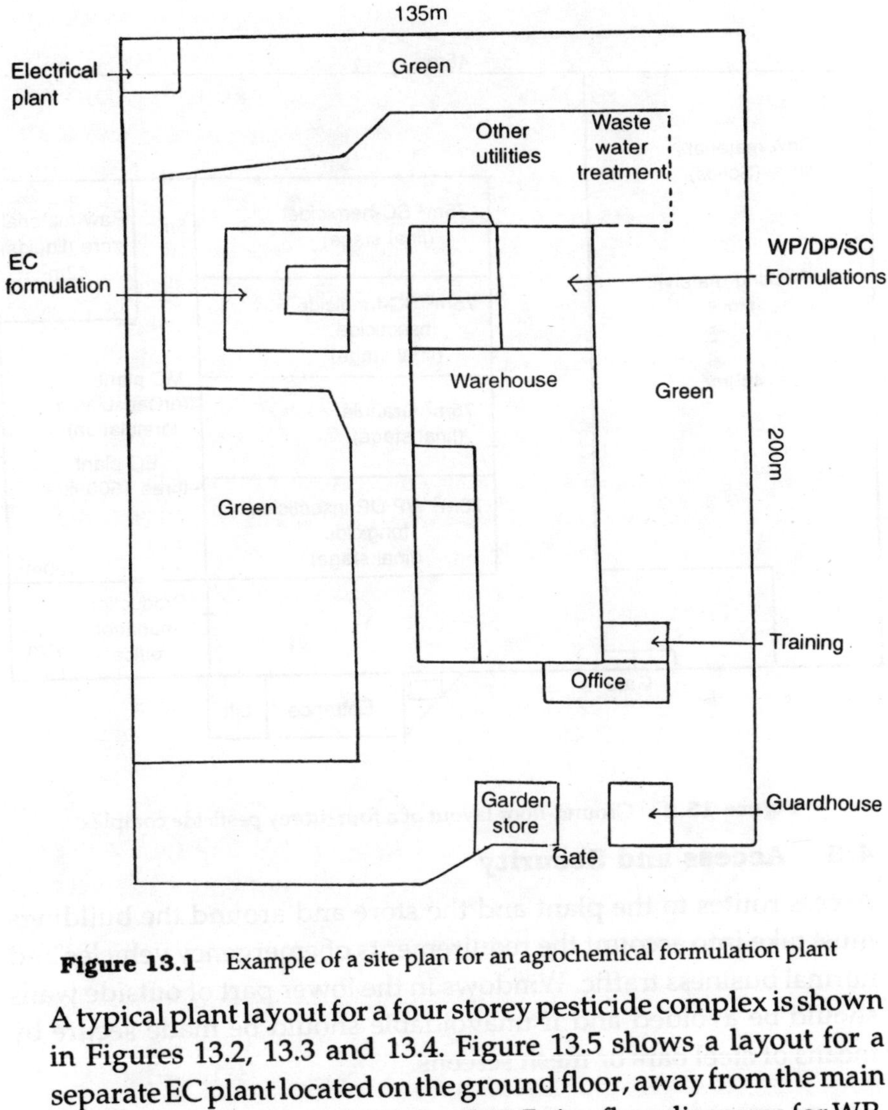

Figure 13.1 Example of a site plan for an agrochemical formulation plant

A typical plant layout for a four storey pesticide complex is shown in Figures 13.2, 13.3 and 13.4. Figure 13.5 shows a layout for a separate EC plant located on the ground floor, away from the main factory building. Figures 13.6 and 13.7 give flow diagrams for WP, DP, GR and WG formulations. Figures 13.8, 13.9 and 13.10 show operating plants for various formulations.

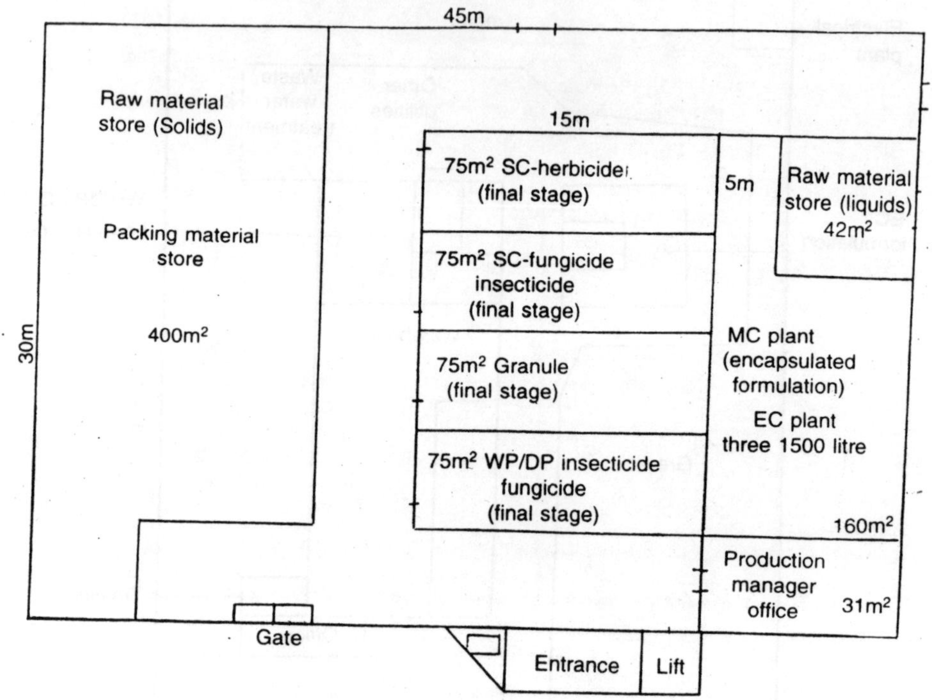

Figure 13.2 Ground-floor layout of a four-storey pesticide complex

4.3 Access and Security

Access routes to the plant and the store and around the buildings must take into account the requirements of emergency vehicles and normal business traffic. Windows in the lower part of outside walls should be avoided and if unavoidable should be made secure by means of steel bars or mesh screens.

4.4 Construction Materials

As far as possible building materials that are themselves combustible should be avoided. These include combustible roof linings, possibly installed for insulation purposes, and bitumen-coated steel sheeting commonly used in older buildings. Reinforced concrete frames are

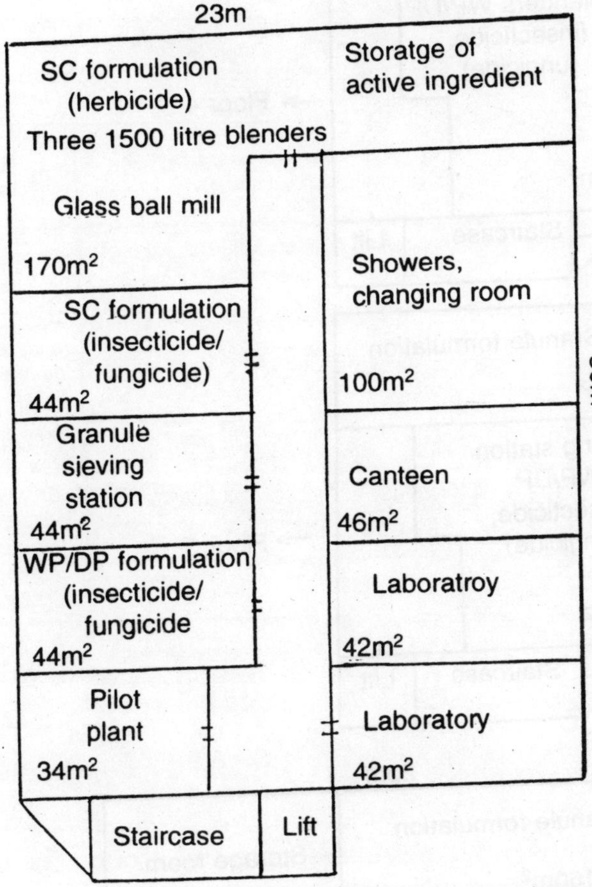

Figure 13.3 First-floor layout of formulation plant

preferable to unprotected steel work, and special treatment of weight-bearing vertical columns is advised. Floors must be impervious and should have a smooth finish for ease of cleaning.

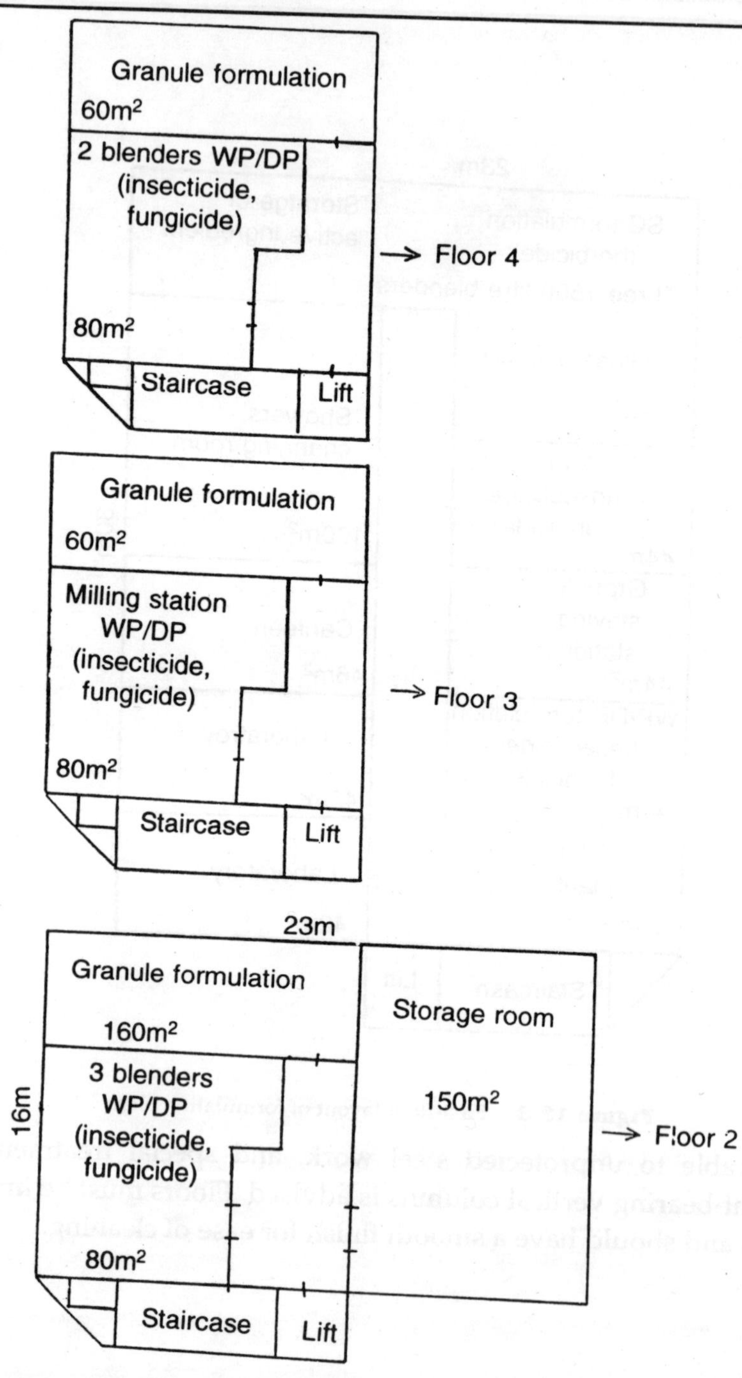

Figure 13.4 Layout for second, third and fourth floor of a formulation plant

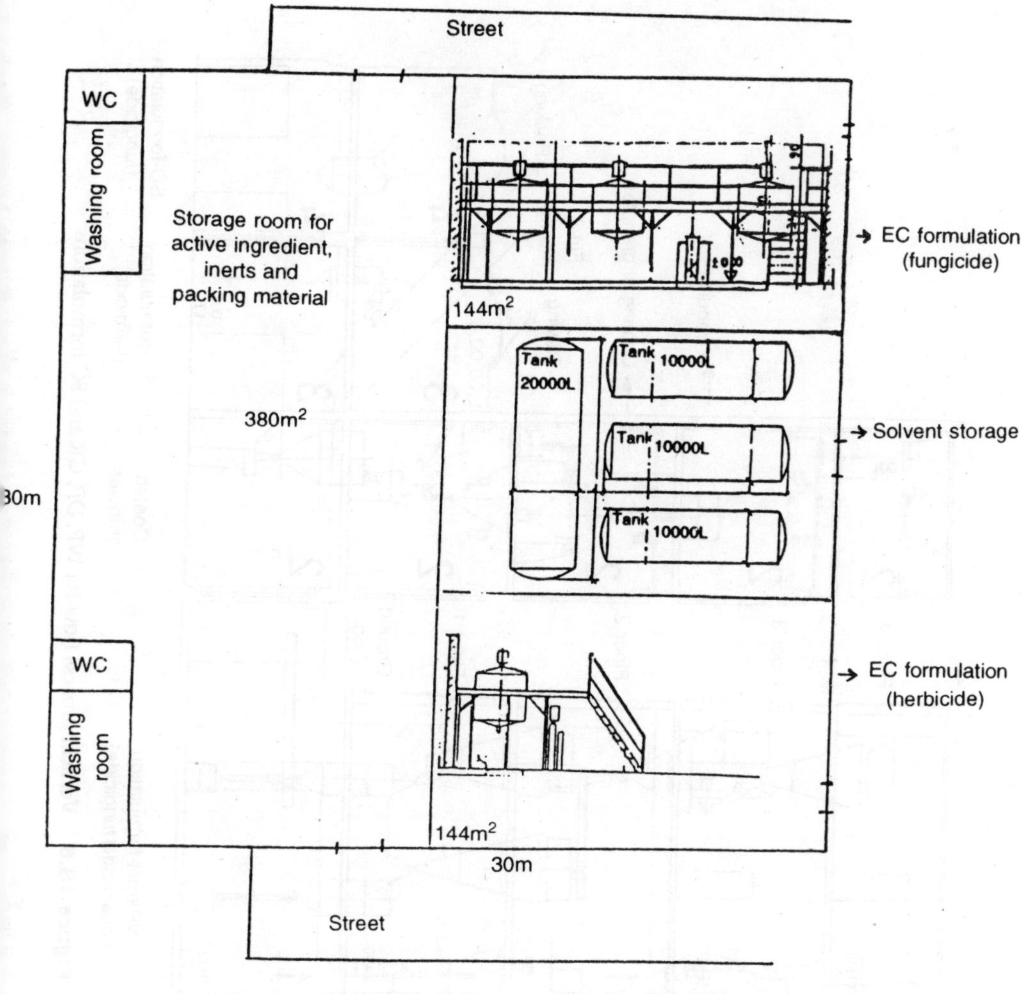

Figure 13.5 Layout for an EC plant on the ground floor

4.5 Lighting

Lighting levels must be sufficient to allow for the routine inspection of stored products and for easy reading of product labels. The location of artificial lighting is important. It should be installed above aisles and other areas where there is routine pedestrian or vehicle traffic.

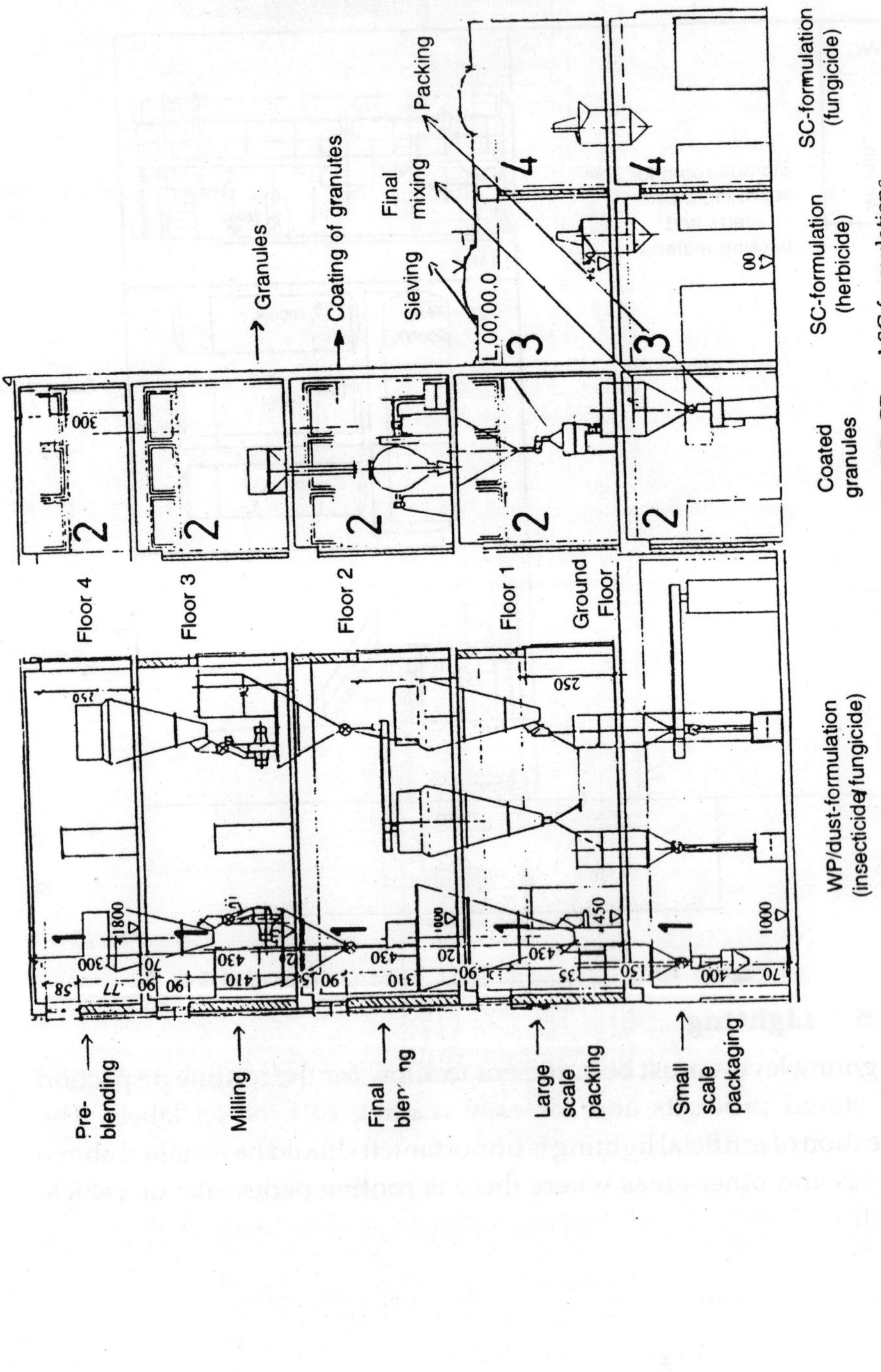

Figure 13.6 Vertical process flow for WP, DP, GR and SC formulations

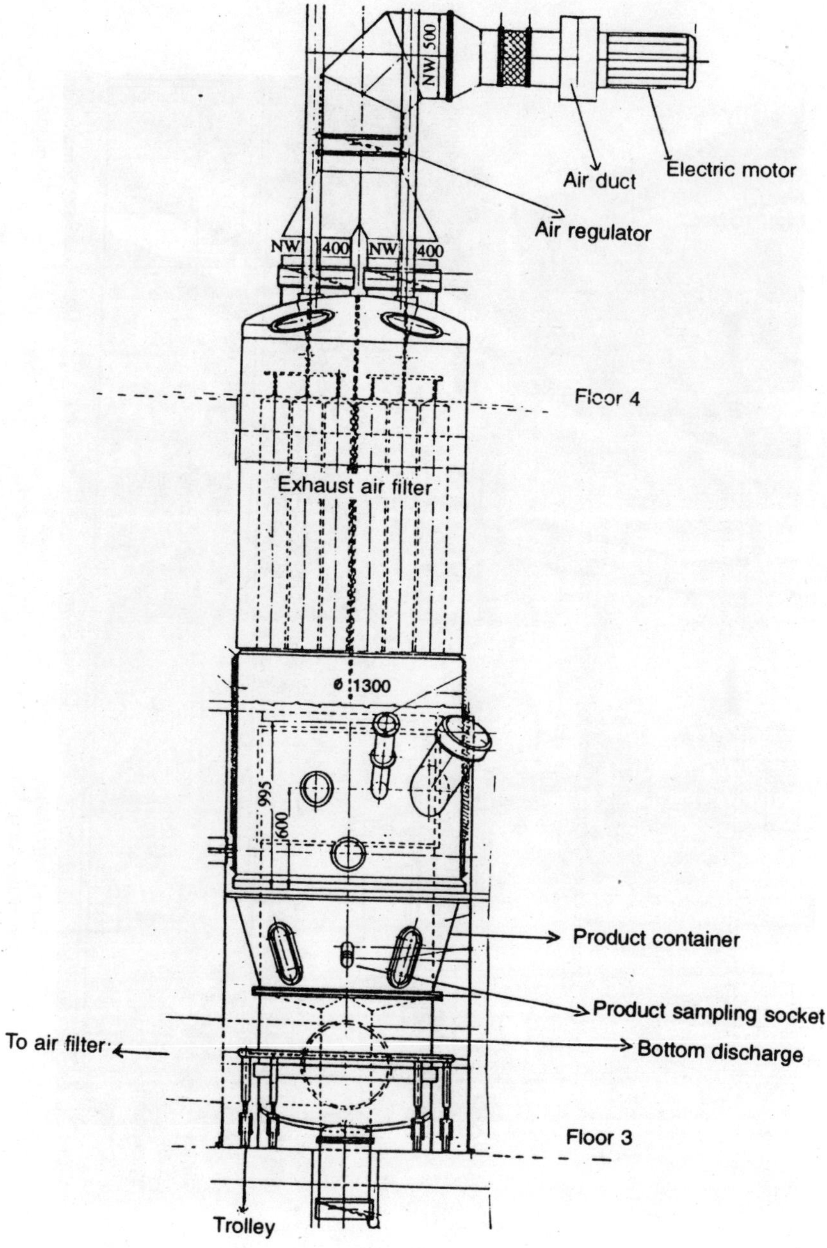

Figure 13.7 Process flow (partial) for water-dispersible granules

Figure 13.8 Plant for suspension concentrate

Figure 13.9 Plant for water-dispersible granules

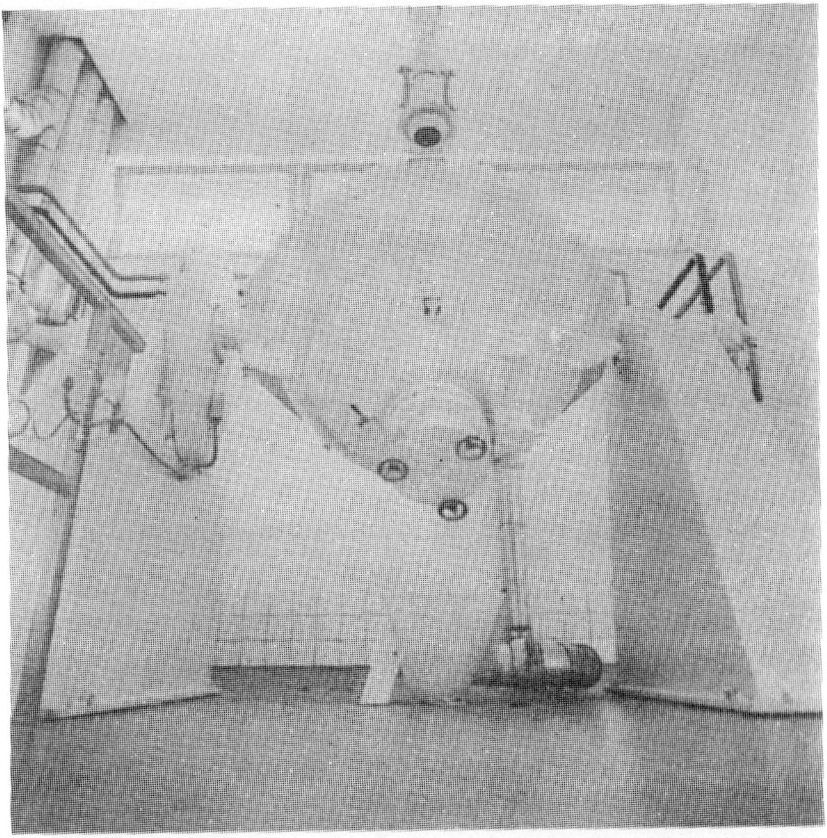

Figure 13.10 Plant for coated granules

4.6 Heating

Generally the large volume of most warehouses does not lend itself to heating by artificial means. However, some water-based products are vulnerable to damage by frost, and where this risk is present, it may be necessary to provide limited heating. This should be carefully planned. Gas- and oil-fired heating are not recommended.

4.7 Lightning

All buildings must have a lightning-protection, which has to be checked every three years.

4.8 Ventilation

Sufficient ventilation must be provided to ensure the removal of smells and vapours from the storage area. In temperate climates, this can normally be achieved by providing suitable ventilation panels in the roof of the building, Figure 13.11.

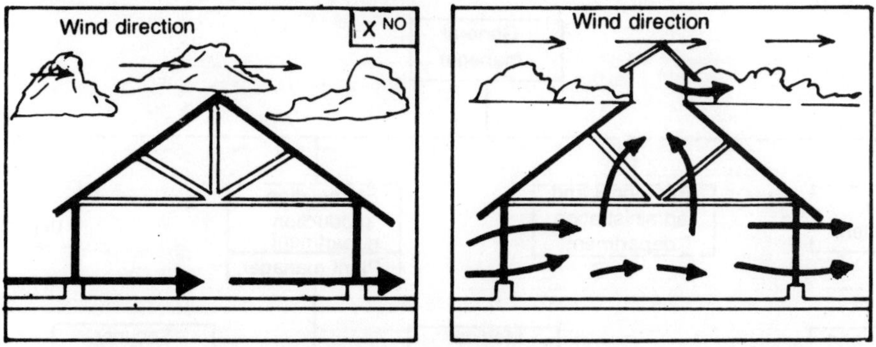

Figure 13.11 Ventilation requirements for a pesticides warehouse

4.9 Office and Amenity Accommodation

Wherever possible, office and amenity accommodations should be located outside and away from the main body of the plant and the store.

4.10 Utility Requirements

The utility requirements also determine the selection of a site. It is important that an uninterrupted, constant voltage power supply and a reasonably good water supply are available. This is especially necessary for sensitive laboratory equipment. In any case a diesel generator and a supply of deep well water should be considered in developing countries.

5 MANAGEMENT AND MAINTENANCE OF PLANTS

Formulation plants are becoming more sophisticated with advances in formulation technology and with the ever higher standards of quality, industrial hygiene, safety and environmental protection. Qualified staff must therefore be selected for managing such operations. Plant operators must be thoroughly trained in the safe and efficient operation of the plant. A typical organogram for a formulation plant is given in Figure 13.12.

Constant training of both shop-floor and management personnel is a vital factor in increasing responsibility and safety awareness of the staff.

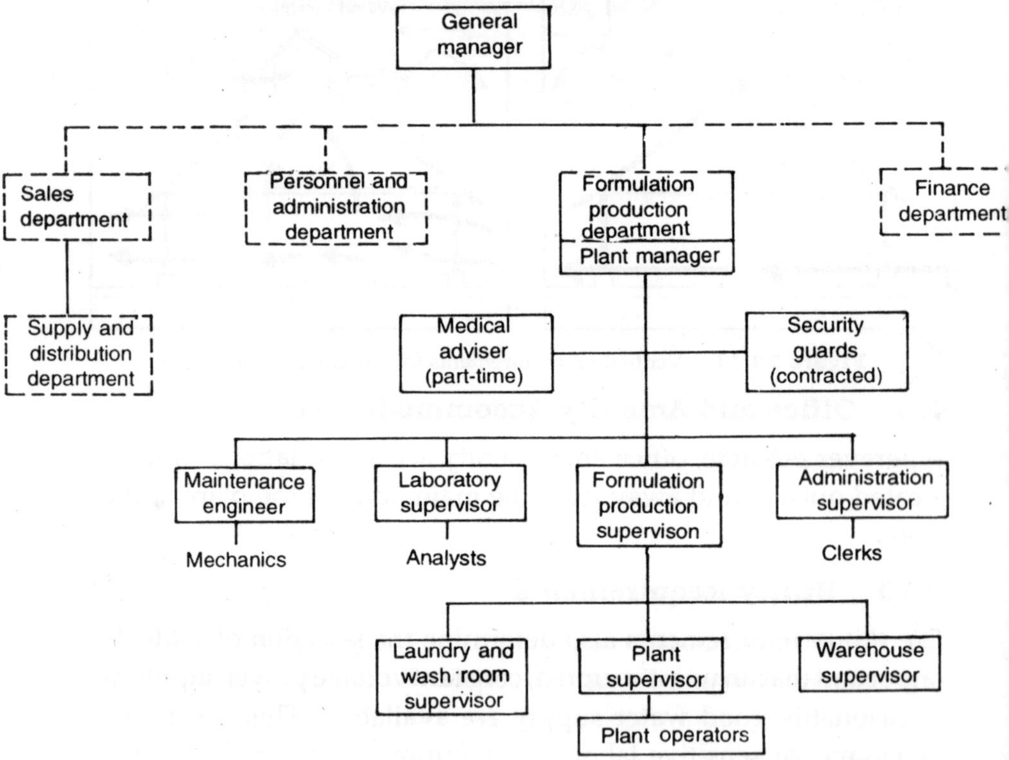

Figure 13.12 Typical organization of a small formulation plant integrated with a sales organization

5.1 Maintenance

Periodic maintenance of equipment is usually carried out by a special maintenance unit using the appropriate maintenance manuals. However, daily maintenance or visual inspection, which is done by the plant operators, is important and useful for the early detection of damage to the equipment before, during and after the operation. The experienced operator is most likely to observe any obvious damage, but he must refer to the inspection manual that has been prepared for each item of equipment. It is important to record the results of the inspection on a check-list and to take the necessary remedial action quickly.

The check-list should include the following: unusual noise, vibration, heat, degree of wear and tear, belt tension, lubricant, trouble in the electric system etc. For maintenance materials used on the equipment, such as lubricants and other consumable articles, the name, type, makers or suppliers and prices should be recorded in case the materials are needed in a hurry.

The results of inspection, maintenance and operating conditions, such as the amount of charged raw materials, mixing time, temperature and electric current, should be recorded on the maintenance and batch sheets. The filling procedure and length of time reports are to be retained should be specified.

Maintenance is of two kinds, breakdown and preventive. Breakdown maintenance is the repair of failed equipment; preventive maintenance consists of daily or periodic adjustment of equipment to prevent breakdown. Some desirable steps would be the following:

(a) Plant modifications will be carried out as required, taking into account quality of the raw materials and the plant design;
(b) Breakdown repair work will be carried out where necessary;
(c) Structural repairs will be carried out on the site, buildings, structures, components and sanitary facilities;
(d) Safety repairs will be done on safety equipment where necessary;
(e) Improvements will be carried out to achieve better performance, to optimize operational procedures and to increase production yield.

Except for daily maintenance, it is usually desirable to perform maintenance work during the shutdown for periodic inspection.

6 OPERATION OF STORES

Operation of stores in conjunction with warehouses is discussed separately because of its importance to the smooth running of the plant. Although this topic is normally not taken seriously in many developing countries, the management of a storehouse (stores + warehouse) is now highly sophisticated in developed countries. It must, however, be pointed out that the direct application of developed country experience may not be appropriate due to supply problems as well as the long time it takes to buy and transport raw materials from abroad. In developing countries, this means planning well ahead of requirements.

6.1 Distribution of Products Within the Store

A clear space should be left between all outside walls and the nearest packs and within block stacks to allow access for inspection, the free movement of air and fire-fighting.

Products must be arranged so that forklift trucks and other handling equipment are not obstructed. Narrow aisles and tight corners will increase the risk that packs may become damaged.

Emergency exits and access to emergency equipment such as fire extinguishers must be kept clear at all times. The use of floor markings to define aisles and walkaways as well as storage areas is highly recommended.

Stored products must be kept well clear of forklift truck parking areas and battery charging points.

6.2 Product Separation

The complete physical separation of products falling into higher risk categories within pesticide stores may not always be possible or justifiable on economic grounds. However, a useful degree of separation can be achieved by using materials of low fire risk (eg. aqueous-based materials and powders) to create breaks between flammable liquids. Where large quantities of flammable liquids are regularly held in stores, physical separation is strongly recommended.

As far as may be practicable, products should also be separated by type, according to their transport hazard category. Transport hazard labels are distinctly coloured according to the nature of the hazard involved and form a useful guide for this purpose.

Thus, flammable materials (red label) should be stacked separately from corrosive materials (black and white labels). Oxidizing materials (yellow label), as already indicated, should be stored separately, as should substances liable to spontaneous combustion (red and white label).

Rodenticide baits require special consideration, for reasons other than fire or general safety. These products are vulnerable to tainting by other strong-smelling materials and this must be borne in mind when selecting an area in which they are to be stored. Herbicides should be stored in a separate sector so that there is no possibility for contamination.

6.3 Rotation of Stock

As a further safeguard against unnecessary pack damage and deterioration, stocks should be used on a rotational basis, i.e. first in, first out. Strict adherence to this rule may not always be practical in warehouses operating a block stacking system, but in every case a regular rearrangement of slow-moving stock is required to prevent undue ageing and consequent deterioration of products and packs.

6.4 Bulk Storage

Bulk storage facilities are likely to be required only at formulation or filling plants. All bulk storage tanks must be located within an impermeable bund (walled area) where the bund is capable of containing 110% of the total contents of the largest single tank located within it. Sufficient space should be maintained between adjacent tanks to provide access for fire-fighting. Also the area with each bund must be kept clear of debris and water. Provision will need to be made to drain storm water from within the bunds, and this can be achieved by means of a small pump located outside the bund wall or a drain valve built into the wall. In the latter arrangement, the drain valve must be chained and locked in the closed position except when required for use.

Outside storage of bulk stock in drums may require overhead protection in hot climates or when the drum contents are particularly sensitive to extremes of temperature. Drum storage areas should have a firm, impermeable base surrounded by a bund. Access for forklift trucks will require the sill to be ramped from either side. The security of any outside storage area will require careful consideration.

7 PRODUCT DIVERSIFICATION

In any formulation plant product diversification is important to satisfy customer demand and to improve the safety and efficiency of pesticide formulation. Products can be diversified in the following ways:

(a) Change the active ingredient but keep the same type of formulation;
(b) Keep the active ingredient the same but change the type of formulation;
(c) Change the both active ingredient and the type of formulation.

To plan product diversification, the plant manager should be aware of market trends and changes taking place in pesticide technology and have access to people who can generate and compile the data needed to get a product registered. Close links with the Ministry of Agriculture are vital for product diversification. In addition, the plant can develop expertise to carry out the basic work but also participate in various other studies in collaboration with other institutions or foreign companies. In developed countries, product diversification is a route for survival while in developing countries it is a slow process. In the latter, product diversification should aim at (a) moving to more active and less toxic pesticides, (b) changing EC formulations to water-based formulations, (c) moving to solvents with higher flash points and (d) changing to pesticides with a different mode of action to avoid the onset of resistance. Finally, any minor or major changes in product line should be tested for bioefficacy, phytotoxicity and shelf-life.

8 CONCLUSIONS

The establishment of pesticide formulation in a developing country demands trained personnel who can judge the market trends as well as get good feedback from users regarding the performance of a product. This should include feed-back relating to any symptoms of the development of resistance in the field. Depending on its classification (refer to Section 1), a developing country might need assistance on a short- or long-term basis for the maintenance and operation of the plant. The workforce should develop awareness of quality and safety in the manufacturing process so that customers are provided with products that:

(a) Give consistent and effective biological control;
(b) Are safe and convenient to handle;
(c) Conform to declared registration and specification requirements;
(d) Are properly packaged and clearly labelled;
(e) Exhibit good and consistent storage characteristics.

CHAPTER

14

Quality in Relation to Pesticide Formulations

Konrad Pavel

In den Birken 85
D-42113 Wuppertal
Germany

1 INTRODUCTION

Quality is a term of considerable importance. It plays a major role in manufacture and trade. It influences prices and in turn influences competition. With the expansion of trade and the intensification of the world-wide exchange of goods the importance of quality continues to increase. This importance is evidenced by the increasing frequency with which the language contains the word "quality". Examples of this are "quality work", "quality product", "quality design", "quality guarantee", "quality control" and even "quality of life". In such combinations the word "quality" often expresses a degree of excellence of a product or a service.

2 DEFINITION OF QUALITY

What does "quality" really mean? The word is derived from the Latin interrogative *qualitis*, which translates as "What is the essential character of matter?" or "How is the matter constituted?" To define the quality of a product — in this case a pesticide — a description of its characteristics is necessary. The more features described, the more precise is the characterization. Many characteristics of a product, especially its physical properties, can be described by numerical values, e.g. melting point, density, vapour pressure and solubility. Such characterizations are independent of the observer and the intended use of the product and are not based on any ratings whatsoever. A listing of a product's measurable properties results in an "objective" describtion of the quality of the product.

Some of these properties may be suitable, others not, for a product that is intended to fulfil certain needs and special functions. Since such a ranking of product properties with respect to the applicability, usefulness and reliability of the product is dependent on the observer and the intended use, it is called a subjective or relative quality assessment. An assessment of the quality of a product carried out by the manufacturer may differ from that made by users, by the authorities or by the public. Thus, quality statements about products generally comprise an objective part (the measurable features mentioned above) and a subjective part rating suitability, safety and significance for the environment.

Quality is not inherent to a product as are colour, density, melting point etc. It cannot be measured. Assessments such as "high" or "minor", "good" or "bad" are relative, so quality is a relative concept. Individual features can be determined, but quality comprises the totality of numerous features and characteristics of a product (or service) that bear on its ability to satisfy stated or implied needs and which can be rated in comparison with similar products or a standard.

3 SPECIFICATION

Quality is an aspect that should be considered from the very beginning of a product. When the designated purpose of the product is known, it is advisable to bear in mind what each step in research, development, production and final delivery to the customer contributes towards quality. A survey of features determining the quality of a product is called a specification. This is not only a list of properties but also defines allowable limits within which the characteristics of a product may vary and remain acceptable. In the feild of pesticides, most of the products needing to be specified are complex mixtures of one or more biologically active substances in combination with inert ingredients such as diluents, carriers, surfactants and other adjuvants. The process of mixing these components is called formulation. The resulting products are called formulated products, simply formulations.

The customer who buys a pesticide expects that the information given by the manufacturer is correct and that the product fulfils the expectations of efficacy. The product should always have the same properties, i.e. it should be of constant quality. The constancy of some properties can be assessed simply by noting appearance, e.g. the state of aggregation, colour, odour and so forth. Other characteristics can only be determined using physical, chemical or biological means, which are probably time-consuming and costly. Constant product quality creates confidence in a manufacturer's ability to provide products of a consistently "high" quality. There are, however, exceptions to this principle. For instance, if technical improvement makes it possible to produce a product of superior quality, the specification should be changed. With increasing experience, the demands of the consumer may grow and "quality" turns out to be a moving target.

It is important, for both the customer and the manufacturer, that the details of product quality should be laid down by establishing written specifications.

As quality should be kept in mind from the beginning, specifications must be developed for (a) the raw materials needed for the production of active ingredients, (b) for materials used for formulations, (c) for the finished product, (d) for packaging materials and (e) for labelling. As a rule, specifications are developed by the manufacturer of the analytical and control department. It is advisable that a committee be formed by representatives of the research, manufacturing, quality control and sales departments to discuss the specifications. For a number of important pesticides, specifications have been issued by FAO and WHO and some other authorities. As already mentioned, a specification should contain a detailed description of the characteristics of a product, together with the tolerable limits for certain properties. In the following Sections, it will be shown which items must be taken into consideration when writing the specification for a pesticide.

3.1 Written Specifications

In the heading, one specifies as a minimum (a) the trade name, (b) the type of formulation and (c) the range of application. For the type of formulation, the two-letter international coding system is recommended. A general description of the appearance, e.g. the physical state, colour and odour, is part of the formulation specification. The range of application indicates whether the product is applied as an insecticide, herbicide, fungicide or plant growth regulator. Some further applications are nematicide, molluscicide, rodenticide and bactericide. The formulation code stipulates how the pesticide is to be applied.

In addition to such purely descriptive features, several measurable properties should also be given to specify a pesticide product. Table 14.1 lists typical characteristics in alphabetical order.

Table 14.1 Typical characteristics of pesticide products

Measurable properties for potential specifications	
Abrasiveness (e.g. of solid carriers)	Flowability
Acidity	Heat stability
Active ingredient content	Particle size
Alkalinity	pH
Attrition (e.g. of granules)	Sieving
Cold stability	Suspensibility
Density	Tap density
Dispersibility	Viscosity
Emulsifiability	Water content
Flash point	Wettability

Not all of the listed features must be measured for each lot of product. Which data are to be included in a specification depends on the kind of product. In each case the active ingredient content is an essential part of the specification, together with the tolerable maximum and minimum limits (GIFAP Monograph No. 1). Not only the absolute percentage determines the quality but the accuracy of the agreement between the analytically determined and the declared content. The declaration of (a) the amount of isomers in proportion to the active ingredient and (b) the amount of impurities originating from the manufacturing process (or from the raw materials used) is often more relevant for the assessment of quality than the active ingredient content itself. Isomers and impurities can influence the toxicity of a product or may be of importance for the environment, e.g. the fact that they are degradable to a very limited extent and tend to accumulate in the soil.

The features mentioned up to now constitute important information for the description, identification and registration of a pesticide, but they are not sufficient to assess whether a product is of "high" quality or not. Quality is determined by the answer to the question, How perfectly is this product suitable for its intended use in comparison with other, similar products or compared to a standard? "Good" or "high" quality of a pesticide product based on the active ingredient content can be confirmed when the active ingredient content is in the upper part of the tolerance range of the normal value. Small analytical deviations and slight degradation due to aging or climatic influences can then be compensated and will not cause rejection of the product.

3.2 Toxicity

Besides its physicochemical properties a pesticide has, of course, biological properties on which the efficacy is based. With few exceptions most pesticides are toxic. After all, their purpose is to kill pests. But, the toxicity of chemicals differs greatly, so it is very useful to know something about the toxicity of pesticides, e.g. the LD_{50} towards warm-blooded animals such as mammals and birds. Therefore toxicity data are also part of the specification of pesticides. This knowledge contributes to the assessment of quality.

3.3 Packaging

Ease and safety in handling and during application are further features that influence the evaluation of the quality of a pesticide. Both are closely related to the packaging. For example, it should be possible to handle solid formulations such as granules without generating dust and to decant dust formulations without covering the surroundings with dust. Packaging should be such that it ensures easy and accurate dosage and use. It should be possible to empty the packages as completely as possible without leaving residues. Packages must be safe against unintentional opening or unauthorized use, e.g. by children. Packages should be subjected to appropriate stacking tests. Such a test is passed when there is no deformation, breakage or leakage even at elevated temperatures. No leakage or corrosion must be observed after storage, even under extreme climatic conditions. This discussion shows that the pesticidal product itself and its package are considered as an entity when to assessing quality. For this reason specifications should also be established for containers, drums, boxes, cans, bags etc.

3.4 Labelling

Similar considerations are also applicable to the labelling of pesticide products. Even the smallest portion of a pesticide, whether an active ingredient or a formulated product in a diluted form, should always be labelled, at least for identification. Both the technical active ingredients intended to be used as raw materials in a formulation plant and the finished products must be labelled. The label has at least two functions:

(a) To provide information about the product. e.g. its identity, its purpose and potential hazard;
(b) To give instructions for its use and measures to be taken in case of accidents.

Labels must be clear and legible. The ink and printing colour should be light-fast and water-proof. The label should be protected against loss or damage by handling, humidity or leakage. The detachment of a label should be ruled out by using suitable adhesives. If pesticides are distributed in regions having low levels of literacy or even illiteracy, the information should also be conveyed using drawings or simple graphs. This is particularly true with regard to the mode of application and to warnings.

The following statements should be included on the label: trade name, common name(s) of the active ingredient(s), chemical name(s) of the active ingredient(s), intended use (insecticide, fungicide, herbicide etc.), percentage of active ingredients(s), instructions for use (shortened form), warnings (poisonous, burnable, inflammable etc.), instructions for first aid, advice for storage and disposal of both the ingredients and packaging, procedure in case of spillage and so forth. Most important is a clear and unambiguous batch number, from which it is possible to identify the origin of the product and the mode of processing. Moreover, it is useful to know the original size and type of a package so that unauthorized fills into smaller packages can be detected. Unauthorized fillings cancel the manufacturer's guarantee of quality. This is important to protect the user from falsifications and perhaps contaminated products. It is quite evident that correct labelling is an integral part of a high quality product. Therefore, label specifications are an additional means of maintaining high quality.

3.5 Expiry Date

It may be hazardous to human health if a medicament is taken after the expiry date. Similarly, the crops may be adversely affected if pesticides are applied after the expiry date. Pesticides of dubious origin or from tapped or unclosed packages should be viewed with suspicion and if possible not used. Decrease in crop yield, unnecessary costs and other harm may arise if a product is insufficiently effective. Damage to crops may occur due to partial decomposition or degradation of the active ingredient when using unsatisfactorily specified products. It is a hallmark of good quality when the stability of a pesticide product has been determined and a relevant statement e.g. expiry date, is made on the label.

The most important feature of a pesticide, its effectiveness against the target organisms, is difficult to define and, in particular, to measure. The efficiency depends largely on all the features and properties that have been mentioned hitherto: active ingredient content; nature and amount of impurities; kind and composition of the formulation; quality of packing; age of product; conditions of transport and storage and expertise in application (timing, weather conditions, mode of application). The detailed and numerous descriptions, specifications and analyses of pesticides are ultimately justified in that they are efficacious to agriculture.

4 QUALITY CONTROL

It is desirable that manufacturer and user, seller and buyer trust one another. However, to trust is good, to control is better. Thus, quality control is important for both manufacturers and customers. Prerequisites for an effective quality control are the specifications, the standards against which all products, raw materials. intermediates and finished products are measured. As already mentioned, quality should be built into a product from the very beginning. Quality control of pesticides does not begin with an examination of the finished product. It begins with the design of the synthesis and the establishment of specifications for the active ingredient. Quality control continues with (a) analytical methodology for the active ingredient as well as the raw materials, (b) on all aspects of the production of the active ingredient and (c) on the solvents, formulation additives and other auxiliary materials used in the finished product.

One cannot overemphasize that all the manufacturing processes, both for the active ingredients and for the formulated products, must be subjected to quality control. In this connection one responsibility task is to prevent cross-contamination, i.e. the presence of one pesticide in the formulation of another. Particularly critical is the presence of an herbicide in an insecticidal or a fungicidal formulation. Cross-contamination can be eliminated by very carefully cleaning all apparatus in the formulation plant, especially when different types of formulations are prepared one after the other using the same equipment. The preparation of a new batch of a formulation should not be started until all traces of the preceding product have been eliminated. This can only be achieved by thorough solvent rinsing of the apparatus. Since highly sensitive analytical methods are available it is not too difficult to test the effectiveness of different amounts of solvents used for rinsing and even to detect traces of a preceding product.

5 QUALITY SYSTEM

Quality control is part of the quality system that should be arranged in each institution that is occupied with purchasing, producing, selling, examining and testing pesticides. A quality system comprises the organizational structure and the responsibilities, procedures, processes and resources for implementing a quality management program. Quality control involves the operational techniques and activities aimed at monitoring a process and at eliminating causes of unsatisfactory performance.

Generally, quality control is realized in two different ways. The manufacturer or seller sets up specifications for a product, carries out the necessary analyses and sells the product if the test results meet the requirements. The buyer or user checks the product specifications that have been laid down by the manufacturer, or that have been issued by neutral authorities, e.g. FAO or WHO, or that are derived from contractual agreements. The extent of examination depends on the contractual or the official requirements. A total system of quality surveillance is expensive because it consists of monitoring procedures, methods, conditions, processes, products, services and records in relation to stated references to ensure that specified requirements for quality are met. It must not become an end in itself but should guarantee a certain degree of safety.

This presentation shows that analyses, tests, checks etc. are the basis for evaluating the quality of a pesticide product. Such tests are carried out on a sample of a pesticide product. The analyst must be certain that the sample is obtained skillfully and that it is representative of the complete batch or amount of product. A great deal of money and effort could be wasted if the sample taken was not a representative one.

It is much more economical to avoid damage to crops and perhaps financial loss by spending on quality assurance than by buying insurance against financial loss due to insufficient and inadequate quality control.

6 QUALITY ASSURANCE

Quality assurance is a management tool to provide confidence that a product will satisfy given requirements of quality. The aim of quality assurance performed by a manufacturer is "the customer should return but not the product". In view of the increasing awareness of quality in the field of industrial production, the International Organisation for Standardization (ISO) has issued a series of standards dealing with quality systems. Where a quality management and quality assurance system is to be established, it is recommended that ISO standards 9000 through ISO 9004 should be consulted.

7 EDITORS' ADDENDUM

The importance of quality to the agrochemical business in the developing world can not be overemphasized. Quality is a route towards survival of the business and towards an ability to compete on a global basis. Without it, a business is doomed to mediocrity, diminished profits and a reduced ability to serve mankind.

Before the Second World War the Japanese were noted for their inexpensive, poorly constructed products. Then Edward Deming introduced his quality concepts to the Japanese people, who in turn introduced quality in all aspects to their research, production and marketing. The results are history. The Japanese have prospered and benefitted by the sale of products and goods that meet customer expectations. The automobile business in the United States was caught napping when the Japanese introduced cars into the United States that were more reliable than domestically produced cars. One result of this is that Japanese Honda Accord is now the best-selling automobile in America.

When Dr. Pavel says that quality is a moving target, he speaks of something that is critical to the developing world. When a new product is introduced into a market, customer expectations have been set by previous experiences and may not be high.

For example, suppose you are introducing a new wettable sulphur fungicidal powder into your country for the first time. You do not have a good grinding mill and the average particle size of your active ingredient in your product is 15 μm. It will take almost 1,000 μg/ml to completely control conidia of barley mildew (1). However, if the average particle size were 1μm, it would require only 10 μg/ml to control conidia, a 100-fold increase in efficacy.

Thus, if your company is selling a product containing the larger particle size sulphur and you are the first to sell such a product, your customers might be satisfied. However, as soon as a competitor introduces a superior product with a smaller particle size sulphur, you will have a problem. Your customers will be dissatisfied with your product because it does not perform as well as that sold by your competitor and you will have to improve the quality of your product if you are to survive in the market place. As Dr. Pavel said, quality is a moving target. As your customers become more sophisticated in the use of pesticides, they will demand better and better performance. Producers must anticipate this trend and must continuously improve their products to meet these expectations.

REFERENCE

(1) W., Maas, *Advances in Pesticide Science*, H. Geissbuhler, ed. (Pergamon, 1978).

SUGGESTED READING

(1) American Society for Quality Control, *Quality Assurance for the Chemical and Process Industries, A Manual of Good Practices*, 1987.
(2) Lawrence S. Aft, *Fundamentals of Industrial Quality Control* (Addison-Wesley, 1986).
(3) Robert C. Camp, *Benchmarking, The Search for Industry Best Practices That Lead to Superior Performance* (Quality Press: Quality Resources, 1989).
(4) Rowland Caplen, *A Practical Approach to Quality Control* (Business Books, 1972).
(5) Philip B. Crosby, *Cutting the Cost of Quality* (Industrial Education Institute, 1967).
(6) Joe, Cullen, *Implementing Total Quality* (Bedford, IFS (Publications), 1987).
(7) W. Edwards Deming, *Quality, Productivity and Competitive Position* (Massachuseits Institute of Technology, Center for Advanced Engineering Study, 1982).
(8) A.V. Feigenbaum, *Total Quality Control*, (McGraw-Hill, 1992).
(9) J. M. Juran, *Juran on Planning for Quality* (Free Press: Collier Macmillan, 1988).
(10) Peter W. Moir, *Profit by Quality: the Essentials of Industrial Survival* (E. Harwood: Halstead Press, 1988).
(11) William W. Scherkenbach, *The Deming Route to Quality and Productivity; Road Maps and Roadblocks* (CEEP Press Books, ASQC Quality Press, 1986).
(12) *Total Quality: An Executive's Guide for the 1990's* (Dow Jones-Irwin, 1990).

CHAPTER

15

Industrial Safety in Pesticide Formulation

M.G. Srivastava

B-168, Niralanagar
Lucknow 226 020, India

1 INTRODUCTION

Safety is a key component of good manufacturing practice in the chemical industry. Irrespective of the size of an operation, good manufacturing practice is specially important in the pesticide industry due to the toxic nature of pesticides and the hazards associated with some of the other chemicals used. Therefore, the key task before management in the pesticide formulation industry is to identify hazard areas and to educate operators in the safe operation of equipment and in dealing with eventualities should any arise.

In the area of safety, awareness leads to knowledge and knowledge leads to skill. This chapter therefore seeks to identify the cause and effect of hazards and lists the ways and means to minimize risks. In doing so it takes into account the plant layout, equipment used for best possible design and construction and operator protection within the constraints of climate.

Government regulations on safety (they may be known by such names as "Pollution Control Act", "Factories Act" or "Fire and Explosives Act"), exist in almost all countries of the world. Regulations cover, for example, boiler maintenance, toxic and inflammable product storage and effluent disposal. However, by their nature regulations are generalized and despite them, accidents still occur. Since the risks in the pesticide formulation industry are product- and equipment-related, in-house safety regulations must be established for each individual plant, bearing in mind that government regulations take precedence over plant regulations. Of even greater importance is the need for commitment at all levels to achieve high standards of safety. Operator training therefore becomes essential for safety.

The sections of this chapter deal with the safety precautions that must be observed in operating pesticide formulation plants.

2 PROPERTIES OF PESTICIDES AND RAW MATERIALS

In considering safety, the first step is to know the characteristics of the products involved.

2.1 Pesticide Concentrates

Not all pesticides are toxic. Indeed some are quite safe to human beings. However, others are toxic on contact or inhalation. Some produce harmful fumes. Still others can be a source of chronic toxicity through prolonged exposure. It is essential to be completely aware of the toxicological properties of the pesticide to be formulated in order to plan for operational safety. In practice, since most plants are used to formulate several pesticides, certain basic precautions are always necessary, with provision for supplemental protection when a highly toxic or volatile pesticide is processed.

2.2 Solvents

A variety of solvents are used in the production of formulations. Organic solvents, in particular, demand care due to their volatility or combustibility. Adequate precautions are needed to minimize fire hazard. Similarly, vapours of certain solvents can be toxic if inhaled, so precautions are essential to prevent the excessive build-up of vapours and minimize their inhalation.

2.3 Solid Carriers

For dry formulations, minerals are used as carriers. The commonly used ones are talc (soapstone), china clay, calcite and bentonite, which by themselves are non-toxic. However, considerable dust is produced in handling and pulverizing them. The inhalation of such dust over a long period of time leads to a variety of respiratory diseases. Furthermore, if the plant atmosphere is heavily charged with dust, it could lead to an explosion. Precautions are thus necessary on both counts.

2.4 Surfactants

These mostly consist of emulsifiers, wetting and dispersing agents. Other adjuvants include stickers and anti-caking and anti-foaming agents. Generally they are considered non-toxic. However, due to their organic nature, some could pose to a potential fire hazard.

Awareness of properties is thus essential for safe handling during storage, plant transport and processing.

3 PLANT LAYOUT

In countries with an advanced chemical industry, the formulation of pesticides is a tail-end operation of basic manufacture. In most developing countries, where the chemical industry is not adequately advanced, formulation may be an independent activity. As the design and layout of formulation plants was described in detail in chapter 13, only a brief reference is made here in relation to safety considerations plant layout.

Formulations can be of different types:

(a) Solids (dusts, dispersibles, granules and microcapsules);
(b) Liquids (emulsifiables and ultralow-volume concentrates);
(c) Aerosols.

All require different types of machinery. Yet there exist some common features: handling of concentrates/technical grade pesticides, fast-moving equipment etc. The critical aspects of safety therefore include the following:

(a) The plant location and orientation must take care to prevent flooding and the venting must be oriented towards the prevailing wind direction;
(b) Adequate movement space must be provided for operators

and transport of goods. Lack of space could lead to cluttering and risk of accidents;

(c) The hoisting and loading areas should be adequately separated to avoid accidents from slipping of bulk containers;

(d) The inflammable solvent storage areas must be kept at prescribed distances from the mixing and packing zone to minimize risks of accidental fire;

(e) Solid and liquid formulation units should be separated, especially the heavy machinery area; such as jaw crushers pulverisers, which could lead to the build-up of an explosive dust;

(f) Areas using volatile pesticides or solvents must be located close to natural ventilation. Generally, it is advisable to place vents close to ground level.

4 OPERATIONAL HAZARDS

Like the chemicals, machinery can also be a source of hazard during plant operation. It is therefore essential that the processes are carefully documented and precautions for the safety of operators listed.

4.1 Liquid Formulations

4.1.1 Formulation equipment

This generally consists of stainless steel or glass-lined vessels fitted with agitators. Depending upon the solubility of the pesticide, the operation may be carried out at room temperature or at a higher temperature through the use of heating coils.

4.1.2 Charging of vessels

Safety precautions are essential in charging the vessel with ingredients. For example, solids should be charged slowly to prevent ignition from static electricity. Solvents and technical liquid and raw materials should be charged separately into the vessel through their respective pump and valve systems. An important precaution is to electrically ground the whole system to prevent a build-up of static charge.

4.1.3 *Ventilation*

Apart from the risk of inhalation, special precautions are also necessary because most organic solvents present a fire hazard. The formulation area must therefore be provided with a proper ventilation system, including exhaust fans. The exhaust should lead to a secondary scrubber or filter and thereafter be harmlessly vented, bearing in mind the environmental regulations.

4.1.4 *Pumps and valves*

Pumps and valves develop leaks. When that happend, pending repair/replacement, it is essential to provide catch pots to ensure against undesirable contamination and risk of fire. Seals on pumps, if excessively tightened, may become a source of heat and lead to a fire. They need regular attention.

4.1.5 *Heat exchangers*

Malfunctioning or faulty operation of heat exchangers is a potential source of hazard through excessive heat build-up and fire.

4.1.6 *Blending equipment*

The agitator of the blending vessel is a fast-moving piece of equipment carefully adjusted so as not to interfere with the vessel walls. Vigorous agitation of the liquid in the vessel may lead to frothing and splashing. Therefore sufficient ullage should be provided and under no circumstances should the agitator operate without securely closing the lid of the vessel. Furthermore, port windows must be provided for visual checking of the inside of the vessel. Operators must be made aware of the hazards of opening the lid during operation.

Accidents can also occur if the agitator is not switched off or is accidentally switched on while the vessel is being scrubbed. It is therefore essential that the start-up switch of the agitator is so located and secured that there is no risk of its being turned on. A security system of isolation of circuits and switches is recommended.

4.1.7 *Fire and explosion hazards*

The minimum requirements for prevention of these hazards are:

(a) Electrical motors and switches must be flameproof;

(b) Spark-producing tools and accessories must be kept away from the formulation area. For on-line repairs, copper hammers should be used instead of steel ones;
(c) Open flames and smoking must be prohibited;
(d) Fire detectors must be installed at appropriate locations and height.

4.1.8 Toxic substances

Apart from the solvent vapours, a formulation may emit fumes of toxic pesticides. Additional precautions are necessary in such cases. Entry into the mixing vessels, which contain toxic material, for repairs or any other reason, must be by management permission. The conditions of entry must be specified and the vessel marked with an "out tag". The processing should not be restarted until authority is given to remove the tag.

As a part of good manufacturing practice, it is desirable to establish a classification system for raw materials based on the handling procedure and type of personal protection required. Under this system, containers, equipment and operating areas are prominently marked with an alphabetical code to indicate particular hazards such as toxicity or inflammability. Individual companies or units might develop their own codes. For example, one company has adopted a labelling system for handling procedures and personal protection that uses the following code:

A Requires use of thin cotton or rubber gloves.
AB Requires, in a addition to A, single-use gloves, respiratory protection or goggles, as appropriate.
ABC Requires, in addition to AB, protective shoes, disposable overalls or plastic aprons.

The documented procedures for handling and operation in such cases must record the code to be used. The system, oncė introduced, must be properly monitored and managed.

4.2 Solid Formulations

In processing solid formulations, different types of equipment (grinding/pulverizing equipment, blending equipment etc.) are used. Each presents a different hazard.

4.2.1 *Jaw crushers*

These operate with heavy hammers or beaters to reduce the large lumps of diluent mineral into a coarse powder. The hammers and beaters require daily checking as they might get loose and damage the machine itself or hurt the operators if a metal projectile gets detached while the machine is running. Uneven wear or damaged rivets often lead to such a hazard.

4.2.2 *Pulverizers*

These are generally roller mills used to grind the pre-mix (mixture of technical-grade pesticide, diluent and additives) to the desired fineness. As they have fast-moving parts, probing to locate and remove blockage should only be done after stopping the machine.

In the grinding chamber, considerable frictional heat is generated. This may lead, in some cases, to the melting of chemicals and the formations of cake, which, in turn, blocks roller movement. In such cases, air vents must be opened to adjust internal temperature. If caking has already occurred, the mill must be fully stopped and only then the observation ports opened to clear the jammed rollers with the help of probes.

It is equally important to install a safe system for start-up to prevent the risk of trapping anyone in the machinery. The system must ensure that the key for the start-up switch is kept with a designated person whose name is on the key tag. Several devices with multiple locks and keys are available for shutting down and preventing restart until every key has been turned or lock removed. The final key, which is most critical, should be used to unlock the starting switch. To ensure safety during the maintenance shutdown, the keys required for unlocking the device may be provided only to the technician in charge.

4.2.3 *Blenders*

The pulverized material is homogenized by blenders. Depending upon the nature of blender used, appropriate precautions must be observed. Generally, there is greater risk in hopper- and worm-type blenders because of continuing thrust and the structure of the machine itself. It is essential that the lid of the blender is securely closed when starting and throughout operation. If attention is needed, the equipment should first be switched off. Moreover, no effort should be made to disturb the pockets of unagitated material by hand or small probes.

4.2.4 Granulators

In the manufacture of extrusion granules, plain or pesticide-containing dough is fed through the extruder. While the precautions listed for blenders apply here, special care is needed to prevent accidents while charging the dough into the feeder end.

At the drying stage of granulation, the wet extruded material is exposed to a hot air blast, with the temperature at the exit point as high as 85°C. Direct exposure of individuals to the blast must be prevented.

4.2.5 Ventilation and dust collection

Dry formulations involving mineral diluents generate dust. Due to static charge on the particles, the dust clouds are prone to explosion. The dust must be kept from spreading in the plant as it may cause serious maintenance problems. The dust-laden atmosphere also exposes operators to inhalation of the particles, leading to respiratory ailments (see operator safety). It is essential, on both counts, to provide exhaust fans in the mill areas. In addition, it is essential to provide dust-catching and explosion protection equipment. If flammable powders are handled, it is absolutely essential to ensure that all equipment is properly grounded electrically.

4.2.6 Filling equipment and operation

The filling point is the tail-end of the formulation activity. The filling of the formulated pesticide into user or bulk packs is carried out with automatic or semiautomatic filling machines (Manual filling of liquids, once common in developing countries, must be discouraged). The filling operation may involve splashing, overflow (due to frothing or otherwise) and contamination of the operators. Several precautions are therefore essential:

(a) Machines should be appropriately adjusted for each individual formulation for correct quantity and speed of delivery;
(b) Machines must be provided with safety guards;
(c) Liquid filling machines should be provided with automatic closure devices to stop product delivery when the container has been filled to the required level;
(d) At tail-end of the packing line (i.e. after filling/closure), a water shower should be provided to wash off the liquid pesticide on the container surface;
(e) Receptacles should be provided for spillage and washings of the filled containers. These must be disposed of safely;
(f) Sufficient ventilation must be provided in the filling area. If toxic fumes are involved, an exhaust hood over the filling machine is required. Similarly, where powders are filled, dust exhausting equipment is essential;
(g) Filling of formulations involving a static charge demands that the metal containers be earthed while filling. This may be achieved by a crocodile clip attached to an earthed lead or by placing the drums on an earthed metal plate;
(h) Finally, it must be ensured that while working on a filling machine, operators wear protective gear (gloves, goggles and overalls) and, where necessary, use a gas mask.

Besides filling equipment and operating precautions, the type and construction of the container is important. Metal and polymer containers are preferred and commonly used the world over. However, particularly in developing countries, the use of glass bottles is not uncommon. Glass containers are fragile and need special care in handling during filling and other operations.

Glass bottles have smooth surfaces and are liable to slip and break during filling, labelling and packaging. To minimize the risk, glass bottles used as pesticide containers should be specially designed to be held firmly. This may be achieved by providing ribs on the surface or suitable depressions. Furthermore, to avoid misuse of the used empties, beer and soft-drink bottles must not be used for packaging pesticides.

4.3 Services

4.3.1 *Hoisting equipment*

The process of formulating a product involves transfer of a pesticide concentrate, solvents, premixes etc. from bulk packs into mixers or pulverizers. Hoists are often used for this. The use of hoists is a safe operation but periodical checking of chains and locking devices is extremely important. Drums should be properly secured for lifting to prevent accidental slippage and injury to the workers.

4.3.2 *Boilers and pressure resistant equipment*

Larger formulation units have boilers for steam generation for warming up the formulation vessel. Where formulation is carried out using pressure-resistant vessels or equipment, the design must suitably provide against bursting of pipes. The delivery lines and equipment must be provided with relief vents and valves to prevent unusual build-up of pressure. Relief vents should be short and straight and directed away from workers. Regular inspection and testing of all pressure equipment is essential and should follow the plant safety rules, statutory regulations and guidelines.

4.3.3 *General maintenance*

Norms of maintenance should be established for each plant and monitored properly. The regular maintenance of equipment, especially those pieces having moving parts, is essential to avoid excessive wear and tear. Since pesticide production pressure is seasonal, checks and maintenance should be completed in the off-season to avoid down time during the peak period. For formulation units using several types of equipment, a maintenance schedule must be planned in advance.

During maintenance, entry into areas where there may be a toxic hazard must be restricted by a suitable entry system. The risk of accidents can be further minimized if the locking out devices (described earlier) are incorporated into the mechanical, electrical and total chemical system. All valves must be clearly marked so that accidental risk of turning on the wrong gas etc. is eliminated.

5 OPERATION MANUAL

The foregoing description of risks and precautions emphasizes the importance of the operating procedures. These procedures should be written in a language that is easily understood by the work-force. The written instructions must incorporate a data sheet on chemicals and equipment used in the plant as well as the precautions that must be observed while operating. As far as possible, the precautions must be supported with suitable illustrations. It is good practice that no formulation plant should be operated without a written procedure and training of the personnel. Copies of the manual must be available to the workers and supervisors. Management must also closely monitor the observance of laid-down procedures to prevent the development of dangerous work habits.

6 SAFETY CONCERNS OF MANAGEMENT

Besides protection of plant and property, it is both a moral and legal responsibility of management to ensure maximum safety of the workers, the environment and the neighbouring community. Management, therefore, must not only provide the necessary protective facilities at site but also educate workers on the potential hazards and train them on procedures and precautions. Only with such conscientious effort can it hope to achieve the desired safety standards.

6.1 Personal Safety

Personal safety is by far the most important ingredient of any safety procedure. In the pesticide formulation industry, the workers are liable to be exposed to undesirable quantities of concentrated chemicals during formulation, packing and storage. A high degree of worker safety can be achieved by the use of very sophisticated and expensive equipment. Experience has, however, shown that even with conventional equipment adequate safety can be ensured by observing precautions commensurate with the risk.

6.1.1 Protective clothing

Workers must put on clean, long- sleeved overalls of material suiting the climate at the start of each shift. The legs of the overalls must fit snugly on the boots to prevent seepage inside the boots. No street garments should be worn under the overall. When working with liquid pesticides or packing powder formulations, suitable water-proof or cloth aprons should be worn. Aprons and overalls should not be taken home.

6.1.2 Protection of hands

Normally, gauntlet-type rubber or plastic gloves of medium weight should be used. In a hot climate, use of disposable cloth gloves under the rubber gloves is desirable. During heavy repair and in the workshops, heavy leather gloves are recommended.

6.1.3 Protection of head

Depending upon the work situation, cloth caps, hard hats or helmets must be used to prevent injury and also settling of pesticide onto the hair.

6.1.4 Protection of feet

Use of rubberized or leather safety boots is essential. Exposed steel toes and studs must be avoided in the inflammable vapour areas.

6.1.5 Protection of eyes

In any operation involving liquids, protection from splashes is essential. Unbreakable eyeglasses, goggles or even face shields must be used.

6.1.6 Protection of respiratory tract

Dust masks or respirators with appropriate filter cartridges should be used depending upon the compound handled. Cartridges must be changed after every work shift. In a high vapour situation, gas masks with canister-type filters are recommended. Respirators with a self-contained air supply must be used when working inside blender tanks or hoppers.

6.1.7 Personal hygiene

Protective equipment alone is not sufficient. The workers personal hygiene is as important:

(a) At the end of work or whenever required, gloves should be removed only after they have been washed;
(b) Soiled overalls and aprons must be removed and deposited in a special bin for laundering. Contaminated work clothing must not be stored in lockers;
(c) Immediately after removing the work clothes, workers should shower, using plenty of soap and water;
(d) Street clothes may be worn only after the worker has showered;
(e) Used waterproof aprons, boots and gloves should be properly washed before re-use. Gloves should be checked for tears and perforations before use;
(f) Workers must not smoke or eat at the work site and when they do smoke or eat they must first wash hands and face with soap and water.

The above are essential but routine precautions. Special hygiene or medical assistance may be required under the following circumstances:

(a) If excessive accidental contamination with dry or liquid pesticide occurs, the worker must immediately strip and take an emergency dousing shower and wash the body thoroughly with soap and water before putting on fresh work clothes;
(b) If the contaminated person had been working with very toxic pesticides (like organophosphorus compounds), after the shower he should be examined for toxicity symptoms;
(c) If the chemical has splashed into the eyes, the worker must wash them with plenty of cool water. Use of eyecups (provided in the emergency shower) is even better. This must be followed by medical attention;
(d) If accidental inhalation of toxic fumes has occurred, the worker must as far as possible be moved quickly into a well-ventilated area, tight apparel around the neck (if any) must be loosened and shoes and nose pad must be removed. The

supervisor or fellow workers should immediately come to the aid of the worker in trouble.

Management has to ensure that the workers and plant staff are made aware of the above precautions, including personnel hygiene. Plant supervisors have a specific duty to regularly monitor and ensure that the codified safety precautions are observed by all workers and defaults corrected on the spot. Habitual defaulting must be penalized.

6.2 Plant Environmental Safety

Workers spend eight hours or more per day inside the plant premises and are therefore exposed to the physical and chemical hazards therein. Plant procedures must therefore include the following important considerations:

(a) *Orderliness*. Bulk containers, empties and packed containers, unless kept properly and out of the way of the normal work area, can cause physical injury to workers. Orderliness in the in-and-out handling and storage of containers must be ensured;

(b) *Decontamination*. Leakages and spillage are a common occurrence during work. Unless these are promptly removed, workers may get injured by slipping on or coming into contact with toxic substances. Liquid spillage must be removed using sawdust, which should be sent for safe disposal. In addition, periodical washing of the plant floor is necessary for hygiene or statutory reasons. For this, the plant floor must be provided with covered drainage leading the washing into the effluent treatment system;

(c) *Detoxification*. The plant atmosphere may become contaminated with toxic or combustible fumes at night. Under such circumstances it is necessary that the atmosphere be detoxified before the start of the morning shift by switching on the exhaust fans for at least a half an hour. Similar precautions are desirable between shifts if there has been a time lapse. The switch for the exhaust system is ideally located just outside the plant entrance.

(d) *Fire fighting*. Fire extinguishers must be provided (see bellow).

6.3 Warehousing Safety

Plants are frequently provided with a raw material and finished product warehouse. Ideally, separate warehouses should be provided for toxic and inflammable substances. They should be located so as to provide easy access into and out of the plant. In addition, materials should be stored in suitable bays with adequate movement space for fork-lifts/trollies etc. Stacking must be done only up to an approved height and each stack suitably labelled. A first-in, first-out procedure is essential from both a material and a safety point of view.

Other warehousing precautions, e.g. decontamination of the atmosphere, removal of leakages and spillage, washing of floors and fire-fighting, are similar to those described for the plant environment. The warehouse supervisor should be made responsible for a monitoring and observing appropriate precautions.

6.4 Safety of the External Environment

Ensuring safety of the external environment, i.e. the areas outside the plants and warehouses as well as outside the factory premises, is essential. The subject of environmental impact of pesticides is dealt with in detail in chapter 19.

Reference has been previously made to dust collection, effluent treatment, disposal of wastes etc. These procedures must be rigorously observed. A few additional important precautions are listed below:

6.4.1 Fire-Fighting

Fire extinguishers of the dust or foam type (as appropriate) must be provided at suitable locations in the plant, warehouses and quality control laboratory for dealing with localized small fires. The workers at all these sites should be trained and periodically updated in the use of fire-fighting equipment. Extinguishers must be regularly serviced as per the manufacturers' recommendations. Fire alarm systems must be provided in the vulnerable areas to draw the attention of management and fire-fighting personnel as soon as a fire is noticed.

To deal with a major fire, factories must provide access for fire trucks and have an adequate water supply. Hydrant points must be provided at appropriate and easily accessible locations. Every effort must be made to contain a fire to minimize loss as well as to protect neighbouring properties. In case of a serious fire, the municipal fire service must be summoned without delay.

6.4.2 Explosions or major leaks

In the event of an explosion due to dust or inflammable materials or a major leak of liquids or toxic gases due to valve or pipeline failure, emergency steps are essential. These include switching off the plant and pressure systems, closure of gate valves on delivery lines and other emergency steps to stop leakage.

Gases may also leak due to sudden power failure. Emergency alarms and safety measures for this eventuality must be made known to all members of the work force.

6.4.3 Effluent and waste disposal

A much smaller quantity of effluents and wastes are generated in the formulation process than in the basic manufacturing of pesticides. Nevertheless, the haphazard disposal of solid wastes, off-specification materials, residues, washings etc. can be a source of serious environmental contamination and health hazard for workers. Waste disposal is discussed separately in chapter 17. The following steps for safe disposal are emphasized:

(a) *Solid wastes*. These include plant floor and warehouse sweepings, collections from dust catchers etc. which are difficult to detoxify. Burning and high-temperature incineration, practiced in some countries, may be inadequate because of incomplete combustion and associated air pollution problems. A commonly used method is to bury such wastes in soil to a minimum depth of 45 cm. Such burial sites, unless determined by the state, must be carefully chosen to ensure that surface or subsurface water contamination does not take place. Addition of lime or caustic soda before burial hastens decomposition. If highly toxic wastes are involved, these must be encased in concrete blocks;

(b) *Effluents.* Fluid wastes must not be drained into the sewer system. Such wastes from the plant, scrubber, laboratory and warehouse must be chemically treated (where necessary) and then channelled into a detoxification tank containing activated charcoal. The detoxified fluid from the tank is still not safe for discharging into a river. It is therefore desirable that it should be drained into an evaporation pond or lagoon of suitable size and allowed to slowly evaporate.

Solids that settle in the evaporation pond and the spent charcoal should be removed periodically and incinerated or disposed of along with other solid wastes. While removing these wastes (as well as any other solid wastes) the workers must use safety gear, e.g. gloves and rubber boots;

(c) *Disposal of empties.* A sizeable number of large and small empty containers are generated during formulation operations. These comprise bulk containers of technical-grade pesticides and solvents and smaller containers of adjuvants. It is a hazardous practice to sell or otherwise dispose of these empties in any way that would lead to their reuse as a container (not uncommon in developing countries). The following procedure should be followed before disposal of such containers:

(i) First, the inside of the empties of pesticide concentrates must be washed with a suitable solvent;

(ii) All empties should then be washed with a caustic soda solution;

(iii) They should then be thoroughly rinsed with plenty of water and, as far as possible, their bottoms perforated to prevent reuse;

(iv) All washings must be collected and fed into the effluent treatment system;

(v) Containers such as bags and cartons should be destroyed by burning, taking adequate care to minimize air pollution.

A system should be introduced to ensure that no uncleaned empty container is taken out of the factory premises;

(d) *Education of the neighbouring community*. Serious fire, explosion or toxic leaks could also expose the neighbouring community to physical or chemical hazards. Safety of the community is a responsibility of the industry. Therefore, without creating undue fear or alarm, it is essential to inform the neighbouring community of the types of chemicals, processes and hazards involved and the precautions that should be observed in case of an unforseen eventuality. Explaining the safety measures already provided in the factory helps to build the trust of the community.

7 SECURITY FOR SAFETY

Apart from routine factory security, special measures are required in pesticide formulation units to ensure against acts of omission by factory personnel that may lead to hazards:

(a) Restriction of free access to the hazardous areas. For example, the solvent storage battery must be barricaded and the gate secured;
(b) Authorized workers at plants and warehouses should be made to wear distinguishing badges;
(c) Personnel from other plants/activities should be allowed entry only against specific permission of the concerned manager;
(d) At the point of entry into the work site every person must be frisked for matchboxes, lighters etc. and inspected for the use of appropriate apparel and equipment;
(e) During the off shifts, all entries and exits must be securely locked;
(f) There should be only one entrance to the formulation and packing plant and it should lead to the changing room and then on to the plant.

8 MANAGEMENT OF SAFETY

Safety is unquestionably the concern of both workers and management. The latter, however, have a special responsibility to ensure that the procedures and precautions discussed earlier in this chapter are observed. The main managerial tasks in this regard are given next.

8.1 Standard Operating Procedures

It is customary that when they design formulation plants, the engineers also write the standard operating procedures (SOP) for each unit. In doing so, they must pay special attention to the safety aspects. Copies of the codified SOP should be made available at each plant for reference by the supervisors and workers. Ignorance of SOP is cannot be condoned.

Because in many developing countries, formulation is a small-scale industry, process/design engineers are not retained. Instead, the equipment and plant are of standard design and construction and are obtained from established suppliers. In such cases, management must demand that SOP documents are provided along with the plans for the plant.

8.2 Safety Manual

Safety precautions are specific for each plant and service. Like SOP, precautions for each unit or activity must be written down in simple and clear language. The safety manual should be made available to all workers and supervisors. It is desirable that the manual include a check-list of the precautions to be observed at the start of every shift and during plant operation or maintenance. Periodical monitoring of safety performance is the responsibility of the senior managers.

8.3 Sign Posting of Warnings

It is essential to install appropriate warning signs in the plants, warehouses and workshops. Some of the more important warnings are given in Table 15.1

The design and message must take into account differences in language and cultural practices.

Table 15.1 Warning Signs

Situation	Sign
Volatile chemicals and solvents	NO SMOKING, NO OPEN FLAMES
Toxic chemicals	DANGER + Skull and crossbones or other readily understood symbol
Power line switch boxes	DANGER + high tension (lightening) symbol
Emergency evacuation	EMERGENCY EXIT
Restricted entry	NO ENTRY

8.4 Training and Motivation

Availability of SOP and a manual is by no means a guarantee of safety. Desirable safety standards are achieved only when workers are committed to translate norms into practice. This is done by suitably training the workers on procedures and precautions. Management must provide suitable training programmes for new employees and refresher programmes for all workers at appropriate intervals. Refresher programmes should also be used to update the safety procedures.

Experience has shown that the initial enthusiasm for safety on the shop floor begins to sag after a period of time. Workers progressively become casual about precautions, which makes them more prone to accidents. It is therefore essential to keep the work- force motivated on safety measures. Companies have achieved this in different ways, *e.g.*:

(a) *Group reward*: All workers in a plant/activity may be collectively rewarded for achieving long accident-free periods. The length of such periods may be fixed according to the work teams' experience and the degree of hazard. It is desirable to progressively increase this period to keep the challenge alive;

(b) *Safety competition*: Some companies organize competitions between plants and services for the highest level of safety standards achieved. Fairness in evaluation and reward is the key to the success of such competitions;

(c) *Individual awards*: Individuals should be encouraged to suggest practical improvements in safety systems. For suggestions accepted and implemented, management must reward the individual.

The objective of these and similar schemes, e.g. safety committees, is to cultivate a sense of belonging amongst the workforce. However, rewards should not be treated as a substitute for good management, including punishment for serious or habitual default on safety measures.

8.5 Medical Care

Medical care for the employees in pesticide formulation plants is essential. The main guidelines for such case are as follows:

(a) *Health monitoring:* Because of the risk of toxic exposure of workers to pesticides, a system of health monitoring should be established. Before an individual is assigned to a formulating or packing plant, a medical examination should be carried out, including blood cholinesterase level and urine tests. Thereafter, blood and urine tests must be carried out at regular intervals for all personnel working with carbamates, organophosphates or other toxic pesticides. Medical data must be recorded on individual health cards. A physician's advice must be sought in case blood cholinesterase activity drops by over 20%. If it drops by 40% or more, the individual must be removed from the work situation until the cholinesterase activity returns to at least 80% of the pre-exposure value and the return to work is approved by the physician. Similarly, a high level of pesticides in the urine is a warning of exposure and calls for proper medical advice.

Size of the workforce permitting, it is desirable to rotate workers from toxic areas to less or non-toxic areas;

(b) *First aid and medical attention:* Despite precautions and supervision, accidents can occur. Besides physical injuries, these may include acute exposure to toxic chemicals or excessive inhalation. To deal with these eventualities, a first aid manual similar to the one published in 1977 by EPA and information on antidotes must be available and accessible. Ideally, a treatment room with paramedical staff is desirable in a formulation factory. Additionally, supervisors and managers must be given training in first aid. First aid boxes must be placed at prominent points in the plants and workshop for immediate attention. In case of serious injuries a doctor should be called or the patient hospitalized. Eyewash facility should be provided in the emergency shower rooms in the plant. In case of suspected or acute intoxication with pesticides, the assistance of a qualified physician or hospital should be sought. The doctor/hospital should be informed of the nature of the chemicals handled and appropriate antidotes, e.g. atropine or pralimdoxine (PAM). Indeed, it is strongly recommended antidotes should be available in the factory itself.

8.6 Emergency Plans

Major disasters are rare in formulation plants. However, serious leaks or fire hazards may occur. The steps to be taken must be codified. Supervisors and operators must be trained in the procedures to be followed should such a need arise. It is desirable to carry out drills at suitable intervals so that the workers remain familiar with the code of practice.

8.7 Supervision

Management must recognize that precaution and safety facilities are no substitute for good supervision. Alert supervision eliminates or considerably minimizes risks. Some other steps may also be taken by management:

(a) *Safety valves and vents*. These must be periodically checked for proper functionality;

(b) *Effectiveness and accident review*. A periodical review of suitability of the procedures and facilities is good management practice. Each accident or accidental exposure must be thoroughly investigated to establish reason. Corrective steps should be initiated to plug the lacunae.

(c) *Hazard and operability audit*. In recent years, a technique to pin-point areas of hazard and minimize risk has been adopted by management. The hazard and operability audit involves a formalized study before commissioning a plant or making modifications at an operating plant. The study is undertaken by a team of three persons, including a worker, an engineer associated with the plant and an independent technician. It critically examines design, equipment, flowsheets and the actual plant and identifies locations involving risk from process reactions, equipment failure, raw materials or any other cause. It also recommends appropriate modifications to eliminate hazards to operators and other plant workers.

To conclude, industrial safety in the pesticide formulation industry is largely achieved by good plant design and procedures, but risk cannot be eliminated without the active involvement of the factory personnel and their commitment to observe precautions. The acid test of the success of management therefore lies in creating an awareness of the hazards and a safety-motivated workforce that recognizes the importance of not only the safety of the plant and personnel but also that of the environment.

REFERENCES

(1) M. Harmer and R.W. Homer, "Safety in storage and transport and disposal of pesticides and containers," *Formulation of Pesticides in Developing Countries* (UNIDO, 1983), p. 175.

(2) K.M. Naef, "Occupational hygiene and industrial health in pesticide formulation plants," *Formulation of Pesticides in Developing Countries*, (UNIDO, 1983), p. 165.

(3) S. Singhal, "Safety in manufacture and application of pesticides, *Pesticide Information* (Pesticides Association of India), 9, No. 3, 1983 p. 85.

(4) M. G., Srivastava, "Acceptable international standards for industrial safety in the operation of pesticides in developing countries, *Formulation of Pesticides in Developing Countries* (UNIDO, 1983), p. 155.

(5) *Manual for Pesticide Users*, Safety with Pesticides, (Pesticide Association of India, 1989.) p. 82.

(6) United States Department of Health, Education and Welfare, *Health and Safety Guide for Pesticide Formulators*, 1977.

CHAPTER

16

Advances in Pesticides Formulations Packaging

Jürgen Hartmann

Bayer AG
PF-Zentrum Monheim, PF-Forschung/FT-TE
Packungstechnologie Geb. 6800
D 51368 Leverkusen
Germany

1 REQUIREMENTS PROFILE

When developing packaging for a crop protection product, 10 basic rules are to be observed. Matters relating to the presentation of the package are not included in these rules.

1.1 Statutory Demands

Very specific legal requirements apply in the following situations:

(a) *Agrochemical legislation:* Country-specific regulations on safety in the workplace, protection of health and the environment and packaging must be satisfied to gain registration. The FAO Code of Conduct (1) provides valuable assistance;

(b) *Directives on quantity accuracy:* National laws are based on two principles. The mean value requirement is the fundamental idea of the directive of the European Community (2), which has been adopted as national law in European and many other countries. This statistically logical concept influenced the development of more accurate dynamically controlled filling and weighing machines.

The alternative concept is the minimum quantity requirement, which does not so strictly demand the economy of product;

(c) *Hazardous goods transport regulations:* United Nations recommendations on the transport of dangerous goods (3) cover the classification of products, packaging requirements, conditions for performance tests and the official registration of packages. They serve as a framework for more specific national and international laws;

(d) *Waste disposal legislation:* Governments in places such as Germany, the Netherlands and several states in the United States have already introduced laws on disposal. The common philosophy is based on a hierarchy of demands:

(i) Avoidance of packaging material,
(ii) Reduction of the quantity of packaging material,
(iii) Re-use of packages and packaging material.

Deposition in landfills or burying is least desirable alternative.

(e) *Laws on product liability:* Responsibility, in the event of damage is governed by such legislation. The tendency in many countries is that, in case of a defect, the manufacturer of the packaging product must prove it made serious efforts to prevent any damage.

1.2 Product Compatibility

The essential prerequisite is that there should be no interaction between the packaging material and contents. A pesticide formulation is said to be stable when the changes in its properties and characteristics over a given period of time vary only within predetermined, acceptable limits (4). Further details will be given in Section 3.

1.3 Pack Size

The packs should contain either the right amount for one use, e.g. the quantity for one hectare (taking into account active ingredient content and concentration) or a multiple of such amounts.

Solid products (WP, WG) have different bulk volumes and hence will require different size packages. When liquids are packaged, a certain space must be left empty (prescribed in the transport regulations).

1.4 Closure Concept

Basic requirements call for a hermetic seal and the possibility of opening the pack without tools and with a gloved hand. If the closure consists of two devices, one of them must be of the screw-type.

Tamper-proof seals are of increasing importance. The aperture must be large enough to allow glug-free pouring and easy rinsing (the minimum internal diameter should be 38 mm). For liquids the recommended design is a neck with a buttress-type thread as per DIN 6063 KS 50 mm, or as per GIFAP Industrial Standard 63 mm, using appropriate caps.

1.5 Design

Packages for liquid and solid products must be designed in order to allow a safe and economical filling process. They must be easy to handle and completely rinseable. Packages for liquid products must not have hollow handles, while those for solids must not have any hidden pockets. Graduations can facilitate the removal of partial quantities. GIFAP will publish detailed design criteria for packaging for crop protection products.

1.6 Filling Process

Packages are filled in semi- or fully automated processes. It is advisable to carry out trial runs before drawing up the definite specifications. Symmetrical packages for liquids, such as cylindrical containers with a concentric opening, facilitate filling. Form-fill and sealing machines and bag-in-box systems are common for powders and granules. Two-stage filling processes (first bulk and then controlled metering) improve accuracy and safety.

1.7 Information on the label

The package has to convey information from the manufacturer to the consumer. It is important also that containers for crop protection products should not be confused with edible food packs and, above all, that the empty packs are not re-used to preserve foodstuffs.

Embossed pictorial information is useful. Quick and precise identification is essential. The following need to be included: product name, manufacturer, quantity, registration marking, statements on risk and safety, hazard symbols, recommendations for application and first aid measures. The FAO Code of Conduct (1) provides valuable assistance in this regard.

1.8 Labelling

Labels must stick firmly to the packaging material; the glue must fulfil its function even under unfavourable conditions. It must not cause corrosion on metal packs. Overlapping paper labels are advisable. If the prescribed texts are too long to be accommodated, folded multiple (double, triple) labels could be used.

1.9 Transportation and Storage

The packs are mainly transported and stored on pallets. Consequently the shipping units must be designed for pallets, i.e. they must allow stacking on pallets without projecting. The material used to secure the load on the pallets must be strong and durable; shrink-wrapping or stretch films must not relax due to temperature or time.

Standard rules for palletization are as follows:

(a) Maximum load on one pallet, 1 tonne;

(b) Maximum height of the unit load, 2 m;

(c) If the individual packs are too small to form a safe load, half height, i.e. 1 m, is recommended.

The compression strength of the shipping cartons must comply with these rules and with customer specifications.

1.10 Disposal

The final task in developing packaging is to guarantee safe disposal. Quantities of packaging materials should be minimized. The selected material should not absorb the chemical. It should, however, allow the recovery of energy or be reusable after having been properly rinsed according to good agricultural practice.

2 SELECTION OF PACKAGING MATERIALS

2.1 Solid Products

There are four typical solid products:

(a) Wettable powders (WP);
(b) Water dispersible granules (WG);
(c) Dusting powders (DP);
(d) Granules (GR).

The universal material of construction is polyethylene. It is thermoplastic, therefore an ideal sealing medium, and it offers very good moisture barrier properties. To preserve the agrochemicals in a humid atmosphere is the main requirement. In monolayer constructions, a thickness of 0.1 mm will be sufficient in most applications.

If the product is highly hygroscopic (which fact would have been established in an accelerated storage test), the only suitable package would be of an aluminium composite foil. Depending on the weight of the product, a thickness of 0.009 mm (up to approx 1 kg) and of 0.012 mm for greater weights should be sufficient. Aluminium foil requires a sealable layer on the inside and, on the outside, protection against environmental influences, a strengthening layer and a pre-treated surface that can be printed on. Ideally, the shape should be symmetrical.

A typical composite foil would consist of 40 g/m^2 low-density polyethylene (LDPE) as the sealing medium, 0.012 mm aluminum, 20 g/m^2 LDPE and one of the following: 30 g/m^2 polypropylene (PP), or 0.012 mm polyester (PET), or 0.015 mm polyamide (PA).

Products with certain properties, e.g. a strong odour or volatile toxic by-products, require a composite film consisting of LDPE and an additional barrier, which should be PET or PA rather than aluminium. These are also an ideal reinforcement for larger packages, e.g. those containing 10–20 kg GR.

A typical construction would be 0.015 mm PA (external, to form the gas-barrier and to give physical strength) and 0.05–0.1 mm LDPE (internal, responsible for moisture protection and for sealing).

The principle in designing multi-layer films is that of additive properties. A comparison of the permeation rates of LDPE and PA in Table 16.1 illustrates this principle.

Table 16.1 Permeation rates of films[a]

	O_2 permeation/ transmission	*Water vapour*
	($cm^3/m^2 \times day \times bar$)	*($g/m^2 \times day$)*
LDPE, 0.1 mm	800	0.7
PA, 0.015 mm biaxially oriented	15	40
LDPE/PA biaxially oriented, 0.05 mm/0.15 mm	25	1.9

(a) 23°C and 75% relative humidity (DIN 53 122).

There are numerous alternative barrier plastics, of which only films based on vinyl might be used for crop protection products. In the author's opinion, the agrochemical industry should concentrate on the four most common films, mentioned above, which are available in almost all countries, as their demand is usually not as high as in other industries. This will result in economical solutions.

Film or composite foil packages for powders or granules, are usually tubular pouches made on form-, fill- and sealing (FFS)-machines. They can also be gusseted square-bottom bags or cartons with an internal lining.

Beside considering the protective properties, any decision will also consider costs. It is essential to calculate the cost not only of the material but also of the filling operation. More expensive packaging material can easily be offset by more economical filling costs.

2.1.1 Water-soluble film

A special packaging option for WP is water-soluble bags of polyvinyl alcohol (PVA). These packages do not protect the product in the usual sense, but they do protect the operator when applying the product. They contain premeasured doses and leave behind uncontaminated outer packages. The packaging material becomes part of the spray liquid, whose properties it could influence.

Tubular pouches made on FFS-machines or pre-fabricated bags for quantities from 0.1 to 25 kg have been introduced to the market. Of course, water-soluble packages have to carry water-soluble or at least water-degradable printing for identification and they need suitable outer protection, which adds to the packing costs. Cartons lined with almunium-composite foil are the usual choice.

Although water-soluble film is ideal for WPs, WGs and gels (GLs) (see Chapter 4), even liquid herbicide have also been packed in PVA bags.

2.2 Liquid Products

Liquid agrochemicals always contain an active ingredient and may also contain solvents and emulsifiers. As all these component chemicals influence the packaging choice, it is almost impossible to make a decision based on the literature alone. Usually compatibility studies have to be performed to prove the suitability of a packaging system. The tests should be carried out on the packaging in which the pesticide is to be sold, with original closing systems.

From the point of view of a packaging engineer, the most important types of formulations are the following:

(a) Emulsifiable concentrates (EC) and soluble liquids (SL), which are usually not compatible with high-density polyethylene (HDPE);

(b) Suspension concentrates (SC) and emulsion concentrates (EW), usually HDPE-compatible.

The most common packaging materials are glass, metal (aluminum or tin sheet), plastics and lined corrugated cartons.

2.2.1 *Glass*

In spite of its universal product compatibility, glass is not recommended for chemicals because of its fragility.

2.2.2 *Metal*

Usually, containers are made from aluminium or tin sheet with or without internal lacquer linings. Aluminium has the advantage that it can be converted in mono- or biblock containers, if a suitable alloy is used. Although its purity should be min. 99.5% (as per DIN 1712.3) an internal protective lining is required for many products. However, because the lacquer has to be applied in a discontinuous spray process, it is not very economical. The latest aluminium containers have standard closures (DIN 6063/KS 50), using seamed lids, diminishing the disadvantage of previous designs, which had small apertures and internal threads.

All tin sheet containers have to be threee-piece constructions, which means the seams must be precise and uniform to prevent leaks. The seaming compound also has to be tested for chemical compatibility.

Internal lacquer linings are generally required. They are applied by roller, which results in a rather uniform coating. Disadvantages include the possibility of pore formation and the resulting effect on adhesion. The longitudinal seam requires a separate protection. Due to limited construction possibilities, it may be difficult to remove product quantitatively.

Closures for metal packages must also have good barrier properties.

2.2.3 Plastics

Blow-moulded containers made of HDPE offer a wide flexibility in construction and good moisture barrier properties. However, they are suitable mainly for water-based products.

Chemical compatibility can be improved by fluorination, which is done either during the blow-moulding of the container or, off-line, after blow moulding in a vacuum chamber, resulting in the treatment of the internal and external surfaces of the containers.

Another means of modifying HDPE to improve its barrier properties, is to blend the resin with, for example, polyamide. The resulting laminar structure (SelarRB) will allow its use for products where HDPE normally is unsuitable.

The best possibilities are offered by coextruded, multi-layer plastic containers, which are available in sizes up to 10 litres. Here, plastics with different properties can be combined to produce multi-layer packaging tailored to the product requirements in the same way as has long been done with laminated films.

A typical 1-litre bottle (weight 85 g) is made of HDPE at least 0.7 mm thick, a bonding agent at least 0.01 mm thick and an internal EVOH or PA layer at least 0.02 mm thick. Reground material from the manufacturing process could be blended with virgin HDPE and used as a fourth layer between the external PE and the bonding agent without increasing the container weight.

Although the initial investment is higher and the process control has to be very strict, coextruded bottles are the most advanced solution for all products for which HDPE is not sufficient.

PET is widely used for the packaging of liquid products. However, its water vapour transmission rate is approximately six times greater than that of HDPE. This deficiency is often offset by its rather good compatibility with many chemicals.

The investment cost for the manufacturer of stretch blow-moulded containers is very high. As the resin is expensive, there is a constraint on bottle weight. For example, a 1-litre bottle is competitive with alternatives only if its weight is not more than 50 g.

There are signs that combination packs (film liners in corrugated cardboard boxes) may find application in the agrochemical industry. They will undoubtedly be successful if the problem of quantitative removal of liquid can be satisfactorily solved. Their advantage is the possibility of removing the contaminated lightweight liner for the clean outer carton and disposing of them separately.

2.2.4 Biodegradable Plastics

In view of the limited availability of landfills and the problem of disposing of packaging, biodegradable plastics are gaining attention. In fact it would be a splendid idea to design packages that just fade away.

We are not as optimistic as some about photo- or biodegradable packaging material, because when they are disposed of in a landfill, they will require as much space as other materials and they will take a long time to degrade. Moreover, they cannot be re-used as packaging material. Despite these reservations, an overview of the state of the art is provided in Table 16.2 (5).

Table 16.2 Photo- and biodegradable plastics

Source	*Product*	*Degradation type*	*Current/ potential use*	*Technology*
Ideamasters Miami, FL	Plastigone proprietary resin	Photodegradable additive system	Used in Israel in mulch film. Wider uses expected.	Additives absorb UV, releasing metal ions in anti-oxidant to act as catalysts for the breakdown of chains.
ICI Americas Wilmington, DE	Biopol resin	Biodegradable aliphatic/ polyester copolymer	Production of bottles in Europe.	Polyhydroxybutyrate polyesters are produced by soil bacteria during fermentation of sugars to produce a natural polymer.
U.S. Dept. of Agriculture, Washington, DC	Technology available for licensing	Biodegradable starch additive	Agri-Tech Industries, Gibson City, IL, plans pilot plant for blown film.	Starch is used to fill ethylene-acrylic acid or PE film, after which microorganisms attack the natural polymer causing film to degrade.
St. Lawrence Starch, Mississauga, Ontario	Ecostar master batch	Biodegradable starch additive	Used in trash bags and bottles.	Starch is silane treated for compatiblity with polymer; when consumed by fungi, the starch leaves porous structures that disintegrate.
Ecoplastics Willowdale, Ontario	Ecolyte master batch	Photodegradable ketone copolymer	Used in mulch film and has potential in bags.	Molecules are altered by interspersing ketone carbonyl groups on polymer chains during synthesis.

(Contd...)

Table 16.2 *(Contd...)*

Source	*Product*	*Degradation type*	*Current/ potential use*	*Technology*
Dow Chemical Six-pack, Midland, MI DuPont Wilmington, DE Union Carbide Danbury, CT	Resin	Photodegradable ethylene/carbon monoxide comonomer	Currently used in 11 states requiring degradablė six-pack yokes. Dow has applied for permission in food use.	Incorporates carbonyl groups by copolymerizing ethylene and carbon monoxide so that UV-absorbing carbonyls snap to form shorter, weaker chains.
Ampacet Mt. Vernon, NY	Poly-grade master batch	Photodegradable additive system	Used in Bespac trash bags made by Webster Industries.	A proprietary additive susceptible to UV light absorption degrades liner LDPE.
Princeton Polymer Labs Princeton, NJ	Additive system for licensing	Photodegradable additive system	No current commercial use. Suited for use in polyolefins and polystyrene.	Combines a photo activator and prooxidant synergistically to break polymer chains.

2.3 Large Returnable Containers

Intermediate between primary packs (Sections 2.1 and 2.2) and transportation packs (Section 2.4) are large multiway containers. In countries where the farms are very large, there is an increasing usage of minibulk multi-way (approx 500 l) and of small volume (approx 50 l) returnable containers.

Multi-way receptacles would be suitable for products required in large quantities, products applied over a long period of time and products that need quick refilling. In addition, the products (EW and SC) must remain homogeneous and should allow volumetric dosing. The advantage, which are primarily for the consumer, are obvious:

(a) Application in closed transfer systems, which reduces the risk of contamination to the operator;

(b) No need to dispose of empty containers;

(c) The containers can be used for the same product without cleaning them;

(d) Tampering with products can be avoided more easily.

However, the disadvantages cannot be overlooked:

(a) High initial investment costs;

(b) Higher requirement profile for the development of formulations.

The marketing conditions have to be adapted to special conditions of logistics and financing. It is recommended to use stainless steel kegs, as is common for beer. A chemical-specific coupling system for the transfer of material could prevent unauthorised re-filling of the container thereby avoiding contamination due to interchanging containers.

2.4 Transportation Packaging

The fundamental task is to protect product and environment by means of primary containers. Individual receptacles are conventionally collected in shipping cartons for dispatch from the formulation and packaging plant to warehouses and customers.

Attention must be paid to the quality of these cartons, since loads are often transported over long distances on bad roads. Faulty or unsuitable packaging can lead to accidental leakage of the product during transport and present a safety hazard. All packages should meet recognized performance standards (3), be able to withstand the conditions normally experienced in transport and provide the desirable level of safety (6).

The main concern is with corrugated cartons, which should have double nodulation and kraft paper at least on the external layer. One set of guidelines gives the very simple Cobb test (3). The moisture absorption should not exceed 155 g/m^2. In humid climats, the stacking stability of corrugated cartons can be up to 50 % less if paper of poor quality has been used.

The more recycled fibres are used, the more important proper specification and stringent testing of the corrugated carton. Corrugated boxes should be safely closed by means of suitable tapes. Staples are not recommended as there is always a risk that irregular hooks will puncture the bottom of the internally packed containers.

It improved safety and economics if the shipping boxes are stacked on pallets and protected by shrink film (min. thickness 0.1 mm) or stretch wrapping. The preferably interlocking arrangement of the boxes on pallets needs thorough planning. When developing primary packages (bottles, folded cartons), the arrangement of the shipping cartons and the utilization of the surface of the pallets must be considered at a very early stage.

Protection of the load with shrink film or with stretch wrapping should take care of the safety requirements in the plant. However, since it is more difficult to operate shrinking equipment in a packaging plant that has to observe explosion-proof conditions, spirally wound wrapping may be the best method. During storage, pallets have to be separately covered with a film hood.

3 MEASURES OF QUALITY CONTROL

3.1 Storage Stability Test (4)

The stability of a pesticide formulation depends on the intrinsic stability of the active ingredient(s), the formulation itself and the protective function of the packaging material.

The quality of a pesticide formulation is not only that quality which the manufacturer measures before the product is released for sale but, more importantly, it is also that quality which is present when the end-user finally applies the product. The end-user therefore expects that the manufacturer will have carried out tests to determine whether the product will be fit for use at its final point-of-use. As the manufacturer cannot follow on a real-time basis all the potential transportation, storage and environmental conditions that a product may be subjected to, tests need to be devised that will simulate such conditions as closely as possible. Such tests are called stability tests.

Based on the knowledge gained from studies of the formulation, important conclusions can be drawn from the stability tests in the selection of the primary packaging configuration. Specifically, stability tests will yield the following information to the formulator:

(a) Whether the commercial pack provides adequate protection from the climatic influences to which the formulation may be subjected;
(b) Whether there is a catalytic reaction between packaging material and the formulation;
(c) Whether the function of the package is unaffected by the product it is supposed to protect (cap impossible to open, stress cracking of the container etc.).

Before initiating any stability study it is essential to plan this carefully. In the planning of stability studies the following points must be considered:

(a) *Storage conditions*: It is essential that the storage conditions selected for stability studies are identical and standardized;

(b) *Test intervals*: Typical time intervals for removing samples for testing are 4 weeks, 8 weeks, 3 months, 6 months, 12 months and 24 months. Table 16.3 shows recommended storage conditions and time intervals for both accelerated and real time studies;

(c) *Tests required*: It must be pointed out that not every recommended test needs to be carried out at every recommended test point and once the stability profile has been established and the stability indicating parameters are known, the more selective one can be in the choice of tests to be carried out. Depending on the type of the formulation, the testing for indicator parameters (Table 16.4) should be carried out. Table 16.5 gives an overview of tests that should be carried out on important types of containers. Irrespective of the type of the final container, the tests in Table 16.5 should always be carried out as the minimum requirement: weight loss or gain, appearance of the container (including internal surface) and appearance of the container/closure system;

Table 16.3 Storage conditions and test intervals

Time		*−10°C*	*0°C*	*25°C*	*30°C 85% RH*	*40°C 85% RH*	*54°C*
Initial				A			
4	weeks		C	C	C	B	C
8	weeks					B	
3	months	D		B	B	B	
6	months			B	B	C	
12	months			B	B		
18	months			B	B		
24	months	D		B	B		
30	months			C	C		
36	months			C	C		

A = Initial tests are carried out at prevalent laboratory temperature and humidity.
B = Recommended test intervals.
C = Optional test intervals
D = Reconfirm "initial" if necessary

Table 16.4 Recommended tests to be carried out on formulations

Formulation	*Test required*	
EC	Acidity/alkalinity Water content Specific gravity	Emulsion stability Oxygen content/head space
SC	pH Specific gravity Sieve test (wet)	Particle size Suspendability Viscosity
SL	Acidity/alkalinity Water content Specific gravity	Emulsion stability Spontaneity
WP	pH Water content Sieve test (wet)	Wettability Suspendability
WG	pH Water content Sieve test (wet)	Wettability Suspendability Dust content
Dusts	pH Water content	Sieve test (dry)

Table 16.5 Recommended tests of final packs

Packaging	*Test required*	
Containers		
Tin sheet	Tin layer thickness Quality of seams	Corrosion
Tin sheet with protective lining	Porosity Adhesion	Elasticity
Aluminium	Corrosion	
Aluminium with protective lining	Porosity Adhesion	Elasticity
HDPE	Deformation/wall thickness Odour barrier properties	Colour changes Drop stability
PET	Internal pressure (head space) Oxygen content	Stress cracks
Coextruded plastics (multi-layer)	As for HDPE; additionally thickness of single layers, homogeneity and porosity of the barrier	
Closure systems		
Caps	Torque Differences in dimensions	Leaks Stress cracks
Gaskets	Weight differences Differences in dimensions	
Sealing disks	Corrosion Adhesion of laminates/ delamination	
Films		
Laminated films/foils	Odour barrier properties Tightness of seals Brittleness Adhesion of laminates/ delamination	Deformation Corrosion Porosity
Water-soluble films	Water solubility Tear resistance Total penetration Energy (free-falling dart)	Brittleness Elongation Spray properties (with product)

(d) *Test methods*: An absolute prerequisite in the choice of the best test method is that it should be stability-indicating. Secondly, it must be capable of giving a true and accurate initial value. Thirdly, all the parameters that might possibly be affected by a change in quality with time/storage conditions must be determined as an initial value at t = 0. Fourthly, one should not rely solely on recognized international methods, e.g. CIPAC methods, for stability testing. These methods should be supplemented with in-house methods if these are better in detecting changes in product characteristics with time/storage conditions. Fifthly, one should not change the test method without due cause and certainly not in the middle of a stability test programme, without running both methods in parallel;

(e) *Major review points*: After each check-point the results should be reviewed to determine whether the product is behaving as expected. For accelerated studies, a major review should be conducted after four weeks and no later than the third month, at which time a graphical presentation of the data is very valuable;

(f) *Presentation of stability data:* As the purpose of stability testing is to determine change with time, it is important to present the stability data in such a way that such changes are immediately visible. A recommendation is to present the results of each test at each individual storage condition in vertical sequence, from the initial value up to the value measured at the longest time. Major changes in stability indicating parameters should be presented graphically;

Apart from the obvious use of determining whether the product will have an acceptable use life under various climatic conditions in the selected containers, the results of stability studies can be used for the following:

(a) Fixing specification limits;
(b) Evaluating changes in the manufacture of the active ingredient, a change in the synthetic route for making it or a new source for it;
(c) Testing the influence of a changed formulation.

If the commercial pack is changed, e.g. if the manufacturing process of the primary packaging configuration is altered substantially or another supplier is selected, it is necessary to investigate, under accelerated conditions, critical parameters of the formulation and package.

3.2 Continuous Quality Checks

As with all manufacturing processes in the chemical industry, it is important for packaging processes to have a continuous and well-organized documentation of quality measurements. Therefore, a system of in-process control has to be established. The minimum requirement would be to identify the product and all elements of the final package configuration and to control the product quantity on a statistical basis, the function of the closing system and the completeness of the labelling.

The regular inspection of ready goods in warehouses, prior to their dispatch, allows making a detached judgement to the packaging development engineer.

3.3 Control of Incoming Material

All incoming packing material must be tested as to whether or not it meets the specifications. This requires the use of statistical analysis techniques such as statistical sampling, test methods for the important properties, designed experiments and control charts (7).

Elements common to many effective quality assurance programmes are outlined in ANSI/ASQC Q 94 (8). The control of dimensions and weights is the minimum requirement. Other important tests were summarized in Table 16.5.

The test programme and the test methods have to be worked out in agreement with the suppliers. These negotiations should take place in an atmosphere when no problems with the quality of the supplied material or the design criteria are disturbing the peace between the organizations. Harmonized specifications, test programmes and test methods are especially needed in critical, controversial situations for an independent evaluation of the quality of a product.

Periodic audits of the suppliers should create a climate of confidence. This is a prerequisite for the acceptance of certificates. But there should be no misunderstanding: certificates do not replace the receiver's own well-organized control of incoming material and the systematic documentation of results. Certificates may help to minimize the need for in-house control measures.

4 SPECIFICATIONS

A correct specification is the first step necessary to ensure the consistent performance of the completed package. It is important that all packaging material components are precisely specified in package specifications ("raw material specifications"), but it is also important to give clear instructions to the refilling plant. The latter task, packaging instructions, should be independent of the first one.

4.1 Package Specifications

The package specification is the main means of communication between the packaging plant and the supplier. It must define all critical types of raw materials, dimensions and physical and chemical properties. Drawings are self-explanatory. Specifications have to be accepted by the supplier as well as the receiver of the material and must be continuously updated to reflect the latest technology.

Typical package specifications are shown in Table 16.6 and in Figure 16.1.

4.2 Packaging Instructions

The instruction for the refilling and packaging plant has to be as precise as any other specification as it is the plant's main source of information. It also is useful for planning and logistics, for quality assurance (in-process and final goods control) and for accountability. The packaging instruction contains information for the warehouse and for marketing.

The kinds of information that might be included in packaging instructions are shown in Table 16.7.

Table 16.6 Package specification example

Requirement	*Package specification*
Definition of the item	HDPE bottle
Nominal volume	1 litre
Code no.	7 123 456
Prescription	Cylindrical bottle, extrusion blow-moulded, neck KS 50/ DIN 6063
Raw material	Lupolen .... Hostalen virgin material reground only from the same process (closed system recommended)
Colour	Natural transparent
Print	Neutral
Total volume	1.1 litre
Weight	85 ± 3.4 g
For further technical details refer to drawing no.	3353 g
Test instruction (programme, AQL, procedures)	03, dated 1.5.1995
Delivery instruction	Palletized 1.2 m × 1.0 m × 1.8 m, each layer packed on trays, securely wrapped
Date/signatures	

Table 16.7 Packaging instruction

Category	*Details*
Parts list	Code numbers of bottle, closure, label, shipping box, tape, pallet, shrink film
Instruction for the packaging plant	Torque, type of glue, type and width of tape
Weight accuracy	Mean weight, tolerance limits
Marking and labels	Numbers and positions of hazard label (3), batch number, date of manufacture, method of application
Stacking plan	Position, size of pallet, arrangement of boxes and number of layers per pallet, means of securing, gross volume and weight per pallet, warehouse instruction on max. height for stacking of pallets
Annual colour code	Annual colour (first in/first out)
Pallet label	Identification of the pallet load

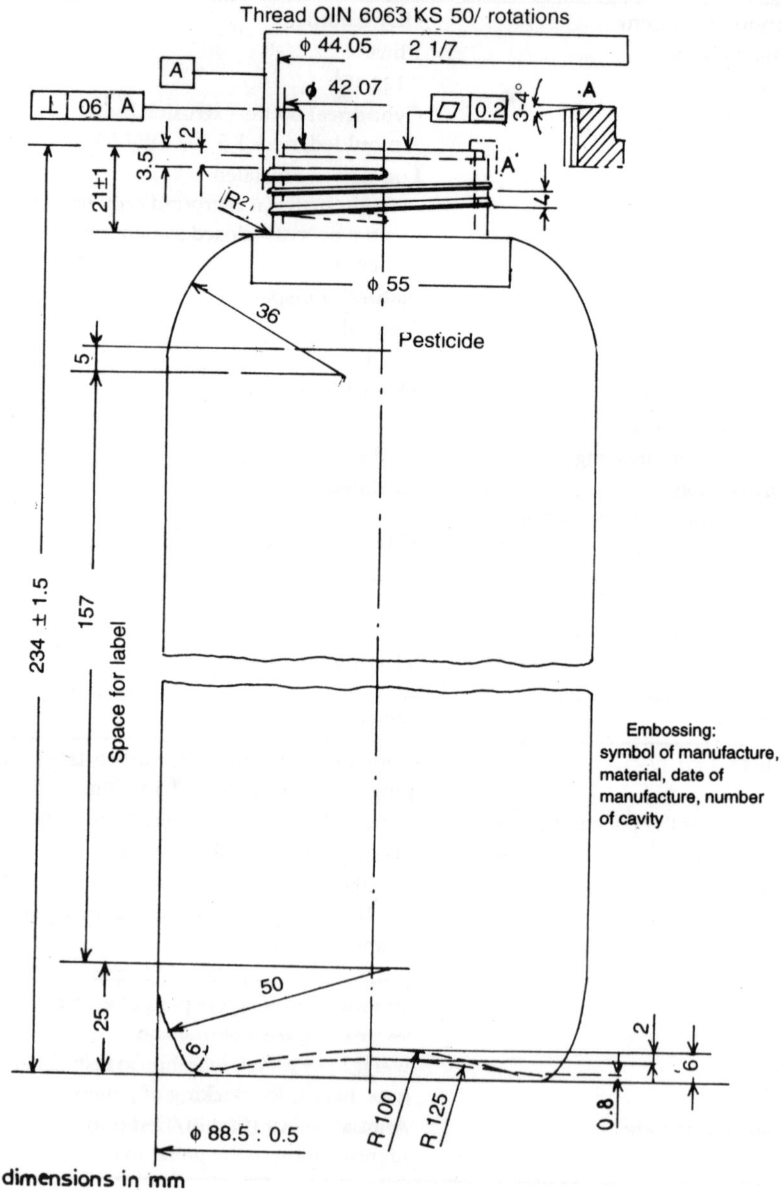

Figure 16.1 Diagrammatical specification

5 DISPOSAL

The agrochemical industry has taken responsibility for the safe, environment-friendly disposal of post-consumer packages. Early in the development stage, this task has to be taken into consideration. The duties of the various parties may be seen as follows:

(a) Industry selects materials of construction, that do not allow intolerable rates of permeation and that allow at least the recovery of energy. Containers are designed so that rinsing is easy; this requires:
   (i) Wide apertures, min. 40 mm diameter;
   (ii) No hollow handles in cans or no hidden pocket in bags;
   (iii) The product can be removed completely (max. 0.01% product residue) after rinsing as per standardized methods with integrated rinsing equipment on the spray rig.

(b) Farmers are committed to rinsing as per the above mentioned procedure and are willing to return the empty packages to the point of sale;

(c) Dealers are prepared to collect the post-consumer packs after controlling for correct rinsing. This makes them responsible for buying goods that can be easily disposed of;

(d) Disposal companies with professional qualifications, take care of the rinsed and clean containers. In a cascade flow they recover the energy content, preferably in cement works or in controlled incineration plants, or re-use the material if possible.

The empty containers must not litter the environment. Farmers should assume that responsibility when they decide to apply crop protection products.

REFERENCES

This chapter has been developed from Jürgen Hartmann, *Technical Report: Findings and Recommendations*, prepared for the Government of India, UNIDO, 1987.

(1) *Code of Conduct: Resolution 10/85*, 23rd FAO Conference, November 1985.

(2) *EG-Richtlinien über Fertigpackungen*, 75/106/EWG, Amtsblatt No. L 42, 2/15/1975, last revision Amtsblatt No. L 398/18, 31/12/1989 (89/676/EWG).

(3) *Recommendations on the Transport and Packaging of Dangerous Goods*, 9th rev. edition (New York, United Nations, 1995).

(4) Norman C. Franklin and Jürgen Hartmann, "Storage stability of pesticide formulations", *Proceedings of the seventh IUPAC-Congress*, 1990.

(5) Robert Leaversuch, *Modern Plastics International*, October 1987.

(6) *GIFAP, Guidelines for the Safe Transport of Pesticides*.

(7) Richard L. Skeehan, "Quality assurance in the packaging industry", *Journal of Packaging Technology*, September/October 1990.

(8) *Quality Management and Systems Elements - Guidelines*, ANSI/ASQC Q 94, American Society for Quality Control, 1987.

CHAPTER

17

Waste Management and Disposal in the Pesticide Formulation Industry

K. S. Johnson

Environmental Consultant
76 Pound Road, East Peckham
Tonbridge, Kent, TN12 5BJ, United Kingdom

1 INTRODUCTION

The disposal of wastes produced during the formulation of pesticides is an area of extreme environmental sensitivity. Strict regulatory controls govern all aspects of waste handling and preparation through to the final process of disposal. Similarly, the requirement for high standards in the waste disposal industry has limited the scope of pesticide waste outlets largely to high technology disposal systems (incineration). Currently these outlets are limited in capacity and involve high costs.

Few pesticide formulation plants have sufficient in-house facilities for the safe disposal of all the wastes produced and must resort to disposal through third-party waste disposal contractors. However, duty rests firmly with the waste producer to ensure that all wastes produced are disposed of safely and responsibly. It is vitally important, therefore, that all formulation units have adequately resourced and funded facilities to support a secure system of waste management and disposal.

This chapter reviews wastes arising at all stages of the formulation process, suggests, where appropriate, methods for reduction at source and recommends a disposal approach for specific waste types. The important aspects of waste identification, documentation and recording are discussed and examples are given.

Necessary standards for the transportation of wastes are highlighted, with particular reference to the use of third-party vehicles. The use of waste disposal contract services is reviewed along with proposals for an audit on the services provided.

Finally, it is acknowledged that situations governing the overall disposal of wastes will vary considerably depending upon location, local regulations and controls. It is hoped, however, that the advice and guidance offered will go some way in providing the basis for a system to manage this important and sensitive area of pesticide formulation manufacture.

2 WASTE MINIMIZATION

Waste minimization, or reduction at the source, is one of the most important elements in a factory system of waste management. Stringent legislative controls, along with public concerns over the deposit of chemicals to the environment, have resulted in restricted outlets for disposal with attendant escalating costs. There are strong incentives for the industrialist to critically review all stages of process manufacture to identify ways of reducing waste outputs at the source.

The benefits are obvious, offering opportunities for savings from recovered raw materials, which otherwise would have been lost, labour and final disposal costs. In planning a waste minimization strategy, consideration should be given to both solid waste and liquid effluent streams arising directly or indirectly from process operations.

In the initial phases of planning of a waste minimization programme, two main areas should be addressed: existing processes and new and modified processes or plants.

2.1 Existing Processes

All stages of the manufacturing process should be examined to identify and quantify the waste streams produced, along with potential losses of raw materials and final product, plus associated costs for waste disposal. This initial stage is important to establish a benchmark against which all future waste reduction and subsequent cost savings and benefits may be measured. During this initial process, objectives and plans for future improvements should be prepared, starting with the high-waste, high-cost elements of the process.

To ensure full commitment to the scheme, the "producer pays" principle should be adopted. This requires the plant producing the waste to pay directly for disposal. Normally, these charges are met from a central budget, which would give the producing plant no incentive to reduce wastes.

Targets for waste reduction for each process should be agreed and set over a sensible time period, e.g. a 50% reduction in waste outputs over 1-2 years. Performance against these objectives should be regularly monitored and reviewed.

2.1.1 *Waste reduction*

Improvements in process efficiency and associated waste reduction require a team effort involving staff at all levels, including senior management. A positive approach is the use of a small, representative task force of staff members to study problem areas and identify solutions.

A number of options exist in the formulation of pesticides for recovery and re-use of waste materials. A common constraint is one of cross-contamination, frequently encountered in the cleaning of equipment in batch-oriented production operations that preclude re-use. Simple operational changes and process plant modifications can often minimize or overcome these problems. For example,

(a) *Liquid wastes.* Considerable quantities of valuable active ingredients can be lost to drains during plant cleaning activities. Every opportunity should be taken to intercept and recycle uncontaminated aqueous and solvent washings back into the process. This may involve, at the end of campaigns, the storage of washings in drums for later use. An investment in dedicated storage vessels, where appropriate, may be worthwhile;

(b) *Solid wastes.* A large volume of solid wastes (powder residues) can come from dust extraction units. The use of dedicated extraction equipment for single-product units enables uncontaminated product residues to be recycled directly back into the process.

2.1.2 *Management and Operation*

The success of any waste minimization exercise is dependent upon the full support and commitment of senior management to the project. An overall strategy must be agreed ensuring that all initiatives are properly resourced and funded, with subsequent benefits identified and recognized. This will ensure that the programme receives attention from the highest management levels and that its importance is understood and accepted by all levels of staff.

2.2 New Plants and Modified Processes

Situations involving the construction of new plants or modifications to existing processes offer opportunities to incorporate waste-reduction measures at the plant-process design stage. Site procedures should be amended, where necessary, to ensure that all aspects of environmental pollution control (along with health and safety) are fully addressed during the early stages of a project.

An environmental statement containing waste and effluent minimization features should be incorporated in the initial project documentation.

3 CLASSIFICATION OF WASTE TYPES

Essentially three classes of wastes regularly arise during the formulation of pesticide products requiring safe disposal: (a) non-toxic wastes, (b) lightly contaminated wastes and (c) toxic wastes.

3.1 Non-toxic Wastes

These comprise discarded packaging materials, for the most part clean and uncontaminated, including cardboard, paper and plastics. Volumes can be large, and for this reason alone, they require regular removal.

3.2 Lightly Contaminated Wastes

Large amounts of these wastes regularly arise from both manufacturing and re-packing operations:

(a) Discarded contaminated packs including paper and plastic bottles and cans;

(b) Assorted metal drums containing residual quantities of raw materials, active ingredients and products.

3.3 Toxic Wastes

These represent a wide range of difficult-to-handle and hazardous waste materials:

(a) Liquid and powder residues, sludge from process operations;

(b) Laboratory wastes, solvents, solids, miscellaneous unwanted chemicals and outdated samples;

(c) Obsolete raw materials and products that cannot be recycled and have no further use.

4 WASTE HANDLING AND STORAGE

In the interests of safety and good working practice, it is essential that a well-organized system of waste handling is established and operated in the factory workplace. Procedures for these tasks must be clearly defined and implemented; adequate resources in the form of equipment and manpower must be provided as part of the factory site service.

4.1 Waste Collection

Clearly defined designated areas in each manufacturing location must be allocated for the deposition and subsequent collection of wastes for disposal. It is necessary to segregate all wastes at source into specific types (Section 3) to ensure these are processed through the correct disposal routes.

A system of colour-coded bins within each waste collection area offers a means of ensuring that each waste type is deposited into the appropriate collection container. A colour-coding system should clearly distinguish among specific waste types: non-toxic, green;

lightly contaminated, blue; and toxic and hazardous, red.

4.2 Waste Removal from the Workplace

These operations must be regulated to comply with the production activity and the needs of good housekeeping and safety. Generally, non-toxic and lightly contaminated wastes are bulky and should be cleared from work areas daily to a central collection point for subsequent disposal. Prior to their removal, toxic wastes, normally more concentrated and of specific composition, require strict controls to ensure good containment, labelling and records. Special authorization from plant management for the clearance of toxic wastes should be obtained before they are transferred to a secure waste storage area.

4.3 Health and Safety

All waste products are potentially toxic and dangerous. The same full protective clothing required for the manufacture of pesticides should also be worn when handling waste products. As a minimum, the following items of protective clothing should be available to operators involved in waste disposal activities: overalls, aprons and arm protectors, rubber boots, safety helmets, face shields, goggles, half-face respirators with organic vapour cartridge and dust masks.

The factory safety adviser must always be consulted on all matters relating to operator safety.

5 PREPARATION OF WASTES FOR DISPOSAL

All waste materials must be safely packed, labelled and, where appropriate, compacted to a minimum bulk before disposal. At most factory sites, adequate supplies of second-hand unwanted drums are available to overpack and safely contain toxic wastes.

Bulky packaging materials and small containers present problems of sheer volume and whenever possible should be compacted on-site before disposal.

A range of proprietary equipment (1) is available for this purpose, and in particular three items for general regular use are recommended for consideration:

(a) *Drum crusher*. Compaction of metal drums of all sizes up to 200 litre capacity to approximately one fifth of the original volume;

(b) *Waste shredder*. Effectively shreds paper, plastic, polyethylene bottles and thin-gauge metal containers into small pieces. Overall bulk is reduced by a factor of 10 : 1 and ensures that containers are not scavenged for secondary use;

(c) *Waste compactor*. Low-density waste paper, polyethylene, small containers etc. are compacted to reduce volume by a factor of 5 : 1 and simultaneously overpacked with a heavy gauge polyethylene bag to eliminate contact exposure risk from dusts etc. to operators.

The overall benefits from the use of this equipment include improved operator safety, cost savings in disposal charges (per unit volume) and conservation of landfill disposal space.

6 DISPOSAL OF CONTAMINATED CONTAINERS

Large numbers of emptied contaminated containers are produced during the manufacture and reprocessing stages of agrochemical pesticide products. The containers vary in size from 1-25 litres (mainly polyethylene) to 50 kg and 200-litre steel drums, used largely for raw materials and bulk supplies of products.

Of necessity the containers are robust in construction and are therefore attractive for alternative secondary uses. Unfortunately it is virtually impossible to completely clean and detoxify them to an acceptable standard for alternative use outside the pesticide industry. Pesticide residues absorb strongly into the walls of containers and will, over a considerable time period, continue to leach out into any liquid medium subsequently placed in them (2).

Once disposed of outside the controls of the pesticide industry, subsequent reuse of the containers is difficult if not impossible to monitor. The prime concern is that the containers could subsequently be used for storage of drinking water, food or animal feedstuffs, with the attendant risks of toxic exposure and poisoning. These reasons alone justify a policy of non-reuse, except within the industry, for all containers that have contained pesticide active ingredients.

6.1 Decontamination and Immobilization

Large numbers of both plastic (polyethylene) and steel containers regularly accumulate for disposal. The sheer volume can present difficulties in disposal, particularly where high-temperature incineration is not available.

Manual methods of rendering containers unusable, by rupturing or crushing, are labour-intensive and result in large volumes of waste for final disposal. A mechanical approach to this problem can be applied using simple shredding and crushing equipment (Section 5).

Necessary washing of containers for the removal of gross-product residues produces contaminated effluent. These can be treated to a high standard suitable for reuse. Raw material containers previously used for adjuvants such as oils, solvents, surfactants and inert fillers are suitable for recycling and should only be disposed off to professional drum reconditioning contractors. This form of disposal, normally carried out on a contract basis, should include regular auditing of the contractor's facilities to ensure good standards of reconditioning are used and to monitor the end-use of the refurbished container.

6.2 Drum Crushing

As a minimum all unusable steel containers should be punctured and flattened before disposal. This can be best achieved by the use of proprietary drum-crushing equipment. Ideally the containers should be rinsed with water and detergent to remove pesticide residues. If this is not possible, the addition of dry sand or other absorbent to soak up liquid residues prior to crushing is recommended.

Washings from the process can be treated for reuse. By placing small containers inside larger drums, it is possible to crush an assorted range of these into compact bundles to approximately one fifth of the original volume.

Final disposal for crushed drums should be either to a steel smelter for recovery as scrap metal or burial in a licensed landfill site.

6.3 Treatment of Contaminated Washings

Aqueous washings from contaminated container washing operations are toxic and polluting. A simple treatment combining chemical flocculation to remove suspended matter and physical adsorption to remove dissolved organic matter will produce a high-quality, detoxified effluent suitable for reuse in drum washing operations.

Small treatment units capable of treating 1, 5 and 10 cubic metres of effluent per day are now commercially available (3). The effluents produced are of high quality and suitable for reuse as wash waters.

7 WASTE STORAGE

Environmental legislation globally will impose strict licensed controls for the storage of all waste materials awaiting disposal. License conditions will vary with different controlling authorities, but the following provisions should be regarded as a basic requirement for a waste storage facility:

(a) Each formulation site should have a formally designated area for waste assembly processing and storage while awaiting future disposal;

(b) The waste area should be contained within a compound having a wall or fence at least 2 m high. Access gates should be of similar height and wide enough to allow the movement of fork-lift trucks and other transport;

(c) All floor areas should be constructed from impervious materials such as concrete or tiles and bunded to contain all spillage. The floor should be either sloped or drained to an integral floor sump, from which intercepted liquid residues may be removed for disposal elsewhere;

(d) Depending upon climate, a section of the storage area should be roofed over to protect operators and equipment from inclement weather;

(e) Clean roof waters should be diverted via sealed drains to an approved surface-water disposal system;

(f) A small office facility should be considered within the secure area to allow the preparation and maintenance of essential records;

(g) The waste storage area must be designated a secure zone, with only authorized personnel permitted access.

8 WASTE DISPOSAL OPTIONS

Four main disposal options are currently available for general use in the disposal of wastes:

(a) *Recycling and reuse*. This is always the preferred option and should be used whenever possible;

(b) *Landfill disposal*. Wastes are co-disposed with domestic refuse, which acts as a diluent, absorbent and ultimately a biomass for the degradation of substances deposited. Strict controls over the use of landfill make it suitable only for the disposal of very dilute pesticide wastes.

(c) *Low-temperature incineration (300°C-500°C)*. This is suitable only for combustible non-toxic and some lightly contaminated wastes. These units operate without secondary gas-stack scrubbing equipment;

(d) *High-temperature incineration (1000°C-1200°C)*. Usually operates in a two-stage combustion process: primary combustion 600-800°C, secondary combustion 1000-1200°C. However, it is expensive to install and to operate, regular feedstock is needed to maintain efficient combustion (capacity 1-10 t/hr) and most large units are required to install full stack-gas scrubbing equipment.

8.1 Disposal of Specific Waste Types

8.1.1 Non-toxic waste

There are three options for non-toxic wastes:

(a) Recycling, reuse;

(b) Combustible waste: By low-temperature incineration or to a landfill facility;

(c) Non-combustible waste, including ash from low-temperature incineration: to a landfill disposal facility.

8.1.2 Lightly contaminated waste

These wastes consist mainly of discarded, empty bags and containers from process operations that are lightly contaminated with residual chemicals and adjuvants. Some containers may have a designated in-house value for reuse. As a rule, however, these wastes should be regarded as unsuitable for re-cycling outside the pesticide industry. In particular this applies to pesticide containers, which has been discussed separately (Section 6).

Caution should be exercised in the low-temperature incineration of these wastes, because, toxic and irritant fumes and vapours may occur due to incomplete combustion. The preferable route would be to seek disposal of compacted, overpacked wastes to an approved, secure, licensed landfill facility.

8.1.3 Toxic waste

These wastes are highly sensitive and are subject to strict legal controls covering storage, movement and final disposal. A critical examination should be made of all waste types to explore possibilities for recycling or reuse at source. Particular reference is made to unwanted, time-barred pesticide products that have no sale value but retain a potential use in field applications, i.e. herbicides. Consideration should be given to offering these materials gratis to carefully selected outlets for use under controlled conditions. This option offers the opportunity to use the product for the purpose it was intended for as well as to avoid expensive disposal costs, which can exceed the market value of the product.

Toxic wastes for which no other outlets exist must be disposed of responsibly and safely. Currently the only acceptable outlet for them is high-temperature incineration. This service is expensive and in certain locations limited or non-existent. Consultation with waste disposal authorities is strongly advised. High-temperature incineration facilities, in the main, are owned and operated by the following groups:

(a) Several large multinational agrochemical companies with in-house facilities; direct contact with these companies may result in limited spare capacity being made available and is worthy of exploration;

(b) Government-owned and -operated facilities, mainly in developed countries. These usually operate at the local or regional level for the disposal of indigenous wastes but may offer resources where appropriate;

(c) There are professional waste disposal contractors, mainly in Europe, the Far East and the United States. They may, subject to prevailing regulations, accept for disposal specified waste types from some developing countries. Wastes sent via this route will require compliance with trans-frontier transport regulations in addition to local disposal regulations in the receiving country;

(d) Cement kilns operating at temperatures in excess of 1500°C with long residence times and having the potential for the thermal destruction of organic wastes, including pesticides. Earlier trial programmes conducted in several countries (4, 5) have demonstrated the effectiveness of this process particularly for liquid pesticide wastes. However, sensitivity surrounding the matter of pesticides disposal, along with the reluctance of the cement industry to accept a possible tier of hazards into a process currently free of such risk, continues to hinder progress on the use of this disposal option.

Currently, a number of international organizations representing the agrochemical industry (6, 7) are in the process of preparing guidelines for the disposal of unwanted pesticide stocks. All available state-of-the-art disposal technology is being reviewed and up dated information is being collected on availability and assistance that may be obtained.

9 DOCUMENTATION AND RECORDS

The handling, storage, movement and final disposal of waste materials is subject to strict legal controls. It is vital, therefore, that every aspect of the waste disposal process should be fully documented and recorded.

9.1 Waste Origin

Records should include information on process, waste composition, quantities produced and details of packages and containers. Attempts at recycling and reuse should also be recorded, even if unsuccessful, to demonstrate that these attempts were made. A simple check-list for the in-house recording of this type of information is shown in Table 17.1.

Table 17.1 Sample check-list for in-house recording

Reference Number:

WASTE DATA RECORD

PRODUCT/MATERIAL DESCRIPTION:
QUANTITY (LITRES/KG):
TYPE/SIZE OF CONTAINERS:
LOCATION:
CONTACT NAME:
REASON FOR DISPOSAL:

RECYCLING RE-USE

CONFIRM THAT THESE ALTERNATIVES HAVE BEEN CHECKED:

Please tick:

RE-PROCESSING ______
RE-CYCLING ______
DISPOSAL BY SPECIAL ARRANGEMENT ______

AUTHORIZATION (HEAD OF DEPARTMENT)
SIGNED: __________________
RECEIVED INTO WASTE STORE
SIGNED: __________________

WASTE DISPOSAL INFORMATION

DISPOSAL CONTRACTOR:
REFERENCE NUMBER/INVOICE:
DISPOSAL AUTHORITY:
REFERENCE NUMBER:
DISPOSAL COST:
DISPOSAL DATE:

9.2 Waste Identification

All wastes must be stored in secure containers and fully labelled with a contents description, appropriate hazard warnings and serial reference numbers, cross-referenced to a waste manifest document. This is a necessary part of the waste disposal procedure and in many countries a mandatory legal requirement.

Copies of these records should be available for in-house records, notification to waste disposal authorities, transport requirements accompanying waste loads and emergency situations.

10 WASTE DISPOSAL CONTRACTORS

Few pesticide formulation sites have resources or facilities for the complete and safe disposal of all waste materials produced. Use is normally made of professional waste disposal contractors to carry out the disposal service. Most waste disposal contractors are highly professional, reputable organizations. Charges for their service will be high, but these reflect the high cost of the technology required to maintain a competent and credible operation. The status of all contractors should be investigated before using their services and, if possible, the operation checked out with the controlling waste disposal authority.

A formal contract must be agreed upon with the contractor to ensure legal acceptance of mutual responsibility and liability. As a standard practice, the contractor's facilities should be checked out before the service is commenced and at least on an annual basis thereafter checks should be made to ensure that standards are being maintained. Occasional spot checks on the operation offer good insurance against any interim lapse in standards. The disposal process and service for each waste consignment must be discussed and agreed with the contractor in advance. It is important to obtain from the contractor a certificate of disposal for each consignment of waste handled, confirming that it has been disposed of in accordance with the terms of the contract. This should be provided by the contractor prior to payment for the disposal service rendered.

11 TRANSPORT

The safe transportation of waste from the producer's premises to the disposer's site is the most vulnerable stage of the waste disposal operation. Loss or insecure handling of the waste in transit could lead to a hazardous incident and an adverse environmental impact, with attendant bad publicity.

In many countries, waste vehicles are registered and licensed. Wherever possible, use should be made of suitably qualified haulers. Where such regulations do not exist, the waste producer, before engaging transport services, should ensure, as a minimum, that:

(a) The vehicle is fully roadworthy;
(b) The loads are secure and not overloaded;
(c) A copy of the waste manifest (or equivalent) and emergency contact references are placed in the driver's cabin;
(d) The driver has received adequate training in coping with and responding to emergency situations;
(e) All legal hazard-warning signs, e.g. TREMCARDS, are prominently displayed on the vehicle.

12 GLOSSARY OF TERMS

Bund: A small raised wall, usually about 15 cm high, erected around the periphery of a work area to prevent the escape of spilled liquids.

Incineration:

Low-temperature: below 800°C (usually 300-500°C).

High temperature: two stages of combustion-800°C and above 1000°C.

Scrubbing of the final combustion gases is normally required.

Landfill site: A name given to worked-out quarries, mineral extraction sites and other holes in the ground that are systematically being infilled with refuse and other waste materials.

ACKNOWLEDGEMENTS

In preparing this chapter the author has drawn upon long working experience with ICI and with Zeneca Agrochemicals (Fernhurst, Haslemere, Surrey) and close ties with the waste disposal industry in the United Kingdom. These relationships are gratefully acknowledged, but the views and advice given are those of the author and are not necessarily shared by the other parties.

REFERENCES

(1) Rodan Engineering, *Drum Cutters/Compactors/Crushing Equipment.*

(2) N. Donegan, *Decontamination Studies on Pesticide Containers,* (London, Overseas Development Administration, 1984).

(3) WMEC Ltd., *Sentinel Effluent and Water Treatment Units.*

(4) G.H. Huden and D.C. Khan, *Pesticide Disposal in Cement Kilns in Pakistan.* (Washington, D.C., Agency for International Development, November 1990).

(5) Yves Guitton-Fumet, *Waste Treatment in Cement Kilns,* Lafarge Coppee Recherche.

(6) FAO, *Technical Guidelines on The Bulk-Disposal of Unwanted Pesticides,* 1991.

(7) GIFAP, *Disposal of Unwanted Pesticide Stocks - Guidelines on the Selection of Practical Options,* 1991.

POST-MANUFACTURING PERSPECTIVES

CHAPTER

18

Application Equipment with Particular Reference to Small-Scale Operations

G.A. Matthews

International Pesticide Application Research Centre
Imperial College, Buckhurst Road, Sunninghill,
Berkshire SL5 7PY, United Kingdom

Most pesticide formulations, ranging from the emulsifiable concentrates to the more recent suspension concentrates and dispersible granules, have been designed to mix with water. Water is regarded as the cheapest and most universally available diluent for farmers to spray and protect their crops. In consequence, the majority of sprayer manufacturers produce a range of hydraulic sprayers, all of which operate on similar basic principles. These principles consist of a container, a pump and one or more nozzles with associated valves and hoses to interconnect the main parts. Such sprayers vary in size from simple hand-operated syringe-type sprayers to those mounted on aircraft.

However, in many semi-arid areas water is extremely difficult to obtain. Thus "waterless" formulations have been developed for ultra-low volume applications. Specialized equipment has been manufactured to apply oil-based ULV formulations at the minimum volume (< 5 l/min) to achieve economic control. And, in some cases, the same equipment has also been used to apply very low volumes (VLV) (5-50 l/min) of water-based formulations. VLV spraying at 10 l/ha has now replaced ULV applications in some countries, as it is possible to use the same formulations that is recommended for knapsack sprayers and to choose the pesticide to suit the pest.

1 LEVER-OPERATED KNAPSACK SPRAYERS

Worldwide, the lever-operated knapsack sprayer is the device most widely used by the small-scale farmer. Traditional metal sprayers have now been largely replaced by polypropylene and other plastic components. In most machines the sprayer tank, usually 15 l capacity, is carried as a knapsack on the operator's back by means of two shoulder straps. A third strap around the waist is now recommended to reduce movement of the container during pumping and to provide a more efficient transfer of energy. Larger tanks are too heavy when full, but in some countries a 10 l tank is preferred. A piston or diaphragm pump, usually mounted inside the container, is operated manually by a lever that is fitted either to the left or right side (underarm action) or to the top of the container (overarm action). It should be possible to change rapidly from left- to right-hand pumping. On some machines the pump is mounted outside the container to facilitate access for maintenance. The flow of liquid from the container to the pump is controlled by an inlet valve; a second outlet valve controls the flow from the pump to a "pressure" chamber. Air inside this chamber, which is compressed by pumping the spray liquid into it, subsequently forces the spray through a hose to the lance. The lance has a trigger valve and one or more nozzles. The pressure chamber should have sufficient air capacity to cushion the variations in pressure with each pump stroke. On some machines a safety or regulating valve is incorporated into the air chamber to release the pressure by venting liquid back into the container if the pressure exceeds a set value. Since adjustment of this valve necessitates insertion of the operator's hand into the container, it should not be used as a pressure-regulating valve.

A filter should be set as deep as possible inside a large filler-hole in the tank to reduce the risk of backsplash of pesticide when filling the tank. An air vent in the filter improves the flow of liquid into the tank. The tank opening should be closed by a tight-fitting lid in which an air vent is protected by a valve to prevent leakage of any liquid that might spill over the operator's back. Also, on some machines there is a flange around part of the tank top to reduce spillage over the back of the tank. A second filter, with a large surface area, should be mounted upstream in the trigger valve in the lance. This is especially important where sprayer manufacturers have failed to fit a filter into each nozzle. Unfortunately many sprayers in the tropics have had lances with a single nozzle unique to the manufacturer so that alternative types of nozzles could not be fitted. The trend, however, is to have a standard nozzle body incorporating a filter into which a range of different nozzle tips may be fitted.

The last filter before the nozzle needs to have a mesh opening that is smaller than the orifice of the nozzle tip, although a similar mesh is useful at all stages to remove foreign matter that might also reduce the efficiency of the pump and valves. A 50-mesh is adequate for most sprays (see Table 18.1).

Table 18.1 Filter sizes

Mesh (wires/in.)	30	50	100	200
Aperture size (mm)	0.6	0.3	0.14	0.08
Nozzle output (1/min)	> 3.0	0.8–3.0	0.4–0.8	< 0.4

Piston pump sprayers (Figure 18.1) are considered more suitable for insecticide and fungicide spraying. With this type of pump, pressures of 3 bar are generally easier to maintain even when there are several nozzles used on a boom. There has been a trend towards replacing the leather cup-washer-type pump seal with an 0-ring of synthetic elastomer. However, relatively little wear of the thin 0-ring reduces pump efficiency, so the preference is now for a moulded elastomer seal similar to a cup-washer. The diameter of the pump and the length of stroke should be designed to pump 90 1/hr at 3 bar using no more than 30 full strokes of the lever every minute. One can avoid excessive wear on one side of a pump if the linkage from the lever that provides maximum thrust is to the centre of the pump. Some sprayers constructed in this manner have operated for over 2,000 hr with minimal attention.

Most wear on the pump seal is caused by abrasive particles that are water contaminants rather than by the constituents of the pesticide formulation. This is expressly true in the newer formulations, which contain extremely small particles to improve their sus-

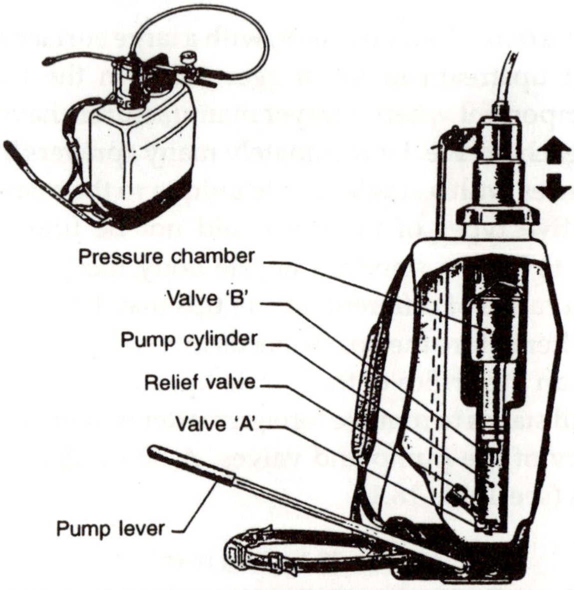

Figure 18.1 Piston pump types

pensibility. A paddle is often fitted to the piston so that agitation is provided in the tank during pumping.

Diaphragm pumps (Figure 18.2) do not have any moving seal and have been preferred for the application of wettable powder formulations in spite of the fact that few of these machines have any agitation of the liquid in the spray tank. The length of stroke and movement of the diaphragm is very much more restricted than for a piston pump. Thus, the output is generally lower and a lower pressure is achieved at a similar rate of pumping. As a consequence, the diaphragm pump is most useful for herbicide application.

The elastomer used in the diaphragm is liable to be adversely affected by the solvents used in some formulations, especially emulsifiable concentrates. Therefore, as with all spray equipment, careful washing by pumping clean water through the whole sprayer at the end of each days's spraying is essential. Irrespective of the type of pump, the design should ensure that all of the pesticide liquid can be pumped from the tank so that no residue of chemical is left to contaminate the next spray.

The lever-operated knapsack sprayer is invariably supplied with a lance, which the operator holds in front of his body while spraying.

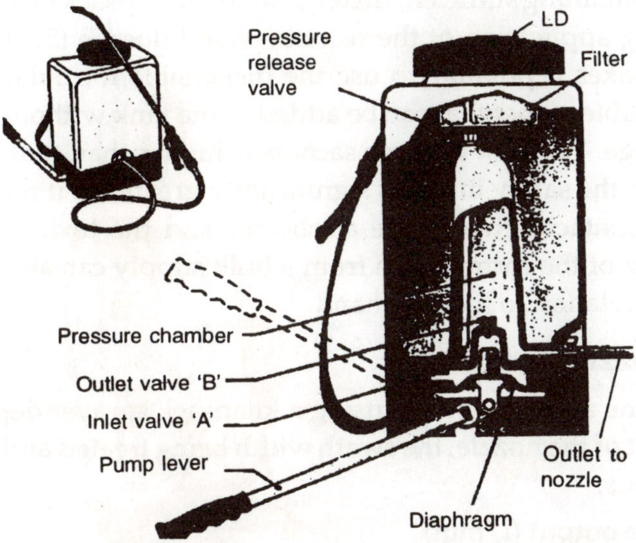

Figure 18.2 Diaphragm pump types

In some crops the lance is waved to either side to treat a wide swath and reduce the number of passes through a crop. The height of the nozzle will depend on the stage of crop growth, but at all stages the operator gets contaminated by the spray as the body brushes against treated foliage (1). Despite the obvious contamination, albeit often at a low concentration of pesticide, many farmers have used highly toxic pesticides, such as methyl parathion, without proper protective clothing. After working for several hours, this chemical is liable to soak through clothing, so none of these pesticides should be sprayed with manually carried equipment. Ideally the operator should hold the lance downwind and where possible fit the nozzle to the rear of the tank. The latter was originally recommended by Fernando (2) for treating rice crops. Cadou (3) extended the idea with four nozzles on a horizontal boom for spraying cotton, while Tunstall (4) fitted a vertical boom on which nozzles were directed up into the crop canopy to achieve some deposition on the undersides of leaves. The latter also allowed the number of nozzles to be increased as the area of foliage increased. The spray concentration remained the same, so the volume increased from 50 l/ha, on small plants to about 200 l/ha on the fully grown crop. The insecticides were packaged in sachets containing sufficient material for one knapsack load, thereby facilitating application of the recommended dosage (5). This technology makes it possible to use the dispersible formulations in a water-soluble sachet that can be added to the tank without opening the package. The cost of these sachets is higher than that of larger packs, but the safety in use is significantly greater as it reduces the physical contact between the applicator and the toxic substance. Repacking of the formulation from a bulk supply can also be done locally in relation to local demand.

1.1 Nozzle Technology

The volume application rate using a knapsack sprayer depends on the output of the nozzle, the swath width being treated and walking speed. Thus,

$$\frac{\text{Nozzle output (l/min)}}{\text{Swath (m)} \times \text{speed (m/min)}} = \text{Application rate (l/m}^2) \times 10^4$$
$$= \text{l/ha}$$

The output of the nozzle will depend on the orifice size and the pressure in the tank. Furthermore, the spray volume is affected by the speed and the way in which the operator moves the pump lever. Short strokes are inefficient as each change in the direction of the lever causes the valves in the pump to open and shut. Thus, it is desirable to have a steady and a higher pumping rate, thereby increasing the output and pressure at the nozzle.

A pressure-regulating valve can be fitted to the lance to provide a constant pressure at the nozzle. A modification of this is a new design of a check valve that can be fitted directly to each nozzle. This valve only opens when the liquid has reached the required pressure but then continues to spray at the same pressure even with an excess of pressure on the input side of the valve.

Use of this spray management valve (SMV) (Figure 18.3) thus eliminates variability of output associated with different pumping rates. A calibrated plastic bottle has been supplied by some sprayer manufacturers to collect spray liquid while treating an area of 25 m^2. The ability of individual operators to measure the spray output without any calculations makes calibration much easier.

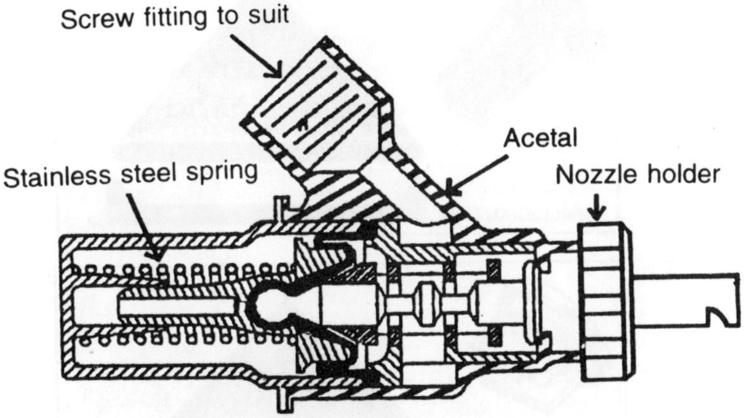

Figure 18.3 Spray management valve

Variable-cone nozzles have been used on many types of small sprayer, but since the angle of the cone is adjusted by screwing the tip in or out from a fixed swirl plate, the operator's hand will invariably be contaminated by spray and their use is not recommended. The main types that should be used on any hydraulic sprayer are deflector, fan or cone nozzles as depicted in Figure 18.4.

All these nozzles produce a wide range of droplet sizes, the average size being larger when the nozzle is operated at low pressure and a large orifice is used to apply high volumes of liquid. Depending on the droplet sizes produced, the spray quality may be expressed as fine, medium or coarse (6). A fine spray may be recommended for insecticide and fungicide applications, while a coarse spray may be essential in the application of herbicides to reduce the risk of droplets drifting to susceptible plants.

A particular nozzle can be recommended by using a code which defines the nozzle type (F, HC or D, for fan, hollow cone or deflector), spray angle (e.g. 110° or 80°) output (l/min) and rated pressure. For example,

(a) F 110/1.6/3 is 110° fan nozzle delivering 1.6 l/min at 3 bar pressure;

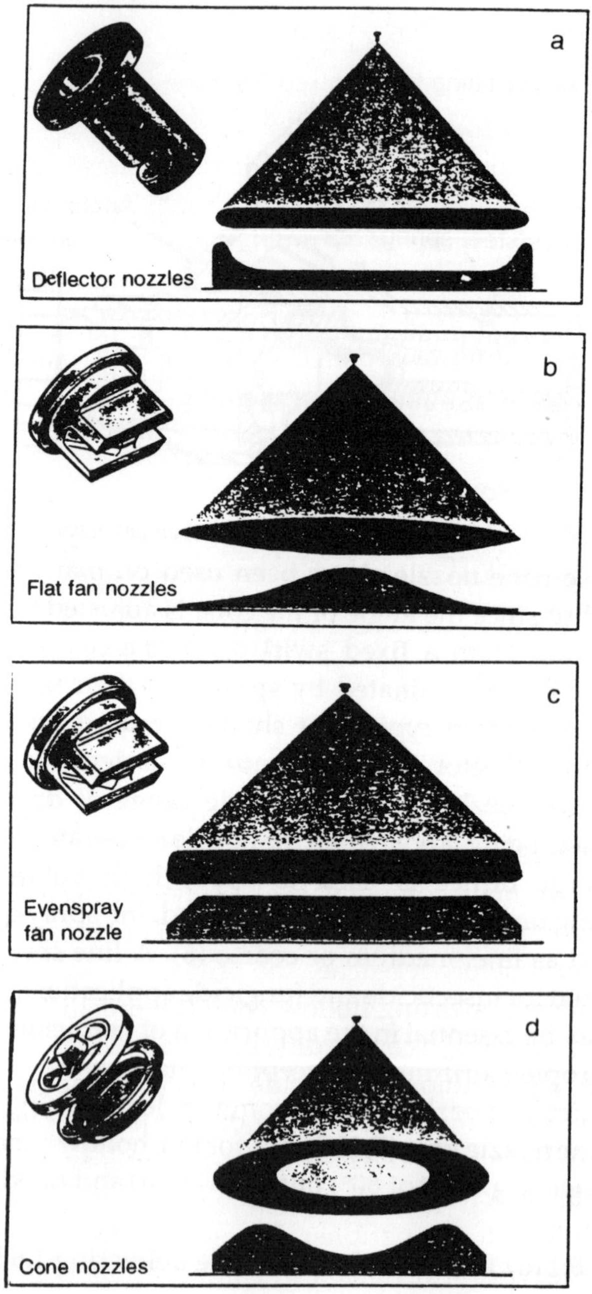

Figure 18.4 Nozzles and nozzle patterns

(b) HC 60/0.39/3 is 60° hollow cone nozzle delivering 0.39 l/min at 3 bar pressure.

The angle is not always indicated for cone nozzles.

1.1.1 Deflector Nozzles

Liquid forced through a circular orifice impinges on a flat surface, which deflects it through an angle of about 90° to form a fan-shaped sheet of liquid (Figure 18.4a). A flat elliptical pattern of spray is produced with a slightly larger volume of liquid at the extreme edges. The spray droplets tend to be large and produce a coarse spray. The position of a deflector nozzle on a lance needs to be checked, especially if the end of the lance is angled, to ensure that the spray is projected in the right direction.

1.1.2 Fan Nozzles

Fan nozzles produce a more accurate elliptical spray pattern than deflector nozzles (Figure 18.4b) but, because the orifice is also elliptical in shape, there is a greater risk of blockage. They are most suitable on a sprayboom as an even distribution of spray across the swath can be achieved by overlapping the patterns of adjacent nozzles. An even-spray fan nozzle (Figure 18.4c) is used when a narrow band of spray is required, for example to apply herbicide along the top of a ridge.

1.1.3 Cone nozzles

There are two main components of a cone nozzle: a circular orifice and a swirl plate with off-centre tangential openings that rotate the flow of liquid before it reaches the orifice, by causing the sheet of liquid to form a swirling hollow cone (Figure 18.4d). For a given output and pressure, a cone nozzle produces a finer spray than a fan nozzle. The distribution of pesticide on a flat surface is more uneven with a fan but, as the cone of spray directs the droplets at different angles, this nozzle is preferred for applying insecticides or fungicides to foliage. A full cone is produced if the swirl plate has a central hole in addition to the peripheral tangential slots.

All nozzle types are available in a range of materials. Most nozzles were manufactured in brass, stainless steel or ceramic, but the trend is towards using injection-molded plastic tips of materials such as Kematal. Resistance to erosion is similar to that afforded by brass tips, but the cost of manufacture is lower. As with all nozzle tips, their output should be checked regularly and a tip replaced when necessary to avoid excessive use or poor distribution of a pesticide. These plastic nozzles are colour-coded, but the colour code systems has varied between manufacturers.

The main disadvantage of the lever-operated knapsack sprayer is that the operator must pump continuously. There are two kinds of manually carried hydraulic sprayer — the compression sprayer or a knapsack with a motorized pump.

2 COMPRESSION SPRAYER

These sprayers consist of a robust container, usually 5-10 l capacity, although some smaller sprayers may be used for localized treatments, as in glasshouses. The container is partially filled with the spray liquid to leave an air space that is pressurized by a pump, which on the cheaper machines forms an air-tight lid to the container. Unfortunately, this design has a relatively narrow opening, so filling requires the operator to use a funnel unless there is a conical flange around the top of the sprayer. The pump is removed to refill the sprayer so there is a channel across the thread to allow any residual air pressure in the tank to escape as soon as the pump is unscrewed. Care is needed to avoid contaminating the spray liquid with dirt that might have collected on the pump while it was outside the tank. To avoid this, on some machines the pump is fixed in position and there is a separate opening for refilling. Sprayers manufactured to meet the WHO specification for compression sprayers (7) have a separate large lid to facilitate filling (Figure 18.5).

Sprayers are seldom supplied with a dial-type pressure gauge having a Bourdon tube as standard. Even when they have one, their accuracy in the field is often very poor. On some machines there is a colour-coded pressure indicator. On others the manufacturers inform the operator of the number of full pump strokes needed to achieve a working pressure when the sprayer is full. If less liquid is

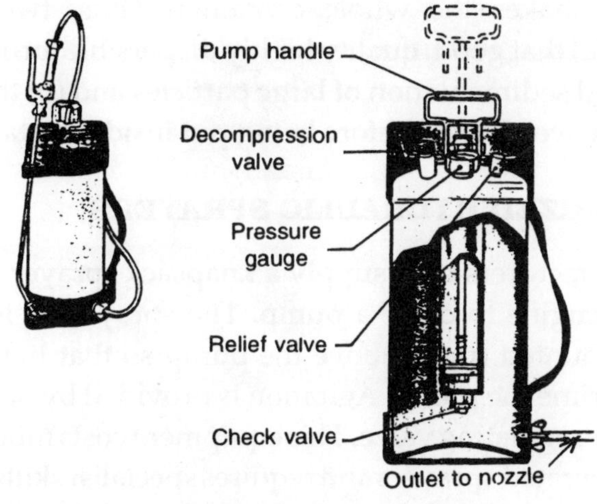

Figure 18.5 Compression sprayers

in the tank, the operator will need to pump more air to achieve the same pressure. All sprayers should have a safety valve to release excess pressure.

The disadvantage of the simplest compression sprayers is that the air pressure, which forces the liquid to the nozzle, drops as the volume of liquid in the container decreases. With this pressure drop, in addition to a change in the application rate, the spray quality becomes coarser, unless of course the operator stops spraying and repressurizes the container. Ideally a pressure regulating valve, such as the SMV described above, is used to ensure more accurate application. In a few countries, the heavier pressure-retaining compression sprayers are still used. Here a pressure-regulating valve must be used and the tank is emptied at a constant pressure. On these sprayers air inside the tank is pressurized initially to a minimum of 3-4 bar before the spray liquid is pumped into the container. Then the pressure inside the container increases to about 10 bar. When all the liquid has been sprayed, a valve retains the air pressure inside the container. Unless there is a motorized pump to refill the container, the pressure-retaining compression sprayer is very laborious to use. As high pressures are used, the strength of the container should be tested regularly.

Agitation is not provided in any of the compression sprayers unless one shakes the whole container. Thus, two things are important: (a) that good, quality highly dispersible formulations are used to avoid sedimentation of large particles and (b) that the spray solutions are well mixed before being put inside the tank.

3 MOTORIZED HYDRAULIC SPRAYERS

A few manufacturers now supply a knapsack sprayer with a small two-stroke engine to drive a pump. The spray tank is fitted on a shoulder-mounted frame above the pump so that liquid flows by gravity to prime the pump. Agitation is provided by backflow from the pressure-regulating valve. This equipment costs much more than manually operated sprayers and requires specialist skills to maintain the engine.

Instead of mounting such a unit on a knapsack, it is more usual to see a motorized unit with a heavier four-stroke engine on a wheeled frame fitted with a larger 25-50 l tank. Unfortunately, these units are often used as a stationary unit at the side of a field, with a single nozzle at the end of a long hose being used to apply relatively high volumes of spray. A good quality hose is required to reduce the risk of stretching and puncturing as it is dragged or carried through the crop. Since the crop can be damaged by the movement of the hose and an increased number of operators per sprayer, their use is not generally recommended.

4 WHEELBARROW SPRAYERS

The manual effort of pumping can be eliminated by attaching the lever to a cam on the axle of two wheels. The unit is pulled through a crop and spray-applied either with a hand-directed lance or with the nozzles fixed to the rear of the spray tank. If the sprayer is pushed rather than pulled, it tends to dig in and is not so easily moved through the crop. Care is needed to avoid mud and weeds collecting on the wheels. Wheeled sprayers, however, are usually fitted with a motorized pump. Thus, the unit is similar to the larger tractor-mounted equipment, albeit on a smaller scale.

5 MOTORIZED MISTBLOWERS

Projection of droplets from a hydraulic nozzle is very limited. So, where the spray has to be directed upwards against gravity, into a tree crop or into dense foliage, it is usual to provide an air stream to propel the spray, make the leaves move and improve impaction of the droplets. Droplets are ideally in the 50-100 μm range as larger droplets will tend to fall out of the air stream and contaminate the soil under the crop.

The smallest knapsack-mounted mistblowers have a two-stroke engine to drive a centrifugal fan from which air is ducted to a nozzle. Some of the air may be used to pressurize the tank to force spray liquid to the nozzle. On the other hand, some machines have a small pump fitted to the engine drive-shaft to pressurize the tank. If the tank is pressurized, the lid must make a good seal, especially when the nozzle is above the level of liquid in the tank. The flow of liquid to the nozzle is controlled by a restrictor. On most sprayers the restrictor has four positions and unfortunately many users always use the maximum setting. This overfeeds liquid to the nozzle so that when atomization is dependent only on the effect of air-shear, too many very large droplets are produced. It is better to have a fixed restrictor that is selected for a particular spray operation.

The standard air-shear nozzle produces a range of droplet sizes, with many very small aerosol droplets. Irrespective of the actual nozzle used on these sprayers, spray is effectively carried in the air stream up to 6-10 m vertically or 12-18 m horizontally. But if the operator attempts to spray against the wind, the small droplets are rapidly blown back due to the fact that the energy of the air is rapidly dissipated as it travels from the air-duct.

As alternatives to the simple air-shear nozzle, a number of different rotary atomizers have been designed to fit the sprayer and produce a more uniform droplet spectrum. A rotary nozzle with a large surface area, such as the Micronair AU 8000, can apply larger volumes more effectively, but it significantly increases the cost of the equipment.

Modern motorized mistblowers with electronic ignition systems are generally easier to start except in hot tropical climates, especially when fuel is left in the engine. Volatilization of the petrol leaves an oily residue that can interfere with ignition. It is better to stop the engine at the end of each spray session by turning off the fuel line and allowing the engine to continue operating until starved of fuel. The correct fuel mixture for a particular engine is essential. Manufacturers should indicate the petrol : oil ratio on the fuel tank. There are principally two engine capacities, 35 and 70 cc. The larger 70 cc engine is able to increase the volume of air emitted and is more suitable when treatment of tall trees is required.

These sprayers have been used to apply ultra-low volumes by using an extremely small orifice in the restrictor. However, care must be taken when applying slightly viscous spray solutions as the actual flow rate will increase as the viscosity decreases with an increase in temperature. Apart from changes in ambient temperature while spraying, heat from the engine can raise the temperature of the pesticide. Particular care is also needed with the choice of other formulations as the small orifice in the restrictor can be easily blocked by particulates.

Mistblowers have also been modified to apply granules and dusts by ducting some air from the fan into the tank and having a larger tube with a variable restrictor connecting the tank to the main air delivery tube. In Japan, microgranules have been extensively applied to rice crops by extending the air tube using an inflatable plastic lay-flat tube in which there are a series of holes along its length. The tube can be carried with an operator on each side of the field. A flap of plastic adjacent to the hole inside the tube increases the turbulence and improves the delivery of the granules.

Motorized mistblowers have been most extensively used on a number of tall crops such as cocoa, bananas and coffee and also on field crops (e.g. cotton). Unfortunately the lifetime of the equipment may be shortened by problems with the engine. The two-stroke engine is designed to operate at maximum revolutions. Thus if the engine is allowed to idle at slow speeds, there is inadequate lubrication. Many users reduce the speed of the engine, especially when walking between fields, or returning to a centralized refilling site, which adversely affects the performance of the engine. If slow engine speeds are also used while spraying, atomization is poor.

6 SPINNING DISC SPRAYERS

The lack of water has severely limited the adoption of chemical control using knapsack or compression sprayers in many areas. However, the initial use of undiluted oil-based sprays to control locusts by vegetation baiting (8) led to the concept of ultra-low volume crop spraying using small hand-held battery-driven spinning disc sprayers (9) (Figure 18.6).

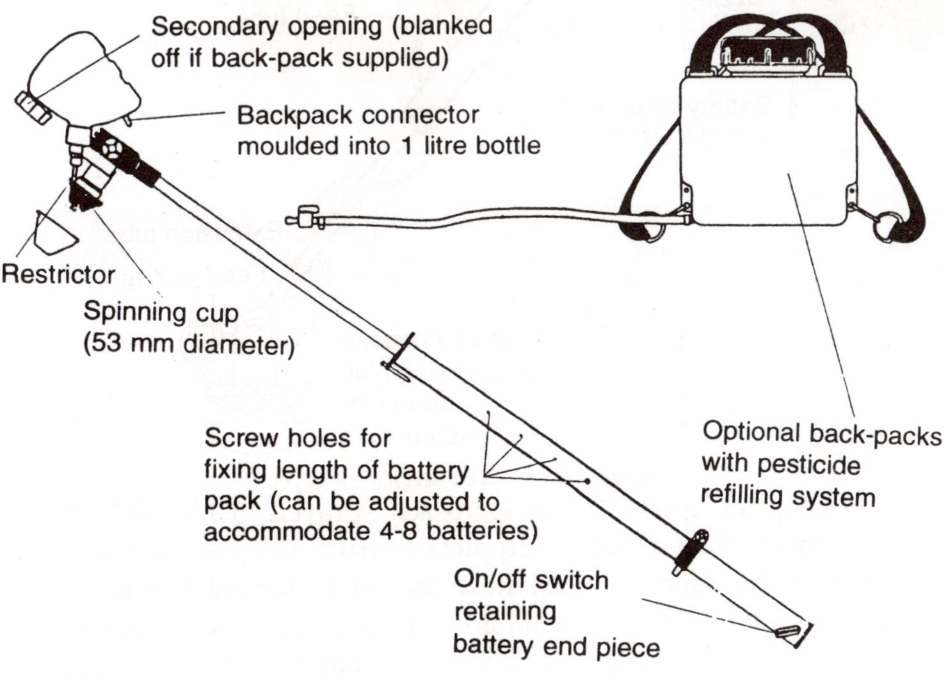

Figure 18.6 Spinning disc sprayer

Effective pest control with a significantly reduced volume depends on maintaining adequate coverage of the target areas where the pest is located. To achieve this, the range of droplet sizes needs to be restricted to avoid wastage of chemical in excessively large droplets and also the loss of the smallest droplets due to drift. This led to the concept of controlled droplet application (CDA) using a narrow droplet spectrum (10) (Figure 18.7). In contrast to relatively

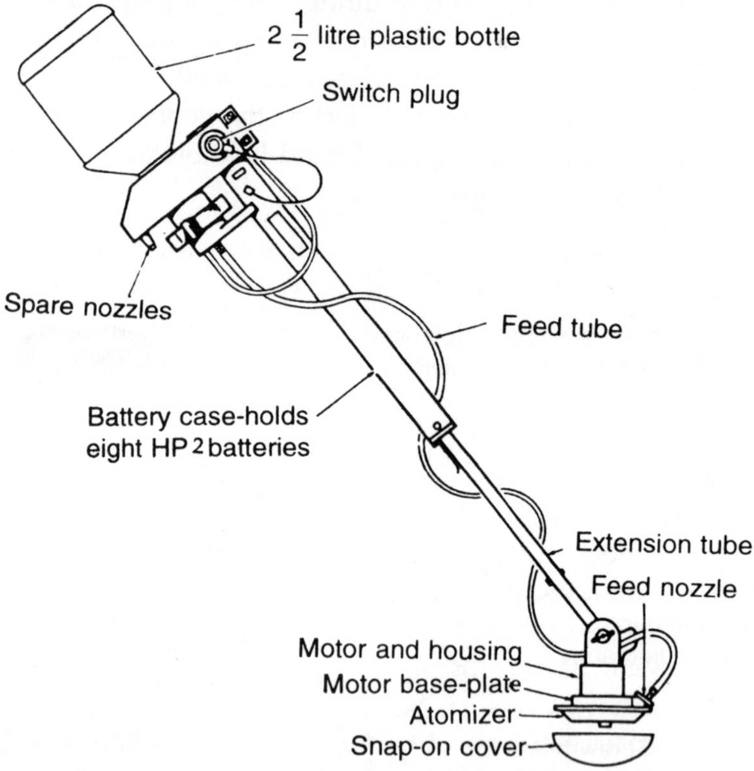

Figure 18.7 CDA herbicide sprayer

high volumes applied with knapsack sprayers, ULV and VLV spraying provides a discrete droplet spectrum. Droplet size must be controlled in order to control the dosage of the formulation in view of evaporation of the carrier liquid. The problem is that evaporation may reduce particle size to such an extent that impaction is reduced. The formulation also must not be phytotoxic despite using a higher concentration of the pesticide. And the deposit of the active ingredient on foliage needs to resist weathering, especially wash-off due to rain. Laboratory experiments have indicated that the discrete pattern of droplets can be extremely efficient in controlling a wide range of pests, e.g. mites (11), whitefly scales (12, 13, 13a), diamondback moth, *Plutella xylostella*, (14).

Under laboratory conditions a reduction in droplet size to as small as 20 μm has increased the efficiency of spray deposits. However, deposition of such small droplets under field conditions is not practical at present and larger droplets have been applied. The principal features of the equipment include a small container, which feeds the spray liquid through a restrictor to near the centre of a small plastic disc, which in turn is rotated by a small electric motor. Liquid is flung from the disc as ligaments by centrifugal force. The ligaments form more uniformly sized droplets than a conventional hydraulic nozzle. Droplets are smaller as the disc speed increases. If the flow of liquid is increased, droplet size can be larger for a given speed. The actual droplet size is dependent upon both the control of flow rate and disc speed. If the flow rate is excessive, separate ligaments are not formed and a wide range of droplet sizes results. At very low rates it is possible to achieve an almost monodisperse spray by direct droplet formation. For insecticide and fungicide application, a high disc speed (6,000-10,000 rpm) is used to produce small droplets (50-100 μm) that move downwind from the disc held 0.5-1.0 metre above the crop. Displacement of the spray downwind is dependent on a wind speed of 2-5 m/sec to produce some turbulence within the canopy. However under normal meteorological conditions wind speed fluctuates. Thus to achieve a uniform distribution, a narrow swath is required to overlap the spray patterns achieved with successive swaths. Many users have adopted a swath of 5-10 m to reduce the time needed to treat a field. In practice, however, better distribution of pesticide (and thus improved insect control) is achieved with a 2 m swath. This will be particularly true when the crop is over 0.5 m tall.

Oil-based formulations have been used at 1-3 l/ha. This programme extensively used was on cotton in francophone Africa, where over a million hectares have been treated. However, the cost of the formulation and in some areas, the lack of control of certain pests, due to the use of wide swaths and negligible underleaf deposits have led to the examination of alternative techniques.

Earlier studies (15,16 and 17) have shown that the same equipment could be used to apply VLV water-based sprays by using a different restrictor. A larger droplet size (ca 150 μm) and narrower swaths (2 m) were used. This technique has now been adopted in West Africa (18) as an alternative to ULV spraying. It permits use of a range of conventional formulations diluted in 10-15 litres of water per hectare. In Zimbabwe, the volume of spray was reduced to 5 l/ha by adding up to 20% molasses, which reduces the effect of evaporation of water on droplet size. The molasses, which also acts as a feeding stimulant to attract moths in a lure-and-kill method of control, is readily washed off the foliage by rain and dew. But in drier areas it could act as substrate for sooty moulds. In these areas, other anti-evaporant adjuvants can be used, such as vegetable oils with an emulsifier.

One criticism of spinning disc sprayers is the poor availability and high cost and/or poor quality of batteries in many cotton areas. Improvements in the sprayer design, e.g. small single disc and selection of motor, have reduced power requirements, resulting in fewer batteries being required. Alternative power sources have been investigated (19), and where electricity is available rechargeable batteries have been used. Ultimately it is hoped that solar recharging systems will be economic, especially if the cost can be spread over the whole year by using the batteries in other equipment.

Another criticism of the technique has been the potential hazard of using pesticides at a higher concentration than that used in knapsack sprayers, especially when the dispersal of droplets is dependent on the wind. In practice the contamination can be significantly less, provided the operator keeps the disc downwind, because the total period needed for spraying is shorter and as the work is less arduous, mistakes are less likely to occur. Where the sprayer container has to be refilled, care is needed to avoid spillage on the outside of the container.

Spinning disc sprayers have also been used to apply herbicides by reducing the disc speed to 2,000 rpm and producing droplets with a VMD of 250 μm. These larger droplets are less affected by the wind when the disc is held close to the weeds/soil. With these larger droplets, the volume of spray is increased to 20-40 l/ha to achieve sufficient coverage of the weeds (droplets/cm^2). Both water-based suspension concentrate formulations and oil-water emulsions have been used, some of which are available in prepacked containers for connecting directly to the sprayer.

The swath width is approximately 1.2-1.6 m, depending on the disc used and the droplet size produced. Narrower swaths are obtained by partially shrouding the disc and respraying the liquid collected on the shroud. The unshrouded disc produces a pattern similar to the cone nozzle, so care is needed to avoid overdosing when adjacent swaths overlap. On some machines the spray liquid can be carried in a 5-10 l knapsack container.

7 ELECTROSTATIC SPRAYING

Experimental techniques for placing an electrostatic charge on agricultural sprays have involved induction, ionic and direct charging of spray nozzles (20). The process has seen very little development into commercial systems. Charging water-based sprays from a hydraulic nozzle was inefficient due to the wide range of droplet sizes in the spray. The largest droplets had a low charge, and deposition on the crop was not improved. But drift of the smallest, highly charged droplets has been shown to be reduced.

Studies on charging spinning disc atomizers using water and oil-based formulations have been made with a range of pesticides and crops (21). A hand-held electrostatic sprayer, the 'Electrodyn' sprayer was developed using oil-based formulations supplied prepacked in a 'Bozzle' applying as little as 1 l/ha (Figure 18.8). Limitations on the choice of pesticides which could be formulated in such a small volume and lack of penetration of dense canopies has constrained its adoption.

8 GRANULE APPLICATION

The most hazardous pesticides, such as aldicarb, have only been

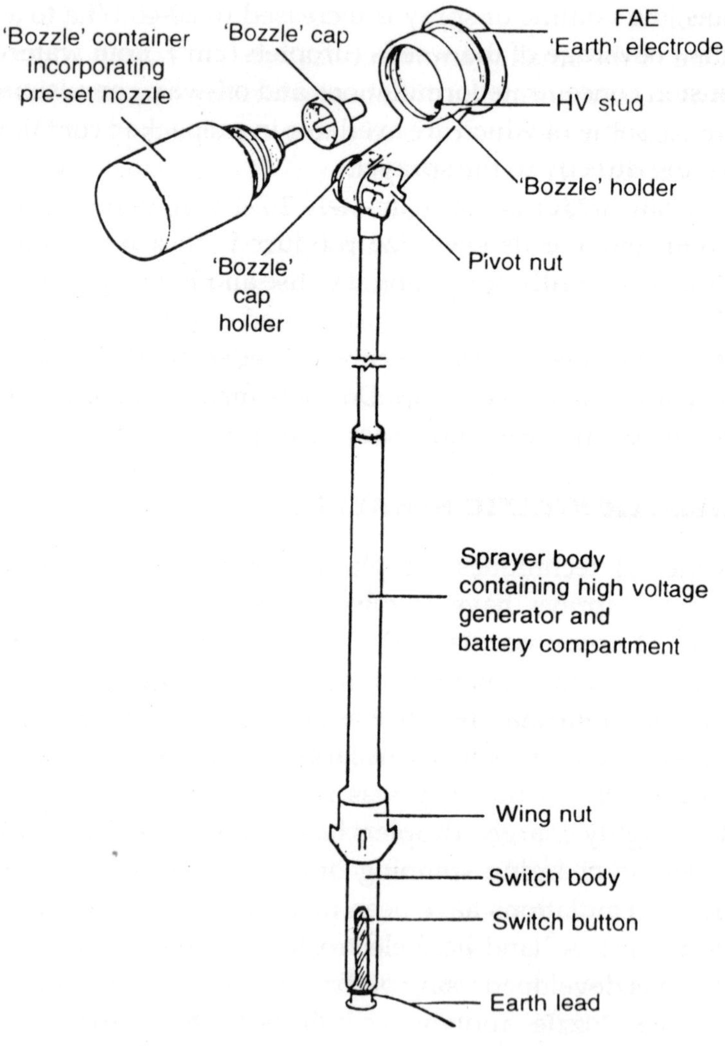

Figure 18.8 Electrodyn Sprayer

formulated as granules and, where possible, their application is confined to well-supervised mechanized equipment, especially in conjunction with sowing crops. Modified knapsack mistblowers, described earlier, have had extensive use in the application of microgranules in Japan. A number of systemic granular products have been applied using simpler equipment. Similarly some granules can be broadcast, e.g. *Bacillus thuringiensis,* for mosquito larval control.

One of the small granule applicators is a hand-held tube, which acts as a container, at the bottom of which is a trigger-activated valve to release a measured dose of granules. This is particularly useful for spot treatments. In some places, containers such as plastic bottles with a few holes (pepper pots) have been adapted to shake granules onto maize for stalk borer control. Care has to be taken to keep the granules dry during treatments. Otherwise they may aggregate and block the apertures.

Broadcast treatments can be made with a simple bag into which a tube is fixed, e.g. a horn seeder. Swinging the tube in a figure eight motion can spread granules over a wide swath. More elaborate and expensive equipment, such as motorized or hand-operated blowers, can be used.

9 WEED WIPERS

Weed wipers (Fig. 18.9), designed by Dale (22), are suitable for localized treatment with a translocated herbicide, such as glyphosate. The herbicide, usually carried in a tube, is metered on to an absorbent surface that is brushed against the weeds. The main

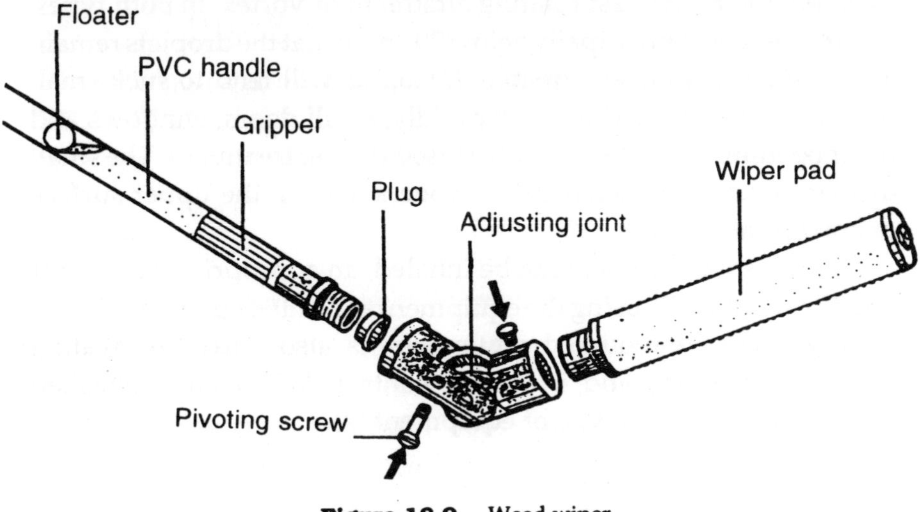

Figure 18.9 Weed wiper

problem is to adjust the flow of herbicide so that the surface remains wet without dripping. Different materials, such as a nylon rope wick, have been used. A disadvantage of this type of application is that the surface can get clogged with dust.

10 FOGGERS/AEROSOLS

Space treatment in glasshouses, stores and other enclosed spaces can be achieved using extremely small droplets. The simplest method to treat a relatively small space involves using a smoke generated by the slow burning of a pesticide mixed with a combustible material. Heat is also used to produce a thermal fog in which a pesticide is vaporized by the hot gases in the exhaust of a pulse-jet engine. The gas-pesticide mixture condenses to form a dense fog. Hand-carried thermal foggers have a pump to pressurize both the spray and petrol tanks. A petrol-air mix is ignited by using an electrical circuit from a set of batteries to a spark plug. Then the hot gases ignite subsequent charges of the fuel entering the combustion chamber. Both oil- and water-based pesticide products have been used through thermal foggers. Manufacturers should be consulted about the operating procedures for a particular product.

A cold fog, or aerosol, is produced when the pesticide liquid is metered into a very fast rotating airstream or vortex. In both types, the droplet size is principally below 30 µm so that the droplets remain airborne for as long as possible. Draughts will tend to suck small droplets out of a building very rapidly, so all doors, windows and other openings need to be firmly closed during treatment. The small droplets eventually sediment, so coverage is on the upper surface of horizontal surfaces.

As such small droplets can be inhaled, an appropriate respirator must be worn while using the equipment and entering a treated area.

Larger vehicle-mounted equipment is also used for treating rubber plantations and in vector control. WHO has published specifications for this type of equipment (7).

11 OPERATIONAL FACTORS

Following are some common-sense approaches to the application of pesticides. Use a pesticide only when it is really necessary and choose the least hazardous chemical when chemical control is needed and suitable equipment is available. Check the sprayer using water to ensure that it is in proper working order and has no leaks before using any pesticide. Check the output of the nozzle and change it if the output is too high or the spray pattern is irregular due to damage or wear of the tip. The sprayer should be calibrated carefully and the amount of pesticide product needed for each tankload calculated. Utilize ancillary equipment for measuring and mixing the pesticide and transporting separate water supplies for (a) spraying, (b) washing (+ soap) and (c) drinking. Drinking water must be kept well away from the pesticide, but it is essential to have it readily available as operators are likely to lose water by sweating due to the manual effort needed to use many types of sprayer . Never eat or drink or smoke while spraying. Always wash well with soap and water before drinking or eating. Read the label carefully to check that the correct quantity of pesticide is used. Use the appropriate protective clothing and avoid contamination with the concentrated pesticide. Wash any contaminated skin or clothing as soon as possible. Wearing a plastic apron will prevent liquid soaking into overalls. Only mix sufficient pesticide for a particular load in case changes in weather prevent the spray programme from being completed. If a pesticide container is empty, rinse it and add washings to the sprayer before completing the dilution. After spraying, wash equipment with water, preferably in the field so that the washings can be sprayed in the field. Never leave any liquid in the sprayer.

12 CHOICE OF EQUIPMENT

Traditionally most small-scale farmers have used very small compression sprayers or a lever-operated sprayer that can be shared by several of them in a village. The choice has been dictated by what was most easily available at the lowest price and because the chemical companies promoted formulations to dilute in water. Understandably these sprayers are versatile and can be used for a wide range of pesticides. However, lack of training in their correct use and maintenance, lack of spare parts, and the absence of information on calibration and choice of nozzle has meant that pesticides have usually been applied incorrectly. Lack of proper repairs has often resulted in leakages, increasing operator contamination.

The advantages and disadvantages of manually carried sprayers are provided in Table 18.2.

Table 18.2 Some advantages and disadvantages of manually carried sprayers

Lever-operated knapsack	*Compression sprayer*	*Motorized mistblower*	*Battery-operated spinning disc sprayer*
Advantages			
Versatility, most pesticides and water-based formulations can be used at low volume (100 1/ha).	Small, lightweight units, suitable for a wide range of pesticides. Intermittent pumping so operator can concentrate on directing nozzle.	Can project spray into trees. Can be adapted for granule application and provide wide swath ULV - 200 1/ha range.	Lightweight separate machines for herbicide application with large droplets. Better control of droplet size. Ideal for ultra low or very low volume application (2-30 1/ha).
Disadvantages			
Requires continuous pumping. Operator usually walks into spray. Water required, so less suitable in arid areas. Pressure to nozzle can fluctuate.	Drop in pressure affects output and spray spectrum unless pressure control valve is used. Low volumes preferred to reduce number of times sprayer has to be repressurized.	Noise, high capital cost. Maintenance of engine. Fuel costs. Heavy to carry. Need to use petrol and oil mixture in correct ratio.	Use of batteries. Higher concentration of active ingredient in spray. ULV requires special non-volatile formulation.

Improved knapsack sprayers are now available but are generally more expensive. Thus their use will depend on governments insisting on minimal safety standards and specifications that should ensure they remain operational for at least two or three seasons. Some training through GIFAP and FAO, is available but much more is needed before pesticides are applied more efficiently, even with conventional equipment.

Where water supplies are inadequate, it has been possible to adopt spinning disc sprayers only where cotton is a major crop and where companies or cotton organizations, such as CFDT in francophone Africa, have provided the sprayers and other inputs. In some areas attempts have been made to group large numbers of small farms into large areas of one crop to facilitate the use of mechanized, tractor mounted sprayer or aerial application equipment. Relatively few schemes have been successful due to differences in agronomic practices and variations in pest incidence between farms, which affects the economic benefits of large-scale pesticide treatment.

In some situations it has been argued that all chemical control is too expensive for the small-scale farmer. However, if yields are to be improved (this is essential where agricultural land is limited) some pesticide use within a pest management programme will be essential. As equipment may be too expensive for the individual farmer, its cost (although not that of the pesticide) should be subsidized and systems of sharing equipment devised. Much of the difficulty in deciding the dosages and other calculations can be minimized by careful collaboration between the agrochemical industry, the scientists and extension staff, so that with appropriate packaging (e.g. water-soluble plastic sachets and squeeze bottles with built-in dispensers) and instructions, the risk of applying pesticides incorrectly is reduced.

Selected manufacturers of pesticide application equipment are listed in the appendix.

REFERENCES

(1) J.A. Sutherland and others, "Effect of application volume and method on spray operator contamination by insecticide during cotton spraying, *Crop Protection*, 9, 1990, p. 343.

(2) H.E. Fernando, "A new design of sprayer for reducing insecticide hazards in treating rice crop, *FAO Plant. Prot. Bull.*, 4, 1956, p. 117.

(3) I. Cadou, "Une rampe portative individuelle pour la pulverisation a faible volume," *Coton et Fib. Trop.*, 14, 1959, p. 47.

(4) J.P. Tunstall, G.A. Matthews and A.A.K. Rhodes, "A modified knapsack sprayer for the application of insecticide to cotton, *Cott. Gr. Rev.* 38, 1961, p. 22.

(5) J. Gower and G.A. Matthews, "Cotton development in the southern region of Malawi," *Cott. Gr. Rev.*, 48, 1972, p. 2.

(6) S.J. Doble and others, "A system for classifying hydraulic nozzles and other atomisers into categories of spray quality" *Proceedings of the British Crops Protection Council Conference - Weeds*, 1985, p. 1125.

(7) WHO "Equipment for vector control", 1990.

(8) Courshee, "Drift spraying for vegetation baiting", *Bull. Ent. Res.*, 50, 1959, p. 355.

(9) E.J. Bals, "The importance of controlled droplet application (CDA) in pesticide applications", *Proceedings of the 8th British Insecticide Fungicide Conference.*, 1975. p. 153

(10) G.A. Matthews, "CDA - controlled droplet application", *PANS*, 23, 1977, p. 387.

(11) D. Munthali, "Biological efficiency of small dicofol droplets against *Tetranychus urticae* (Koch) eggs, larvae and protonymphs", *Crop Protection* 3, 1984, p. 327.

(12) J. Wyatt and others, "Activity of discrete permethrin droplets against whitefly scales", *British Crops Protection Council Conference*, 1984, p. 1045.

(13) A.J. Adams and others, "The relative influence of the factors which determine the spray droplet density required to control the glasshouse whitefly, *Trialeurodes vaporarorium*", *Aspects of Appl. Biology*, 14, 1987, p. 257.

(13a) D. Munthali and T.J. Wyatt, "Factors affecting the biological efficiency of small pesticide droplets against *Tetranychus urticae* eggs", *Pesticide Science*, 17, 1986, p. 155.

(14) D. Omar and G.A. Matthews, "The rainfastness of permethrin deposits of emulsifiable concentrate (EC) and ultra-low volume (ULV) formulation on brussel sprout leaves", *Trop Pest Man.*, 36, 1990, p. 159.

(15) M.D. Mowlam, G.K. Nyirenda and J.P. Tunstall, "Ultra-low volume application of water-based formulations of insecticide to cotton", *Cotton Gr. Rev.* 52, 1975, p. 360.

(16) W.J. King, "Ultra-low volume application of insecticides to cotton in the Gambia", *COPR Misc. Rep.*, 27, 1976.

(17) G.K. Nyirenda, *Studies on the Effects of Insecticide Application on Cotton in Malawi*, Ph.D. Thesis, University of London, 1986.

(18) J.P. Deguine and P. Asfom, "Traitements insecticides a bas volume (BV) en culture cotonniere au Cameroun", *Proceedings of the First African Cotton Research Conference, Lome*, 1989, p. 261.

(19) A.C. Arnold, "Power sources for spinning disc sprayers and their economics", *Tropical Pest Management*, 31, 1985, p. 3.

(20) G.A. Matthews, "Electrostatic spraying of pesticides: a review", *Crop Protection*, 8, 1989, p. 3.

(21) G.R. Cayley and others, "A review of the performance of electrostatically charged rotary atomisers in different crops", *Annals of Applied Biology*, 105, 1984, p. 379.

(22) J.E. Dale, "A non-mechanical system of herbicide application with a rope-winch, *PANS*, 25, 1979, p. 431.

APPENDIX

Selected manufacturers of pesticide application equipment[a]

Manufacturer	*Lever-operated knapsack*	*Compression sprayer*	*Motorized knapsack mistblower*	*Spinning disc sprayer*	*Other*
Brazil					
Maquines Agricolas Jacto s.a., Rua Dr. Luiz Miranda, 1650, PO Box 35, CEP 17.580, Pompeia SP	X	X	X		X
Industria Comercio Guarany s.a., Av. Imperatriz Leopoldina, 112, CEP 05305, Sao Paulo, SP	X	X	X		X
Colombia					
Progen, Urbanizacion Industrial Cazuca, Calle 12 No. 6-64, Soasha, Kilometro 8, Autopista Sev. Apartado 12357, Bogota	X				
Triunfo Fabrica Cra 4a Norte No. 52-137, Cali	X	X			X
Denmark					
Harvig Jensen & Co A/S 5 Farverland DK Glostrup	X	X	X		X
France					
Berthoud 69220 Belleville suSaone.	X	X	X	X	X
Technoma 51.321, Epernay	X	X			X
Germany					
Gloria-Werke GmbH Postfach 1160 4724 Wadersloh/Westf.	X	X	X		X

Manufacturer	*Lever-operated knapsack*	*Compression sprayer*	*Motorized knapsack mistblower*	*Spinning disc sprayer*	*Other*
Solo Kleinmotoren GmbH Postfach 20 7032 Sindelfingen 6	X	X	X		X
India					
American Spring & Pressing Works Ltd. P.O. Box 7602, Malad, Mumbai 400 064	X	X	X	X	X
Italy					
Carpi spa Officine Meccamide, via Romana 82 R.E. 42028 Povidlia	X	X	X		X
Volpi & Bottali via F. Alto bello 2, 26034 Piadana (CR)	X	X	X	X	X
Japan					
Marunaka Sprayer & Duster Co. 11 Mukaida Nishimachi, Kisshoin, Minami-ku, Kyoto	X	X	X		X
Maruyama Mfg. Co. 4-5 Uchi-Kanda, 3 Chome, Chiyoda-Ku, Tokyo 101	X	X	X		X
Malaysia					
Crossmark, Syarikat Jun Chong, Son Bho, P.O. Box 51, Kluan Johore, W. Malaysia	X				X
Mexico					
Swiss-mex-Rapid s.a., Car Panamericano Km 445, 47400 Logos de Moreno Jalisco	X	X			

Manufacturer	*Lever-operated knapsack*	*Compression sprayer*	*Motorized knapsack mistblower*	*Spinning disc sprayer*	*Other*
Netherlands					
Dofra NV PO Box 19, Horst (L).	X				X
Spain					
Goizper S. Coop, (Matabi), 8 Antigna, S/N Anzuola, Guipuzcoa	X	X			X
Switzerland					
Birchmeier CIE 5444 Kunten	X	X	X		X
United Kingdom					
Cooper Pegler Hardi Limited 4, 5 Watling Close Sketchley Headows Industrial Estate Hinckley Leicesteshire LE10 3EX	X	X	X		X
Micron Sprayers Ltd., Three Mills, Bromyard, Herefordshire HR7 4HU.				X	
United States					
B & G, Applebutter Road PO Box 130 Plumsteadville PA 18949-0130.	X	X			X
Hudson Mfg. Co Suite 500, Coal Building, 1130 17th St. NW, Washington DC 20036.	X	X	X		X

	Nozzles
Spraying Systems Co. Ltd., North Avenue at Schmale Road Wheaton, Illinois, United States	

Manufacturer	*Nozzles*
Lurmark Ltd. Longstanton, Cambridge CB4 5DS United Kingdom	
Delavan 811 Fourth Street West Des Moines Iowa 50265, United States	
Hartvig Jensen & Co. A/S 5 Farverland DK Glostrup, Denmark	
Lechler Apparataban KG Postfach 232, D-7000 Stuttgart 1, Germany	
Sprays International 35 Paul Street London EC 2A 4 JU United Kingdom	

(a) The list does not include all manufacturers or all countries in which sprayers are made.

CHAPTER

19

Pesticides: Environmental Impacts

Jack Plimmer

ABC Laboratories, P.O. Box 1097
Columbus, Missouri 65205, United States

1 INTRODUCTION

The potential benefits of the chemical control of pests may be accompanied by adverse impacts. Many of these effects are quite subtle and have been recognized only after extensive experience of pesticide use. Therefore, it is important for pest control specialists and other users to become fully aware of the consequences of the injudicious use of pesticides. Some risks are readily apparent. Several classes of pesticides, particularly those that are powerful cholinesterase inhibitors, may be acutely toxic to man. This toxicity can be readily predicted by mammalian testing. However, among pesticides of low acute mammalian toxicity, it may be very difficult to identify the potential for subtle harmful effects to the environment. Testing for such potential has now achieved a high degree of sophistication through the efforts of regulatory agencies in the developed countries. However, in the face of pressures to increase crop yields, some countries have continued to use pesticides without serious and diligent efforts to understand their potential for adverse effects.

After the introduction of organic pesticides on a largescale in the decade following the Second World War, their spectacular benefits were quickly recognized and the technology was rapidly adopted on a worldwide basis. Malaria and other insect-borne diseases were rapidly brought under control by the introduction of DDT and other long-lasting insecticides. The use of herbicides (or weed killers) in the United States freed many agricultural workers from backbreaking labour and changed the economics of crop production.

Crop production benefitted from the reduction in losses caused by insects, weeds and other pests before and after harvest. Through pesticide use, it became practical to grow crops where because of the prevalence of destructive pest species they could not previously be cultivated profitably.

However, it was soon clear that the benefits of pesticide use could not be sustained if they continued to be used in an indiscriminate manner. Greater demands for chemicals to control agricultural pests and disease vectors resulted in substantial growth in the production of chemicals and also the quantities released into the environment. Their injudicious use led to contamination of the environment and the adverse implications were widely publicized in books such as Rachel Carson's *Silent Spring*.

The extent to which the environment, food, feed, meat etc. had become contaminated was revealed by major improvements in analytical chemistry in the late 1950s and early 1960s. New methods of residue analysis came into widespread use. In particular, gas chromatography made it possible to detect and routinely measure residue at concentrations of parts per million or less on a routine basis. These new analytical techniques were important not only for determining levels of residues in food for human consumption and thus ensuring consumer safety, but also for revealing that pesticide residues were present in many environmental samples, including tissues from fish, birds and other wildlife.

Some pesticides, particularly the organochlorine insecticides, were extremely resistant to environmental breakdown and were termed "persistent". Others, such as the chlorinated phenoxy-alkanoic herbicides, were readily degraded in soils. From a utilitarian point of view, the persistence of the organochlorine insecticides appeared to offer an advantage to the farmer, who could maintain control of pests for long periods without the necessity for repeated applications. However, it was realized that one consequence of such practices was the appearance of insect resistance to insecticides and an unacceptable burden of chemical residues that could potentially contaminate water or crops. Undue persistence is undesirable in herbicides because residues may carry over to damage susceptible crops.

As a consequence of their exceptional stability, organochlorine insecticides were dissipated in the environment but not substantially degraded. They were replaced by short-lived compounds such as the organophosphates and carbamates. These compounds are generally more toxic to mammals but their relatively rapid degradation in the environment reduces their potential impact on non-target organisms. In recent years, these have been supplemented or replaced by highly active compounds, such as the pyrethroids, that are effective at rates of a few grams per hectare and also rapidly undergo degradation in soils.

The problem of resistance is serious, and it is important to adopt measures to avoid its onset by careful mangement of chemicals. In some situations, the continued, repetitive application of insecticides resulted in failure to maintain control of insect pests because resistance developed in the insect population, rendering further treatment ineffective. Consequently, insect pest control programmes were unable to rely exclusively on chemical methods of control. Alternative techniques for insect control were introduced and chemicals were applied only when predetermined management criteria relating to extent of crop damage and insect infestation were met. The combination of alternative pest management techniques and conventional chemicals, known as integrated pest management (see Chapter 3), has effectively reduced the quantities of insecticides used.

Progress in developing new herbicides has paralleled that in insecticides. Initially, weed control placed heavy reliance on esters or salts of chlorinated aryloxyacids. The phenoxy herbicides have generally been supplanted in the United States by more effective compounds. Triazines are still widely used, but newer herbicides that are effective at much lower rates, such as the sulfonylureas, imidazoles and glyphosate, now have well-established markets.

The concern over environmental issues resulted in fundamental changes in the way in which pesticides were registered. Until 1970, this process had been the responsibility of the United States

Department of Agriculture and in the case of those pesticides that were to be used on crops for human consumption, the Food and Drug Administration. Major responsibility for registration was shifted to the United States Environmental Protection Agency (EPA) after it was created in 1970. Since that time, EPA has been responsible for promulgating guidelines and regulations to protect the environment by prescribing the studies that must be submitted before a pesticide can be registered. These regulations cover the chemistry of the product, its mode of use, studies that must be conducted to determine its potential toxic effects and studies that must be conducted to determine its environmental fate and effects on wildlife. These studies must be conducted under precisely defined protocols and the quality and integrity of the data submitted must be assured by compliance with the Good Laboratory Practice Standards prepared by EPA (1). Because older pesticides may have been approved when requirements for these data were less stringent, the registration process calls for periodic review and re-registration to satisfy up-to-date criteria for environmental and human safety.

As part of the submission of data, manufacturers are required to furnish details of the process of synthesis and to provide information on impurities that might arise during the manufacturing process. This information is important because toxicologically significant impurities or contaminants may be introduced during manufacture or formulation. One example is the contamination of the herbicide 2,4,5-T by the extremely toxic compound 3,3,4,4-tetrachlorodibenzo-p-dioxin (TCDD). The manufacture of 2,4,5-T involves the condensation of monochloroacetic acid with 2,4,5-trichlorophenol under basic conditions. The latter is produced by base hydrolysis of 1,2,4,5-tetrachlorobenzene in methanol at elevated temperature under pressure. Strict control of temperature is necessary to avoid the formation of TCDD as an undesirable by-product. Before standards for TCDD content were imposed, stocks of 2,4,5-T were produced that were shown to contain several ppm of TCDD. Polychlorinated benzodioxins (higher homologues of TCDD) were also found to be manufacturing impurities in pentachlorophenol, a fungicide and wood preservative. Such impurities may pose a threat to non-target organisms and, because their physical and chemical properties differ from the precursor, their environmental fate will also require separate investigation.

The pace of development has been slowed by increasing costs of bringing new pesticides to the market, particularly by the high costs of obtaining safety data needed for registration. This is a major factor. But, another important factor is that stringent environmental criteria are an obstacle to the discovery of new biologically active compounds. Formerly, the predominant approach to revealing new leads for pesticide structures was to screen as many as possible synthetic organic chemicals selected at random. Many thousands of compounds were screened annually. Not only did the likelihood of success decrease with the passage of time as existing successful leads were exploited or protected by patent, but the criteria for selection also became more stringent to encompass environmental safety as well as biological activity.

These additional considerations, in recent years, have brought about modifications in the approaches to discovery. Newer pesticides may be synthetic or derived from natural sources. The newer synthetic compounds are active at rates that are only a fraction of those of earlier generations of pesticides. The newer pyrethroid insecticides and the sulfonylurea herbicides are examples of this. Microbial products, such as *Bacillus thuringiensis* preparations, and natural products, such as neem oil extracts, are now marketed. These complex materials of biological origin are relatively short-lived in the environment and their impact is often negligible. There is now an aggressive search for biopesticides. However, industrial operations on a large scale are at present better suited to the manufacture and application of conventional chemicals, so the latter will continue to represent a substantial portion of the market.

The major focus of environmental studies must be the effects of chemicals on soil functions and their toxicological effects on aquatic and terrestrial organisms. This entails not only studies of the parent compound but also of its major metabolites and transformation products that might remain as residues. The presence of bound or conjugated residues that cannot be extracted from the matrix to be analyzed is also a matter of concern. It is necessary for the manufacturer or registrant to establish the magnitude and significance of such residues and their potential consequences.

An approach to assessing "significance" is through the risk assessment process. This process involves identifying the hazard, assessing the dose-response, assessing exposure and characterizing risk. "Hazard" implies the presence of a danger and "risk" may be interpreted to mean the probability of an effect. The process is valuable if reliable data are available and realistic exposure scenarios are adopted. Without these prerequisites, the results may be extremely unreliable. Difficulties in the dose-response assessment lie in extrapolating the effects to humans or ecosystems from data that may have to be obtained in experiments using rats, mice or other test species. Difficulties also occur in the extrapolation of data obtained for specific compounds in a well-defined environment to more complex situations.

In practice the presence of degradation products complicates the toxicological situation. The most likely situation is that the parent compound will be accompanied by its major transformation products and there will be considerable uncertainty concerning their interactions and their availability in the environmental matrix. Here, assessment of the effects of these metabolites requires a knowledge of their physical and chemical properties as well as toxicological information.

2 QUANTITIES OF PESTICIDES USED AND THEIR DEGRADATION PRODUCTS

Assessment of the significance of pesticides in the environment should begin with a consideration of the amounts released into the environment as indicated by pesticide use patterns. Table 19.1 shows the most widely used pesticides in the United States and amounts used by farmers in the United States in 1995 (2).

The metabolites of the compounds shown in Table 19.1 will also occur in the environment. They are generally more polar and water-soluble than the parent compounds. Because few data are available on the occurrence or environmental behaviour of metabolites, predictions must be based on a knowledge of the relevant properties and their probable occurrence in association with the parent compound. It must be assumed that similar environmental transformation and transport processes will influence the fate of the metabolites and that their physical properties will determine their partitioning into environmental compartments.

Table 19.1 Major pesticides used by farmers in the United States in 1995 (metric tonnes)[a]

Herbicides		*Insecticides*		*Fungicides*	
Atrazine	32,802	Oil	23,181	Sulphur	37,596
Metolachlor	26,937	Chlorpyrifos	6,697	Chlorothalonil	5,246
Alachlor	23,402	Terbufos	3,942	Copper	3,751
2, 4-D	19,023	Methyl parathion	2,704	Mancozeb	3,657
Cyanazine	14,601	Carbofuran	2,314	Maneb	1,599
Trifluralin	11,651	Carbaryl	2,073	Captan	1,451
Pendimethalin	9,200	Phorate	2,020	Ziram	1,272
Glyphosate	7,617	Cryolite	1,839	PCNB	754
EPTC	6,558	Aldicarb	1,825	Benomyl	544
Propanil	4,143	Propargite	1,646	Iprodione	396
Total applied	206,034	Total applied	67,594	Total applied	58,634

(a) Top ten compounds and total amount of active ingredient applied in each category.

3 PATHWAYS BY WHICH PESTICIDES MAY BE DISSIPATED IN THE ENVIRONMENT

Pesticides are normally used as formulations in which the active ingredient is diluted by inert materials. These inert additives serve not only as diluents but also improve stability, spreading ability, penetration etc. The formulation may be applied as a spray, incorporated in soil as granules or applied to the target in other ways. In considering its effect on the environment it is important to know the type of formulation that will be used, the method of application and the type of target, because these will strongly influence the fate of the applied chemicals and losses to the environment.

Many different processes may influence the fate of a pesticide applied to soil. These include photodecomposition, adsorption, leaching, volatilization, plant uptake, chemical reaction and microbial metabolism. To predict the fate and behaviour of a pesticide in the environment, it is important to understand the relative importance of these processes. For each class of pesticide, these processes will differ in relative significance, and they will also be influenced by the climate and the soil environment.

The reactions of pesticides in soils have been studied in detail because the data are necessary for approval or registration of the pesticide. Most important among the routes of degradation are those due to soil micro-organisms. The structures of products formed by biochemical transformations have been studied extensively and the major routes of degradation can be predicted. However, studies with sterile soils, isolated microbial cultures and field soils have shown that chemical reactions may also play a role in the transformations of pesticides in soils (3).

Reactions that may occur include hydrolysis, oxidation, reduction and free-radical processes. For example, chloro-s-triazine herbicides are hydrolyzed chemically in strong acid or base solutions to the hydroxy analogue. But, the hydrolytic product of simazine, atrazine and propazine may be recovered from soils treated with these herbicides, and it has been proposed that the reaction is catalysed by protonated forms of montmorillonite clay (4). Hydrolysis of organophosphate insecticides in soils may also proceed by purely chemical processes, and there is evidence that soil fractions may catalyze the reaction.

The presence of free radicals in soils may be responsible for some oxidative reactions. The observed degradation of 3-aminotriazole in sterile soils suggests that radical processes may be occurring because the products are the same as those formed by the action of free-radical oxidants such as Fenton's reagent (a source of hydroxyl radicals).

4 TRANSPORT PROCESSES: MOVEMENT OF PESTICIDES TO WATER AND AIR

The portion of the applied pesticide that is not directly taken up by the target organism will subsequently be degraded by chemical or biological action unless it is chemically very unreactive. This may occur at the site of application or the residues may be transported. The possibility that pesticide residues will be found at a distance from the site of application will depend primarily on their resistance to chemical and biological degradation and on their physical properties (solubility in water and organic solvents, vapour pressure, polarity etc.). For example, although the vapour pressures of the organochlorine insecticides are quite low, residues may be detected in atmospheric samples over wide areas of the globe because the compounds volatilize slowly but are resistant to degradation by sunlight.

Most organic compounds possess measurable vapour pressures. Loss of pesticides to the atmosphere occurs during and after application and if residues deposited on soil or foliar surfaces remain exposed to the air. If degradation is slow, there may be potential for the pesticide to be washed off in solution or adsorbed to soil when heavy rainfall occurs. This is termed run-off. Alternatively, in some soils, the more persistent, water-soluble pesticides may leach or move vertically through the soil profile to considerable depths to contaminate underground water.

Run-off, leaching, drift during application and volatilization all serve as mechanisms for transporting pesticide residues from the site of application. Such effects call for caution in the use of pesticides because of the potential for contamination of surface and well water.

4.1 Run-off

Run-off, or contamination of surface waters, occurs when pesticide or soil is washed into adjacent surface waters by flooding or storm events. The amount that is lost may represent as much as 100% of the amount applied. The major factor will be the frequency of storm events during the period immediately following pesticide application.

In a review of this topic (5), Wauchope uses the term run-off to describe both water and associated sediment lost from the surface of fields. Data for seasonal losses of pesticides from agricultural fields show that the greatest long-term losses occur with herbicides formulated as wettable powders. Losses may total up to 5% from fields of moderate slope or 2% from fields of low slope. Higher losses may be caused by a large run-off-producing event that occurs within two weeks of application.

Water-insoluble pesticides rank second. Generally, these are foliar-applied emulsions of insecticides. Losses may be 1% or less. Losses of water-soluble pesticides applied as aqueous solutions or soil-incorporated pesticides may be 0.5% or less. Storm events are a major influence on losses due to run-off.

Metabolites and transformation products may be considered in relation to run-off-available-pesticide half-lives. The decline in concentrations in run-off water decreases exponentially for successive events. But the decline is much more rapid than that of the decrease in soil concentrations with time. The value for atrazine is 7-10 days. The combination of this information with rates of transformation (metabolism, photolysis etc.) may provide an indication of the potential quantity of transformation products that can enter water in run-off.

4.2 Leaching

During the last decade, the potential for contamination of surface waters or underground water by agricultural chemicals has become an important issue. Initial studies of pesticide behaviour in soils indicated that leaching was unlikely to result in the contamination of subsurface water (groundwater), because the more mobile pesticides would degrade before they had penetrated to sufficient depth to be detected in water. This generalization was shown to be limited, and monitoring studies of groundwater and water from drinking water wells show that there are notable exceptions that depend on both the class of compound and the soil environment.

Several herbicides, insecticides and nematicides have been detected in well water. Some contamination may arise through faulty practices that include such things as pesticide mixing operations in close proximity to well-heads, washing equipment near well-heads, spillage and improperly sealed or unsealed disused wells.

Although integrated pest management and regulatory actions have reduced the use of insecticides, a number of insecticides have been detected in groundwater. This occurrence is dependent on the class of compound, usage patterns and hydrogeological and climatic factors in the area where they were used. The increase in the use of herbicides has also been accompanied by increasingly frequent reports of their detection in groundwater.

The contamination of groundwater by herbicides and nitrates is of particular concern in the north-central and eastern regions of the United States, where more than half of the country's N, K and P fertilizers use is concentrated and where pesticides applied to corn and soybeans accounts for over 80% of all herbicides and nearly 60% of all insecticides used on row crops, small grains and forages (excepting plantation crops and forests) (6). Hallberg (7) has stated that agricultural chemicals can be found in many of Iowa's groundwater aquifers and these commonly exhibit nitrate concentrations above the maximum contaminant limit of 10 mg/l NO_3-N (45 mg/l as NO_3).

4.2.1 Vulnerability

The vulnerability of groundwater to pesticide contamination in specific areas is linked to the management and regulation of pesticide application. The main determinants of leaching potential are the chemistry of the active ingredient and the type of formulation. Whether leaching will actually occur depends on the time of application, soil type, climatic factors, cultural system etc.

It is claimed that the DRASTIC system permits the underground water pollution potential of any area in the United States to be evaluated. This system incorporates hydrogeological settings upon which are superimposed several relative ranking factors. The term "hydrogeologic settings" describes mappable units that possess common major geological and hydrological features controlling groundwater movement.

The factors included as a basis for ranking are the depth to the water table; net recharge to the aquifer; soil media; topography of the land surface; impact of the vadose zone and the hydraulic conductivity of the aquifer. The vadose zone is the unsaturated zone below the soil surface above the water table. The acronym DRASTIC is derived from the initial letters of the system components:

Depth to groundwater
Recharge of groundwater by rainfall and irrigation
Aquifer media
Soil media
Topography (slope of land surface)
Impact of vadose zone
Conductivity (hydraulic conductivity of the aquifer)

Each component is scored from 1 to 10. The ratings are multiplied by weighting factors and the total obtained is used to indicate areas vulnerable to groundwater pollution. The higher the score, the greater the vulnerability of the area (8).

4.2.2 Practical Measurements of Movement in Soil

The soil thin-layer chromatographic technique has proved useful for the study of the effect of soil properties on the relative mobility of pesticides (9). Thin-layer chromatograms were prepared by coating glass plates with different soil types. Relative mobilities of pesticides were then assessed by applying the appropriate radioactively labelled pesticide and obtaining an autoradiograph after the chromatogram had developed.

Soil columns may also be used and leached until the pesticide is observed in the leachate. Alternatively, they may be cut into sections for analysis after a predetermined volume of water has been applied. This technique may be utilized in the field by isolating an intact soil core *in situ* by means of a steel cylinder and measuring the amount of pesticide at a given depth. In this case, irrigation or simulated rainfall may be used. It is customary to include inorganic ions, such as bromide, as reference markers in these experiments. Field studies of movement have been conducted on a much larger scale to assess both movement and potential for contamination of groundwater. This is necessary because there is considerable spatial variability in the factors to be measured. Both movement and degradation (and the potential for movement of metabolites to groundwater) must be measured in large-scale studies such as that conducted at a number of locations in California and other western states to determine whether aldicarb was likely to enter the drinking water in major agricultural areas. This study required the analysis of approximately 2,700 soil samples and 450 water samples. It established that most aldicarb residues degraded in the upper part of the unsaturated soil zone and that aldicarb did not enter the drinking water (10).

4.3 Volatilization

Loss of a pesticide by volatilization may occur during or after application. Some considerations that relate to application are discussed in a later section. During application, there may be a movement of material from the application site due to drift (of spray droplets) or there may be movement of pesticide vapour. This volatility loss during and after application may involve a significant proportion of the applied material. However, the techniques for study and measurement of these losses are complex and time-consuming. The effects of these losses are possible damage to sensitive crops in the vicinity; deposition of residues on crops or plants that may be subsequently classified as illegal residues, if no specific tolerance exists; low-level effects occurring in sensitive organisms at a distance and a general contribution to organic chemical pollutants in the atmosphere.

The vapour pressure of the pesticide and the nature of its immediate environment are the most important factors in determining the extent of loss through volatilization. Losses from moist soil following surface application may be estimated using an empirical rate expression (11):

$$K_v = Q(P/K_\alpha S)$$

where, P = Vapour pressure (mm Hg)

K_α = Soil absorption coefficients ($\mu gg^{-1}/\mu gml^{-1}$) organic carbon basis and

S = Solubility ($\mu g/ml$).

This equation gave volatilization rate constants within a factor of about 10 for six of eight pesticides measured in field experiments. To take into account all factors associated with volatilization from soil, a more complex model would be needed.

Pesticides and other pollutants may be deposited at the surface of the earth in rainfall (also in snow, fog, hail, dew or clouds), in association with dry particulate matter or directly from the vapour phase (dry deposition). Some data are available on the magnitude of pesticide deposition and its impact.

Pesticide concentrations in rainfall appear to be several orders of magnitude lower than injury thresholds for most non-targeted plants. Some aquatic and algae and simple vascular plants would show effects at or near these concentrations, but physiological effects are more likely to result from surface run-off because of the probability that airborne residues would be very widely dispersed in the atmosphere.

The significance and potential magnitude of loading of materials from the atmosphere may be large (12). Estimates of atmospheric loadings in rain to Lake Superior in 1983 showed considerable potential for deposition of organochlorine insecticides, i.e. α-BHC, 860 kg/yr; lindane, 290 kg/yr; heptachlor epoxide, 17 kg/yr; dieldrin, 28 kg/yr; DDTs, 17 kg/yr; methoxychlor, 120 kg/yr and PCBs, 300 kg/yr.

The powerful reactivity of oxidants and free radicals generated by solar irradiation is responsible for extensive degradation and transformation of many organic compounds in the atmosphere. Persistent organochlorine compounds are an exception. Many of the high molecular weight organic compounds transported in the atmosphere are chlorinated compounds associated with pest control application. Experiments conducted globally show that hexachlorocyclohexanes (HCHs) and chlorinated benzenes (13) are among the most abundant organochlorine pesticides in the atmosphere.

In an eleven-year study of the long-range transport of pesticides, α-HCH was the most abundant species present among the organochlorine insecticides (13) . Levels of HCH and DDT isomers were measured in water and atmospheric samples collected in survey cruises of the Pacific and Indian oceans. The highest concentrations of these species were in the mid-Northern Hemisphere. The HCHs are derived from technical BHC and lindane.

The ratios of α- to γ-HCH measured in a 1986 Pacific survey cruise were about 10-15 (14). The two isomers accounted for over 50% of the total high-molecular-weight organochlorines in the North Pacific atmosphere.

Dry deposition is a significant mechanism for removing organic chemicals from the atmosphere. It is estimated that each year about 89,000 tonnes of organic compounds are deposited in the Netherlands by this route (15) . Pesticide concentrations in air were measured and are given in Table 19.2.

Table 19.2 Pesticide concentrations in air

Compound	*Air concentration (pg/m^2)*	
	Mean	Max.
α-HCH	250	1,200
Lindane	360	3,400
DDT residues total	148	–
Heptachlor	37	190
Heptachlor epoxide	32	360
α-Endosulfan	168	1,130

5 ALTERATION OF PESTICIDES

Alteration is defined as a process in which the parent molecule is modified to form photoproducts, oxidation products, metabolites etc. or in which it undergoes total mineralization. It includes the chemical processes of hydrolysis, oxidation and reduction, photochemically induced reactions and metabolic processes involving micro-organisms, plants, mammals and other biota.

5.1 Photochemical Degradation of Pesticides

Sunlight energy is an important factor in degrading organic compounds of synthetic and natural origin. Sunlight wavelengths of less than 295 nm are almost completely absorbed by ozone and are thus not involved in reactions at the earth's surface. Although the shorter wavelengths of the solar spectrum are greatly attenuated by passage through the atmosphere, this radiation is responsible for direct and indirect photolytic processes that may transform pesticides exposed on soil, foliar surfaces or in shallow waters. Because shorter wavelengths (295 nm - 400 nm) are attenuated more strongly than the longer visible wavelengths as they pass through the atmosphere, the rates of photolysis of pesticides will be highly dependent on latitude, season etc. Therefore, in the tropical regions photochemical processes may play an important role in determining the environmental fate of pesticides.

It is essential that light energy be absorbed before reaction occurs. Consequently photochemical reactions are mainly on surfaces or in media that can transmit sunlight. Energy absorption is followed by excitation of the absorbing molecule. The loss of energy from the molecule as it returns from the excited state to the ground state may be accompanied by processes such as reaction, radiation of energy or transfer of energy to a second molecule during collision. Pesticides and their metabolites may undergo direct or indirect photochemical reaction. Indirect (sensitized) processes may occur by energy transfer from a second molecular species that has absorbed light to produce an excited molecule without undergoing chemical reaction. Thus, pesticides or metabolites that do not absorb energy in the solar region of the spectrum may react because they acquired sufficient energy by collision with an excited sensitizer molecule.

Products of photochemical reactions of pesticides have been well investigated, and regulatory agencies require information on the products of photolysis in solution and on soils. A common photochemical process includes the fission of covalent bonds with the formation of free radicals. Because pesticide molecules occur in association with many environmental matrices and radical species are very reactive, the subsequent processes and products may be complex, depending on the molecular environment. For example, irradiation of a halogenated molecule may give rise to a halogen radical and a hydrocarbon radical. The photochemically generated radical may then abstract a hydrogen atom from the solvent to give a reduced product. Alternatively, oxygen may add to the photochemically generated radicals to form a variety of oxygenated products.

Model experiments are usually conducted to indicate the rates of reaction and the types of process that might occur for a particular type of molecule (16). Irradiation conditions used will depend on the purposes of the study. Regulatory authorities prescribe the use of sources such as xenon lamps whose emission spectrum approximates to that of natural sunlight more closely than filtered high- or medium-pressure mercury lamps. The latter, however, proved extremely useful in many earlier studies of pesticide photolysis. Through such studies an understanding of products and mechanisms was obtained. Natural sunlight has been used as a source, but the variability of this source at a single location may give irreproducible results (17).

5.1.1 Photochemical Processes in the Aquatic Environment

Materials present in natural water may increase or decrease rates of photolysis of contaminants. Dissolved substances or suspended particles may absorb light and thus reduce the light available for photolytic processes (18). On the other hand, natural humic and fulvic acids act as sensitizers and probably facilitate the degradation of xenobiotics (19). The absorption of light by humic substances may produce a number of active species, including the following:

(a) Aqueous electrons (20);
(b) Singlet oxygen (19);
(c) Organoperoxy radicals (19);
(d) Hydroxyl radicals (21);
(e) Superoxide (22);
(f) Hydrogen peroxide (22);
(g) Phototransients derived from humic substances (18).

These active species will be responsible for further reactions of xenobiotics in natural waters wherever solar radiation can penetrate.

5.1.2 Photolytic Processes in the Atmosphere

Radiation of wavelengths 290-400 nm comprises about 4% of the solar spectrum. The more energetic region (290-320 nm) is sometimes known as the UV-B region. Many pesticides absorb light in this region and may undergo direct photolysis in sunlight in the vapour phase. As in the case of aquatic photoprocesses, a variety of indirect processes can also occur and compounds that do not undergo direct photolysis may be rapidly transformed by photochemically generated oxidants. Hydroxyl radicals are probably the most important among the species responsible for the oxidative transformation of organic compounds that enter the atmosphere. But other species may be involved, including ozone, singlet oxygen, nitric oxide, hydroperoxy radicals and other radical species. Organic compounds in the atmosphere may be rapidly transformed or degraded by the combined action of free radicals. For example, nitrophenols may be formed by the action of hydroxyl and NO radicals.

The investigation of the volatilization of trifluralin in the field showed that it is exceptionally volatile and is also susceptible to photolysis. It was clear from these experiments that unless trifluralin was incorporated in soil the rate of loss would be very rapid indeed.

Subsequently, in an investigation of the photochemical behaviour of trifluralin in the field, trifluralin and its photoproducts were detected in the air above both surface and soil incorporated fields (23), suggesting that these products probably arose primarily from photolysis of trifluralin on the soil surface followed by volatilization. However, the contribution of vapour-phase photolysis probably increases with air residence time of trifluralin.

The half-life of trifluralin in the atmosphere was estimated at about 20 min (25). The vapour-phase photolysis products include mono- and di-dealkylated trifluralin, a benzimidazole formed by a cyclization process involving one N-propyl group and a nitro group and the corresponding dealkylated benzimidazole.

5.2 Metabolism

The soil is the recipient of a large portion of applied pesticides. It was long believed that soil had an almost unlimited capacity for transforming organic pesticides and other man-made organic chemicals into innocuous materials that were ionic in character (carbon dioxide, carbonates, chlorides, nitrates, phosphates, ammonium and other simple fragments). Prior to methods of analysis that could detect and measure organic compounds at the parts per million level, it was difficult to challenge this convenient assumption. Because pesticides are applied at one kilogram (or less) per hectare, the analytical methods necessary to measure the amounts that will be present in the plough layer following application must be capable of measuring residues at the parts per million level. These limitations presented a major barrier to quantitative residue studies until the widespread introduction of gas chromatographic techniques in the early 1960s. Although experiments in the 1950s demonstrated that unrealistically high applications of organochlorine insecticides would persist in soil for a prolonged period, it was felt that at the levels normally used, residues would dissipate more rapidly.

Experience in the field has shown that these assumptions were incorrect. During the 1950s, sensitive methods of measurement became routinely available and the analyses of soil and other environmental matrices gave a more accurate picture of the rates of degradation of pesticides in soil.

6 ENVIRONMENTAL EFFECTS

6.1 Prediction of Environmental Effects

For each compound, specific tests are necessary to establish whether it should be recommended for use on a particular crop in the environmental situation under consideration. It is important to examine the environmental behaviour of a pesticide in the location where it will be used and also to understand its effects on local ecosystems. The tests are prescribed by national or local regulatory officials and, at the international level, by organizations such as FAO or OECD.

In the introduction to this chapter, the role of EPA was mentioned and the nature of the information required for registration of a pesticide in the United States. To predict environmental behaviour, regulatory agencies require submission of data concerning methods of manufacture of the pesticide and the nature of the impurities and inert ingredients in the formulated product. The physical and chemical properties of pesticide ingredients must be obtained by prescribed test procedures. The manufacturer must also submit information on the environmental behaviour of the active ingredients, significant metabolites and impurities, including their persistence and rates of degradation in soils, mobility, rates and pathways of photolysis in soil, air and water and other data, depending on the type of use proposed.

In addition to toxicity data for mammals, laboratory tests of toxicity to species of microflora, invertebrates, fish and plants must be conducted to indicate whether special hazards exist. Sublethal effects can also be predicted by such observations. Experimental systems that approach natural conditions more closely may be necessary for understanding pesticide distribution and its effects in relation to the variety of species present in the ecosystem. Such experimental systems are described as model ecosystems or mesocosms, depending on their scale.

6.2 Reduction of Environmental Impact through Application and Formulation of Pesticides

Pesticides are commonly applied by spray techniques employing fixed-wing aircraft, helicopters, ground-spray rigs, knapsacks or other portable equipment depending on plot size, location and resources available. In many developing countries, a backpack spray is often used, particularly in intensively cultivated areas. The spray operation presents hazards to the environment and to the operator if it is conducted without strict observance of safety precautions. For the spray operation, protective clothing is necessary to safeguard both the operator and family members who may subsequently come into contact with contaminated work clothing. Mixing and diluting pesticides and washing equipment may result in contamination of soil, wells or other sources of water unless precautions are taken to prevent all forms of spillage. Used pesticide containers must be disposed of in such a manner as to eliminate the possibility of environmental contamination.

The spray operation may be a source of contamination through drift or it may be conducted inefficiently so that considerable amounts of pesticide fall on non-target sites. Aerial spraying must be conducted under favourable atmospheric conditions to avoid drift and damage to adjacent areas. Wind velocity and droplet size are critical factors in determining the extent of drift. Drift is strongly dependent on application equipment. Formulation additives are commonly used to reduce drift. Smaller droplet sizes are more prone to drift, but droplet size is also a factor in efficacy. Insecticidal efficacy may be much greater if smaller droplets are applied as these may more efficiently penetrate to and permeate the microenvironment inhabited by the insect. By contrast, many herbicides act by leaf contact, and larger droplets may more effectively wet the leaf. These considerations underscore the need for spray application equipment to provide a homogeneous spectrum of droplet sizes. To control this factor, equipment must be well maintained and frequently calibrated. It is important to maintain equipment, check nozzle size and replace worn nozzles.

Sprayable formulations may be emulsifiable or water-soluble types, wettable powders or flowable suspensions. Other formulations of pesticides are dusts, granules and baits. Dusts were used extensively in the past, but they present considerable potential for environmental contamination. Although granular formulations may be very useful, some types of granules may be ingested by birds if the granules remain on the surface and are not incorporated in the soil. Water-dispersible granules may offer advantages in minimum-till systems because they may be used to reduce the interception of applied herbicides by surface trash.

Correct placement of pesticides may result in a considerable reduction in the amount reaching off-target sites. For example, seeds can be protected by seed coatings; pesticides may be applied in bands during seed planting; spot treatments may be more economical and a number of devices have been described that provide improved application to the target.

7 CONCLUSIONS

The well-established role of pesticides in crop protection is sustainable if they are managed wisely with an understanding of their effects and the interrelationships among crop, pests and environment. To minimize their environmental impact, the following strategies have been proposed (25):

(a) Use of products with higher biological activity to reduce application rates per unit area;
(b) Use of less persistent chemicals in crop production;
(c) Use of the biologically active isomer only (e.g. lindane, the gamma isomer of BHC, is an effective insecticide, but the remaining mixture of BHC isomers obtained by chlorinating benzene is a potential environmental pollutant);
(d) Use of improved formulation techniques that are environmentally safer, such as controlled release formulations;
(e) Use of improved application methods that result in more specific placement;
(f) Use of chemical techniques in an integrated pest management system.

These strategies must be combined with an understanding of the specific environment in which the pesticide will be used. When selecting chemicals, management should take into account not only the suitability of a chemical for pest control in a particular crop but also its environmental behaviour and effects under specific soil and climatic conditions. In the tropics, soil types, rainfall and sunlight may be very different from the conditions under which experience of the agricultural use of a pesticide was previously gained. It is therefore essential to conduct studies of environmental behavior under local conditions.

REFERENCES

(1) *Federal Register*, 54, No. 158, 34067, 17 August, 1989.
(2) L.P. Gianessi and J.E. Anderson, *Pesticide Use in U.S. Crop Protection, National Summary Report* (Washington, D.C., National Center for Food and Agricultural Policy, 1995).

(3) P.C. Kearney and C.S. Helling, "Reactions of pesticides in soils", *Res. Rev.*, 25, 1969, p. 25.

(4) J.D. Russell and others, "Mode of chemical degradation of s-triazines by montmorillonite", *Science*, 160, 1968, p. 1340.

(5) R.D. Wauchope, "The pesticide content of surface water draining from agricultural fields, *J. Environ. Qual.* 7, 1978, p. 459.

(6) N.L. Hargett and J.T. Berry, *1982 Fertilizer Summary Data*, National Fertilizer Development Center, TVA, 1983.; *Handbook of Agricultural Charts, Ag. Handbook No. 619.*, Washington DC., U.S. Department of Agriculture, 1983.

(7) G.R. Hallberg, "Agricultural chemical and groundwater quality in Iowa, *Proceedings of the 37th Annual Fertilizer and Agricultural Chemicals Dealers Conference*, Iowa State University, Coop. Extension Series CE2158 1985, p. 110.

(8) L. Aller, J. Bennett, J.H. Lehr and R. Petty, "Drastic: a system to evaluate the pollution potential of hydrogeologic settings by pesticides", *Evaluation of Pesticides in Ground Water*, W.Y. Garner, R.C. Honeycutt and H.N. Nigg, eds. *ACS Symposium Series No. 315*, (Washington D.C., American Chemical Society, 1986).

(9) C.S. Helling, "Pesticide mobility in soils. I. Parameters of soil thin-layer chromatography", *Soil Sci. Soc. America Proc.*, 35, 1971, p. 732; C.S. Helling, "Pesticide mobility in soils. II. Applications of soil thin-layer chromatography", *Soil Sci. Soc. America Proc.*, 35, 1971; C.S. Helling, "Pesticide mobility in soils. III. Influence of soil properties", *Soil Sci. Soc. America Proc.*, 35, 1971.

(10) R.L. Jones, "Aldicarb environmental studies conducted in the western United States", *Fate of Pesticides in the Environment*, J.W. Biggar and J.N. Seiber, eds. University of California Agricultural Experiment Station, Publ. No. 3320, 1987, p. 105.

(11) R.L. Swann, P.J. McCall and S.M. Unger, "Volatility of pesticides from soil surfaces", *Handbook of Chemical Property Estimation Methods*, W.J. Layman, W.F. Reehl and D.H. Rosenblatt, eds. (New York, McGraw-Hill, 1982).

(12) W.M.J. Strachan, *Bulletin of Environ. Chem.*, 309, 1985.

(13) R. Tatsukawa and others, *Long Range Transport of Pesticides*, D. Kurtz, Chelsea MI Lewis, eds. 1990, pp. 127-141.

(14) E.L. Atlas, and S. Schauffler, *Long Range Transport of Pesticides*, ed. D. Kurtz, Lewis, MI Chelsea, 1990, pp. 161-183..

(15) R. Guicherit and F.L. Schulting, *The Science of the Total Environment*, 43, 193, 1985.

(16) B.D. Cavell, "Methods used in the study of photochemical degradation of pesticides", *Pestic. Sci.*, 10, 1979, p. 177.

(17) N. Burkhard, D.O. Eberle and J.A. Guth, "Environmental quality and safety", *Third International Congress of Pesticide Chemistry*, (Oxford, Pergamon, 1975), p. 203.

(18) G.C. Miller, and R. G. Zepp, *Water Res.*, 13, 1979, p. 453.

(19) R.G. Zepp, in *Humic Substances and their Role in the Environment*, F. H. Frimmel and R. F. Christman, (New York, Wiley, 1988), pp. 193-214..

(20) R.G. Zepp and others *Env. Sci. Tech.*, 1987.

(21) W.R. Haag and J. Hoigne, *Chemosphere*, 14, 1985, p. 1659.

(22) W.J. Cooper and R. J. Zika, *Science*, 220, 1983, p. 711.

(23) C.J. Soderquist and others, *J. Agric. Food Chem.*, 23, 1975, p. 304.

(24) J.E. Woodrow and others, *J. Agric. Food Chem.*, 36, 1978, p. 1312.

(25) Geissbuhler and others, 1982, cited by N.J. Drescher, *Proceedings of the Fifth International Congress on Pesticide Chemistry*, 1982.

CHAPTER

20

Environmental Toxicological Considerations in Pesticide Formulation

Yong-Hwa Kim

United Nations Industrial Development Organization
Vienna, Austria

1 PROLOGUE

There is no doubt that pesticides have contributed to increasing the yield of crops by controlling insects and plant pathogens, to reducing the labour required of farmers by controlling weeds and to enhancing public health by controlling rodents and other vectors of disease. Therefore, it is natural that the pesticide industry has become one of the major divisions of the chemical industry worldwide, especially of the speciality chemical industry. However, recent global concerns for the environment put significant pressure on the pesticide industry in many ways: several pesticides have been banned, the use of some popular pesticides has been restricted and stricter regulations for the protection of the workers in the manufacturing plant, the users and the environment have been enacted. Even for the registration of a new pesticide, more data are needed for safeguarding the human health and the environment.

Apart from the direct economic impact of environmental concerns on the pesticide industry and society, distrust among the various parties - government, industry and the public - has caused complications in many cases of environmental pollution. For example, pesticides are one of the suspects in fish kills in streams, rivers and coastal waters. In environmental incidents that happen in the field, the direct causes of the accident are often difficult to identify, and pesticides have always become a prime suspect. Of course, there have been several accidental impacts on natural organisms caused by the use of certain pesticides (1, 2, 3). This situation has forced both governments and the pesticides industry to consider every aspect of the adverse impact of a pesticide on the ecological system before introducing it to the market. In this regard, preventive measures such as more stringent data requirement for the registration of a new pesticide or for the re-registration of existing pesticides have come into force, mostly in the developed countries.

The pesticide industry has been trying to meet the requirements of the government proactively. At one time this was enough to satisfy the public and government that human health and the environment were being adequately protected (4), but now they are putting more and more pressure on the industry to address other toxicological and ecological concerns. Research on environmental toxicology by the scientific community needs to go a long way before the three sectors can understand each other and compromise. One of the latest demands of the public has been for conservation of the ecosystem (biodiversity), and the legal requirements in this area are expected to become more rigorous.

This chapter will review the global trends of environmental protection and their implication for the pesticide industry, especially for the registration of pesticides and pesticide formulation. A direction for pesticide formulation will be suggested from the environmental toxicological point of view.

As terminology in this field is still being harmonized, it would be useful to define two terms as they are used in this chapter. "Environmental toxicology" is the study of the movement of toxicants and their metabolites in the environment and in food chains and the effect of such toxicants on individuals and populations (5). "Ecotoxicology" is the study of the toxic effects on non-human organisms, populations and communities (6).

2 GLOBAL CONCERN ABOUT CHEMICALS AND THE ECOLOGICAL SYSTEM

There have been several international efforts to ensure the safety of pesticides: the Prior Informed Consent (PIC) procedure, the Codex Alimentarius Commission of FAO and WHO, the Guiding Principles of the Organization for Economic Cooperation and Development (OECD) and the International Register of Potentially Toxic Chemicals (IRPTC) of UNEP (7). These efforts concentrated more or less on compiling and disseminating information on the human health aspects of pesticides. These efforts were limited by the fact that they were non-binding, and compliance was up to individual countries, even though some of the countries had no resources or infrastructure to regulate the pesticides. The United Nations Conference on Environment and Development, held at Rio de Janeiro in 1992, integrated many of the fragmentary efforts on the environment that had been made in different parts of the world since 1960s (8). The Conference adopted Agenda 21, an action plan for sustainable development of the world. Many international organizations and participating governments are mandated to implement the actions specified in the Agenda at the national policy level. The parts of Agenda 21, that are relevant to chemicals and the ecosystem are as follows:

(a) Chapter 19: Environmentally sound management of toxic chemicals;
(b) Chapter 15: Conservation of biological diversity;
(c) Chapter 17: Protection of the oceans;
(d) Chapter 18: Protection of the quality and supply of freshwater resources;
(e) Chapter 20: Environmentally sound management of hazardous wastes.

All these chapters emphasize the management of the causative agents that are detrimental to the biodiversity of the various climate zones of countries, regions and the entire earth, and most places on our planet have become the subject of conservation efforts. Chapter 19 specifically emphasized the involvement of industry in every action to be implemented by the government authorities: expanding and accelerating the international assessment of chemical risks; harmonization of classification and labelling of chemicals; information exchange on toxic chemicals and chemical risks;

establishment of risk reduction programmes; strengthening of national capabilities and capacities for management of chemicals. Most paragraphs dealing with management-related activities in Chapter 19 begin with the words "Government, through the cooperation of relevant international organizations and industry....".

United Nations organizations have achieved significant progress on launching or strengthening various programmes: the Global Biodiversity Assessment, global monitoring of fish stock, prevention of marine pollution, and the Global Environmental Monitoring System (GEMS) (9). These programmes are documenting the unprecedented impact of human activities on the natural ecosystem. Pesticides have been always one of the human activities monitored by those programmes (10).

3. EFFICACY VERSUS ENVIRONMENTAL TOXICOLOGICAL RISK

Since the main function of pesticide formulation is to enhance the efficacy of the active ingredient of the pesticide, the main task of the formulation chemist has been to prolong the lifetime of the active ingredient at the target site or in the field and to facilitate its action mechanism. The only consideration given to the risk aspect has been the health of human being, the user.

The emphasis of the industry on efficacy has stood in contrast to recent government and public pressure to preserve the ecosystem. Prolonging the lifetime of the pesticide could also lengthen the exposure time to it of non-target organisms in the environment, which might increase their risk. Facilitating the action mechanism could also have an effect on the non-target organisms, jeopardizing their survival in the habitat.

These concerns may not be applicable to every pesticide or every field situation. For instance, technology involved in prolonging the lifetime of the pesticide could serve to decrease its initial application rate. That would decrease its concentration at the site; and a low concentration of the compound is linked with less hazard to the organism.

These aspects of pesticide formulation were summarized by Trought (11):

> "Trends in the development of pesticide formulation has been towards 'safer' formulations at least from the point of view of the operator. Regulatory pressure from environmental and consumer areas is expected to continue to increase, with the result that registration of new pesticides will become more and more difficult, despite new formulations generally being safer. The role of the formulation chemist in developing safer, more revolutionary types of products will become increasingly important if increasingly stringent registration requirements are to be counterbalanced."

As mentioned above, more emphasis should be given to the ecological aspect in the development of the pesticide formulation.

4 REGULATION ON THE ACTIVE INGREDIENTS OF PESTICIDES

In order to gain an insight into the demand of society for preserving the health of the ecosystem, it would be best to review the current regulations on the pesticide registration. Most countries regulate pesticides by evaluating the toxicity of the active ingredients to the organisms in the environment. Some minor additional data could be required on pesticide formulations to ensure that there is no synergistic effect between the inert ingredients and the active ingredient.

4.1 Developed Countries

Environmental toxicological data required for the registration of pesticides in the United States are shown in Table 20.1 (12). The government requires every possible data to understand the fate of the active ingredient in the field and the effect on several representative organisms in the ecosystem. As shown in Table 20.2, European Union countries require almost the same data as does the United States and some more data on various organisms in the ecosystem (13).

Table 20.1 Data requirements on environmental toxicology for the registration of a pesticide in the United States. (12)[a].

Type of testing	*Data required*	*Strength of requirement*	*Test substance*
Environmental chemistry			
Physical and chemical characteristics	Solubility	R	T or P
	Vapour pressure	R	T or P
	Dissociation constant	R	T or P
	Octanol/water partition coefficient	R	P
	pH	R	T and F
Residue chemistry	Residue analytical method	R	T and M
	Magnitude of the residue		
	Crop field trials	R	F
	Processed food/feed	CR	F
	Meat/milk/poultry/eggs	CR	T or M
	Reduction of residue	CR	(T or M)
Environmental fate			
Degradation studies - lab	Hydrolysis	R	T or PR
	Photodegradation		
	In water	R	T or PR
	On soil	CR	T or PR
	In air	CR	T or PR
Metabolism studies - lab	Aerobic soil	R	T or PR
	Anaerobic soil	R	T or PR
Mobility studies	Leaching and adsorption/desorption	R	T or PR
	Volatility - field	CR	F
Dissipation studies - field	Soil	R	F
	Soil, long-term	CR	F
Accumulation studies	Rotational crops		
	Confined	CR	PR
	Field	CR	F
	In fish	CR	T or PR
Ecotoxicity			
Wildlife and aquatic organisms			

(Contd...)

Table 20.1 *(Contd...)*

Type of testing	*Data required*	*Strength of requirement*	*Test substance*
Avian and mammalian testing	Avian oral LD_{50} - mallard or bobwhite	R	T
	Avian dietary LC_{50} - mallard and bobwhite	R	T
	Wild mammal toxicity	CR	T
	Avian reproduction - mallard and bobwhite	CR	T
	Simulated and actual field testing - mammals and birds	CR	F
Aquatic organism testing	Freshwater fish LC_{50} - rainbow and bluegill	R	T
	Acute LC_{50} freshwater invertebrates - *Daphnia*	R	T
	Acute LC_{50} estuarine and marine organisms	CR	T
	Fish early life stage and aquatic invertebrate life cycle	CR	T
	Fish life cycle	CR	T
	Aquatic organism accumulation	CR	T, P or D
	Simulated or actual field testing-aquatic organisms	CR	F
Plant protection			
Non-target area phytotoxicity	**Tier I**		
	Seed germination/ seedling emergence	R	T
	Vegetative vigour	R	T
	Aquatic plant growth	R	T
	Tier II		
	Seed germination/ seedling emergence	CR	T
	Vegetative vigour	CR	T
	Aquatic plant growth	CR	T
	Tier III		
	Terrestrial field	CR	F
	Aquatic field	CR	F

(Contd...)

Table 20.1 *(Contd...)*

Type of testing	*Data required*	*Strength of requirement*	*Test substance*
Non-target Insect pollinators	Honey bee acute contact LD_{50}	CR	T
	Honey bee - toxicity of residues on foliage	CR	F
	Field testing for pollinators	CR	F

(a) Registration of pesticides intended for terrestrial food crops. R, required; CR, conditionally required; T, technical active ingredient; P, pure active ingredient; PR, radiolabelled pure active ingredient; F, formulation; M, metabolite; D, degradation product.

Table 20.2 Data requirements on environmental toxicology for the registration of a pesticide in the European Union (13)[a]

Type of testing	*Data required*	*Strength of requirement*	*Test substance*
Environmental Chemistry			
Physical and chemical properties	Vapour pressure		P
	Volatility (Henry's law constant)		P
	Solubility in water		P
	Solubility in organic solvent		P
	Partition coefficient		P
	Stability in water		
	Hydrolysis		P
	Photodegradation		P
	Dissociation constant		P
	Stability in air		
	Photodegradation		T and P
	Stability in organic solvent		T and P
Analytical methods	Active substance		
	Breakdown products		
	Impurities		
	Additives		
	Residues		
	In Plants, plant products, food stuffs, feeding-stuffs		

(Contd...)

Table 20.2 *(Contd...)*

Type of testing	*Data required*	*Strength of requirement*	*Test substance*
	In soil		
	In water (including drinking water)		
	In animal and human body fluids and tissues		
Residue	Treated plants or products		
	Behaviour of residue from the time of application until harvest or outloading of stored products		
	Overall material balance		
	Feeding and metabolism studies in livestock		
	Effect of industrial processing and/or household preparation		
	Succeeding or rotational crops		F
Fate and behaviour in the environment	Soil		
	Rate and route of degradation		
	Adsorption and desorption (3 soil types)		
	Mobility (3 soil types)		
	Bound residues		
	Water and air		
	Rate and route of degradation in aquatic systems		
	Adsorption and desorption in water		
	Rate and route of degradation in air		
	Testing for distribution and dissipation		
	In soil		F
	In water		F
	In air		F
Ecotoxicology			
Effects on birds	Short term toxicity - 8 day		
	Reproduction		
	Acute oral toxicity		F
	Assess risk under field conditions		F

(Contd...)

Table 20.2 *(Contd...)*

Type of testing	*Data required*	*Strength of requirement*	*Test substance*
	Acceptance of bait, granules, or treated seeds by birds		F
Effects on aquatic organisms	Acute toxicity to fish		T or P and F
	Chronic toxicity to fish		
	Fish reproduction and growth rate		
	Bioaccumulation		
	Acute toxicity, *Daphnia magna*		T or P and F
	Reproduction and growth rate, *Daphnia magna*		
	Algal growth		
	Overspray study to assess risks to aquatic organisms under field conditions	CR	F
Effects on other non-target organisms	Acute toxicity to honey bees and other beneficial arthropods		T or P and F
	Toxicity to foraging bees under field conditions		F
	Beneficial arthropods other than bees		F
	Terrestrial vertebrates other than birds		F
	Toxicity to earthworms and to other soil non-target macro-organisms		T or P and F
	Soil non-target micro-organisms		T or P and F
	Other non-target organisms (flora and fauna)		
	Effects on biological methods for sewage treatment		

(a) T, technical active ingredient; P, pure active ingredient, F, formulation.

All the required data specified in the regulation are generated by the pesticide industry and their associated laboratories. The tendency is for routine tests dealing with ecotoxicity and environmental chemistry to be undertaken by the contracting laboratories and complex tests of a problematic nature to be done by the companies (14). In addition to the ad hoc generation of data, tiered testing approaches allow decisions to be made on proceeding to the next set of tests, which are more expensive. This practice is called environmental risk analysis and has been utilized at multinational companies such as Dow Chemical, Unilever, Eastman Kodak and Monsanto (4).

The submitted data are evaluated by a team of the government authorities mainly composed of personnel with strong scientific and technical experience in the field, supported by research scientists working in government-affiliated institutes (15). The government institutes undertake research to facilitate the process of evaluating the pesticides: upgrading the test methods, field monitoring, developing estimation models and gathering information. Universities and contract laboratories are also actively, if indirectly, involved in method development and evaluation. Therefore, it is very natural that competent scientists from the industry, the government and the public work together on the risk assessment of pesticides. Of course, the general public is given a chance to express its opinions on the registration of pesticides in public hearings or by writing to the responsible officials who announce the decisions in government gazettes (12). Non-governmental organizations may also know a lot about the submitted data and the risk assessment process.

4.2 Developing Countries

Unlike in developed countries, where the pesticide registration process is a proactive way of protecting the ecosystem, the control and regulation of pesticides at the government level in developing countries is advancing relatively slowly (2, 16). By the late 1980s or early 1990s, governments had barely started to enact the basic laws and regulations concerning environmental protection and to create responsible environmental agencies. One of the main constraints is lack of government and industry resources. Public awareness of the issue is also lacking. The industry is not well developed enough to invest in environmental matters proactively. Ecological considerations are mostly not taken into account when evaluating imported pesticides. The government authorities who make the final decision may have little experience in risk assessment and limited support from government scientists. The prime source of manpower may come from the universities. However, rarely do universities have staff qualified in this emerging discipline of environmental toxicology and risk assessment. Specialists in chemical analysis, environmental chemistry, ecology or biology are potential candidates, but they need additional training in testing and risk assessment from persons in those disciplines. The public are probably barely aware of the importance of proactive decision-making procedures for the import of pesticides, partly due to the limitation of manpower and available resources.

The interactions betweeen the sectors of the society involved with environmental problems, including the issues surrounding pesticides, are shown in Figure 20.1. Whenever accidents, complaints or scientific findings concerning the adverse impact of pesticides are reported in the mass media, the public usually are in a panic. The government has to resolve the issue with command-and-control measures to the industry. The process often needs the involvement of the scientific community, which can provide a possible interpretation of the cause and effect of the incident. Lack of manpower and resources in government, industry and the general public, as well as distrust among the sectors, complicates the normal process of problem solving and often leads to more serious complications in the developing countries than in the developed countries.

4.3 The Relationship between Developed Countries and Developing Countries on Pesticide Issues

Most active ingredients for pesticides are supplied by several large multinational companies. Only a few developing countries have their own formulation plants using the active ingredients imported from the multinational companies. Many developing countries import formulations from neighbouring countries.

Most distributors or suppliers of pesticide formulations are concerned with the efficacy of the product, so their training programmes seldom deal with safety of the user and rarely with the environmental impact of the formulation.

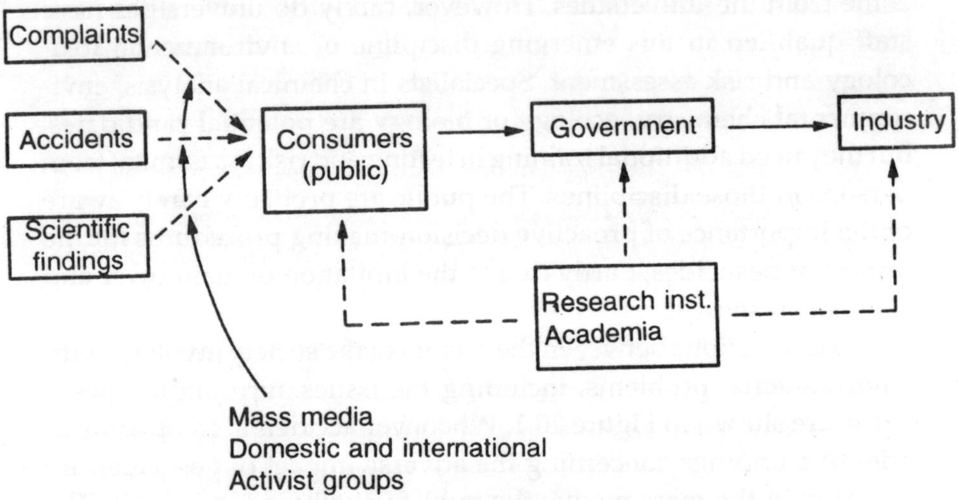

Figure 20.1 Dynamics of environmental problem

The technology for enhancing the efficacy of a pesticide may have spread rapidly from the North to the South, but its safety aspect has spread much more slowly, partly because the governments of the developing countries were not prepared to enact and implement the relevant laws and regulations, partly because of the ignorance of the manufacturer, the local formulator or the distributor and partly because of the low level of education of the farmers and the general public in developing countries. This situation may be causing serious environmental problems, even with formulations that are safe in the developed countries, where the users are certified and well educated. One good example of this is that most farmers in the North may follow the instructions on the label, but we can not assume that this is true in the developing countries of the South. This situation is aggravated by the unpreparedness of the developing countries to protect the environment, as reflected in Table 20.3 (17). Many developed countries require residue data generated in their territories or under similar climatic conditions for the registration of pesticides. Several even require ecotoxicity data and environmental

Table 20.3 Countries requiring residue data generated in their territory or under similar climatic condition(17).

Region	*Country*	*Remarks*[a]
Europe	Czech Republic	
	Denmark	Field trial; include residue (?)[b]
	EEC	Similar climatic condition
	France	Follow EEC directive
	Germany	+Hazard to honeybees
	Greece	Similar climatic condition
	Hungary	
	Ireland	U.K. and similar field trials
	Luxembourg	Accept data from Belgium, France and Germany
	The Netherlands	Similar to the local conditions
	Poland	Similar climatic conditions
	Rep. of Slovakia	
	Switzerland	Local field trial; include residue (?)
	United Kingdom	
Africa	Senegal	Local field trial; include residue (?)
	South Africa	Local field trial; include residue (?)
North Africa and Middle East	Tunisia	Local trial; include residue (?)
Asia-Pacific	Australia	+ Environmental fate
	China	Local field trials; include residue(?)
	India	+ Acute toxicity data
	Japan	+ (advisable) soil, fish, bird study
	Korea, Republic of	
	Malaysia	Field trials; include residue(?) under similar climatic conditions
	New Zealand	
	Pakistan	Local data; include residue(?)
	Taiwan	
South America	Brazil	
	Chile	Local data; include residue(?)
North America	Canada	Local field trials; include residue(?)
	U.S.A	All outdoor studies to be done locally; include residue!

(a) Absence of any remark means local data on residue required.

(b) not clearly stated in the survey result whether residue data is included in the requirement for the local trial.

data generated in the country. However, very few developing countries require such data. The difference in the requirements of developed and developing countries may be considered as a non-tariff barrier to the trade of pesticides and may speed up the movement of less environment friendly pesticides from the North to the South.

It is evident that most developing countries will soon enact or reinforce their laws and regulations following the control and management tactics of the developed countries. This will be accelerated by fast economic growth and public awareness of environmental protection. In the process, the requirements for environmental toxicological and ecological testing and assessment are expected to be strengthened.

The enactment and enforcement of laws concerning pesticide registration that mainly address the preservation of the environment in the developing countries may conflict with practical economic constraints at the national and consumer levels. A manufacturer of an active ingredient or formulation could refuse to supply an item if it was no longer profitable to do so, i.e. if sales revenue in the country did not cover the additional cost of satisfying the new laws or regulations of the importing country. The company must have made the decision based on profitability but the decision maker in a developing country should balance the costs and benifits of introducing the new pesticide against those of the existing pesticide, including hazards to users. This is the point where a conflict of interest between government and industry has arisen, with such conflict being expected to escalate in the future. In this regard, the government official should be capable of assessing the risks and benefits of the pesticides of concern (18).

5 ECOLOGICAL RISK ASSESSMENT AND ECOTOXICOLOGICAL END POINT: PROJECTION FOR THE FUTURE

Even if our discussion is focused on the current regulatory requirements for environmental toxicological data, we need to understand the gap between these regulations and the ideal regulation for the protection of the ecosystem. This is necessary to predict future requirements for data and to be prepared to respond to the anticipated demand of the public for minimizing the adverse impact of pesticides on the ecosystem.

One of the difficulties the government and the pesticide industry encounter in marketing a new product is that they have to predict the impact of the product on the ecosystem before it is actually used in the field. The ultimate goal is to preserve the natural ecosystem as it is. The new pesticide product should not disturb the natural processes or beauty of the world: the tropical forest and fish, the coral reefs of the South Pacific, elephants and their habitat, Alpine forests, the bald eagle of North America, polar bears, seals and so on.

Clearly, it is technically impossible to test all natural organisms in every climatic situation around the globe (19). Therefore, the current requirement is only a minimum attempt to predict the impact on diverse global organisms. This is the limit of the ecotoxicological end point which are measurable. However, ecological risk assessment methodologies are intended to fill in the gap between the actual impact of a pesticide in the environment and the measurable ecotoxicological end points (6). There are two main approaches to filling in the gap: modelling and experimentation in the field. The former approach is intended to predict the behaviour of a chemical in a real field situation, to extrapolate the toxicity, or to integrate both. The latter approach achieves its goal by conducting actual field experiments: mesocosm tests, field tests and environmental monitoring. The modelling approach has the advantage of low cost, but it still needs to overcome the uncertainty of prediction by verification under the local conditions. The field experiments approach should inspire great confidence in the concerned industry and the governmental decision maker but it requires an enormous investment of time, manpower and financial resources. One of the critical limitations of field study is that scientists cannot control the climate and the study can be done only at a few locations. Considering the limitations of the existing approaches, the risk assessment exercise should adopt safety factors or uncertainty factors. This is one evidence of the existence of a gap between the real impact of a pesticide and the ecotoxicological end point. It should be our understanding that the present technology of assessing the risk of pesticides to the ecosystem can never be complete. This is one of the reasons why the industry will be under consistent pressure to supply endless data to the government and the public concerning the ecological risk.

Under these circumstances, it is the direct responsibility of the government and the decision makers of the developing country to ensure the ecological integrity against the hazards of the pesticides, because they are in-charge of evaluating the pesticides being imported and managing their fate in the local environment. Indirectly, the scientific community should support the decision maker technically. It is also the responsibility of the local formulators of the pesticide and the public to ensure that correct decisions are made by demanding that up-to-date risk assessment techniques are used, taking into account of the local climate, biodiversity and human factors.

The pesticide industry in the developing country should take the initiative, at either the individual company level or the industry association level and, with the assistance of the manufacturers of the active ingredients, to build a capacity for evaluating the ecological risk of its products and to identify and phase out the any potentially harmful products. The technology involved in pesticide formulation should mitigate any adverse ecological impact of the active ingredient. The government should encourage the efforts of the industry by regulatory provisions or financial incentives, taking into account the production-oriented economic situation of the industry. This initiative will nearly solve the conflict between the industry and the public on the environmental issues. It will also relieve some of the pressure from international organization, such as the World Trade Organization (WTO) and ISO on environmental matters.

6 ENVIRONMENTAL TOXICOLOGICAL SUGGESTIONS FOR SAFER FORMULATION

Recognizing that the industry itself must take a proactive approach to preventing any adverse impacts on the ecosystem, those involved in pesticide formulation need to do the following:

(a) Adopt the environmental toxicological aspect at the planning stage or when screening the formulation type and the inert components, as is done when inventing new active. It is important to use every possible bit of data on the toxicological behaviour of the component and its effect on the environment. If there is data gap, it should be much more economical to generate data earlier rather than later. In this respect, formulation scientists should be actively involved in the tiered decision-making process for product development. They should be able to suggest the range of ecological risk reduction that could be achieved by applying current formulation technology. This process will keep potent candidates from being eliminated at the early stage of screening.

(b) Emphasize data generation in three areas:
   (i) Toxicity data on marine organisms;
   (ii) Environmental fate and residue data in the climatic conditions prevailing in developing countries;
   (iii) Toxicity data for organisms indigenous to developing countries;

(c) Apply a greater margin of safety to allow for the educational level of users in the developing countries.

The efforts of the industry to produce safer formulations should be fully supported and encouraged by the government and the public.

7 CONCLUSION

This brief review has shown that public concern about protection of the global ecosystem is widespread in both developed and developing countries and expectations that the government and the industry will take action are growing fast. Accordingly, legal requirements for the environmental toxicological data that are critical to the decision-making process of registering new pesticides and re-registering existing pesticides are expected to become more rigorous, especially in developing countries. The proactive approach to this situation is, to be aware of the advances in environmental toxicological methods and ecological risk assessment of chemicals and to apply the latest methodology to the development of pesticides in the industry. Understanding the limitations of current requirements is one way of predicting the future expectations of the society and of preparing for new developments. The role of those involved in pesticide formulation is to resolve the technical conflict between efficacy and ecological risk. They should be involved in the internal decision-making processes of product development from an early stage to ensure that no candidates are unnecessarily eliminated and that the final product is eco-friendly. The industry's contribution to the conservation of the global ecosystem should be supported by the government and the public.

REFERENCES

(1) B. Freedman, *Environmental Ecology* (San Diego, Academic Press, 1989).

(2) *Pesticide Problems, Legislation and Consumer Action in the Third World - the Malaysian Experience*, Consumers Association of Penang, 1985.

(3) R.C. Muirhead-Thomson, *Pesticides and Freshwater Fauna* (London, Academic Press, 1971).

(4) R.A. Conway, "Introduction to environmental risk analysis", *Environmental Risk Analysis for Chemicals.*, R. A. Conway, ed. (New York, Van Nostrand Reinhold, 1982), pp. 1-30.

(5) E. Hodgson and P.E. Levi, *A Textbook of Modern Toxicology* (New York, Elsevier, 1987).

(6) G.W. Suter II, *Ecological Risk Assessment* (Boca Raton, Lewis Publishers, 1993).

(7) G. Goldenman and S. Rengman, *Problem Pesticides, Pesticide Problems* (Penang, International Organization of Consumer Unions, 1988).

(8) N.A. Robinson, *Agenda 21: Earth's Action Plan* (New York, Oceana Publications, 1993).

(9) *The United Nations: Protecting the Global Environment*, New York, Department of Public Information, DPI/1814/Rev. 1- May 1996.

(10) *Water Quality Monitoring in the Asian and Pacific Region*, Economic and Social Commission for Asia and the Pacific, New York, United Nations, 1990.

(11) K. Trought *Trends in Pesticide Formulation* (Surrey, PJB Publications Ltd., 1989), pp. 5-6.

(12) *Code of Federal Regulations*, 40 CFR 158 (Washington, D.C., U.S. Government Printing Office, 1990).

(13) "91/414/EEC: Council Directive of 15 July 1991 concerning the placing of plant protection products on the market", *Official Journal of the European Communities*, No. L 230, 1991.

(14) Personal communications with research staffs of several multinational chemical companies, 1995.

(15) Committee on Pesticide Decision Making, *Pesticide Decision Making* (Washington, D.C., National Academy of Sciences, 1978).

(16) IRPTC and GEMS Monitoring and Assessment Research Center, *Chemtcal Pollution: A Global Overview* (Geneva, UNEP, 1992).

(17) *Pesticide Regulation Compendium* (Cheserex, Switzerland, Editions Agrochimie, 1993).

(18) Personal communications with governmental officials of Sri Lanka, 1995.

(19) S.J. Maund and others, "Ecological considerations in pesticide risk Assessment for Aquatic Ecosystems", *Pesticide Science*, 49, 1997, pp. 185-190.

Author Index

Affiliation Index

ABC Laboratories, Columbus, Missouri, United States
Accessible Technologies Inc., St. Paul, MN, United States
Bayer AG, Leverkusen Germany
Department of Agriculture, Bangkok, Thailand
Hindustan Insecticides Limited, New Delhi, India
Imperial College, Sunninghill, Berkshire, United Kingdom
Instytut Przemyslu Organicznego, Warszawa-Zeran, Poland
Kwizda G.m.b.H., Leobendorf bei Korneuburg, Austria
Maryland Department of Agriculture, Annapolis, MA, United States
Rhone-Poulenc Agriculture, Fyfield Road, Ongar, Essex, United Kingdom
Sandoz Crop Protection Corporation, Palo Alto, CA, United States
S.C. Johnson Wax, Racine, WI, United States
United Nations Industrial Development Organization, Vienna, Austria
United States Department of Agriculture, Peoria, IL, United States
Zeneca Agrochemicals, Bracknell, Berkshire, United Kingdom

Consultants

L.G. Copping, Saffron Walden, Essex, United Kingdom
K.S. Johnson, East Peckham, Tonbridge, Kent, United Kingdom
K.l Pavel, Wuppertal, Germany
M.G. Srivastava, Lucknow, India

Subject Index